The Remix Manual

Simon Langford

Amsterdam • Boston • Heidelberg • London • New York • Oxford
Paris • San Diego • San Francisco • Singapore • Sydney • Tokyo
Focal Press is an imprint of Elsevier

Focal Press is an imprint of Elsevier
30 Corporate Drive, Suite 400, Burlington, MA 01803, USA
The Boulevard, Langford Lane, Kidlington, Oxford, OX5 1GB, UK

Library of Congress Cataloging-in-Publication Data

Langford, Simon.
The remix manual/Simon Langford.
p. cm.
ISBN 978-0-240-81458-2
1. Sound recordings—Remixing. I. Title.
MT723.L36 2011
781.3'4—dc22 2011006660

British Library Cataloguing-in-Publication Data

A catalogue record for this book is available from the British Library.

For information on all Focal Press publications
visit our website at www.elsevierdirect.com

11 12 13 14 5 4 3 2 1

Printed in the United States of America

The Remix Manual

In memory of Clive Button... rest in peace my friend.

Contents

Companion website: www.theremixmanual.com

The Remix Manual: Acknowledgments and Thanks

Writing this book has been an amazing experience for me on so many levels, and a huge number of people have contributed to it—directly and indirectly—and they all deserve recognition and thanks for their part.

First and foremost I have to thank Catharine Steers for her seemingly endless patience, encouragement, kindness, and understanding through what has been, undoubtedly, a very challenging time for me. Catharine, you have been involved in this project from the very beginning and you had the vision to see what I could see and get me started on this path to where I am at now. At some points I felt like I might have "bitten off more than I could chew," but with your guidance, everything worked out just perfectly in the end, so thank you so much. As well as Catharine I would also like to thank Carlin and Melissa at Focal Press/Elsevier for their involvement in the project, and all of the other people who have played a part, behind the scenes, in taking this idea of mine and turning it into a reality. I wish I knew all of your names, but sadly, I don't. Nonetheless you have my wholehearted thanks and gratitude.

Next up is my very good friend, colleague, collaborator, and most recently, my "second opinion guy," Alex Sowyrda. We have worked together a lot over the years and it has always been a blast. I am glad you have been involved in this, my latest project. I have to apologize for some of the crazy deadlines I hit you with, but you always came through when I needed you. Next time around I promise that I will be more organized and won't give you such a hard time!

There are a lot of people in the music industry who have helped me get to where I am today, both professionally and creatively. I can't possible write my thanks here without paying my respects to those who have taught me what I know and enabled me to pass on my knowledge and experience. There are too many to mention all of you, but my greatest thanks must go to: Alex, Matt, Paul, Julian, Phil, Clive, Billy, Andy G, Nigel, Mark, and Danny H. Without all of you (and all of the others) I wouldn't be doing what I am doing and my life would be much less interesting.

I certainly wouldn't be here without Matt Houghton and the rest of the team at *Sound On Sound* magazine, who gave me my "big break" in writing and ignited the flame that now burns within me. I need to thank you for the chance you took on an "unknown" writer for your magazine, and ultimately, for planting the seed for the idea of this book.

I also have to thank my family and friends for getting me started making music, for putting up with all the noise I have made over the years, for smiling reassuringly through my dreadful playing when I was learning, and for anything else I may have forgotten.

To my son, Ben, I hope this book proves to you that anything is possible with effort and determination, and, perhaps most importantly, belief: belief in yourself and the belief of others *in* you. I also hope you realize that, in years to come, I will support and encourage you to follow your dreams in the way others have done for me.

And, finally, I can't write my thanks here without mentioning someone who has, for the longest time, been nothing short of my ultimate source of confidence. Hayley, the way you convince me that I am capable of anything I set my mind to has had an incredible effect on my life. Here I am, writing thanks for a book I could never have written were it not for your kindness and encouragement. I remember, not long after we first met, you telling me that I had "a way with words." Well I guess there may have been some truth in that after all! So here, many years later, is the result of all you have done for me, and I simply couldn't have done it without you. I hope that one day I may be able to repay the kindness.

SECTION 1
The Art of Remixing

CHAPTER 1
The History of Remixing

Welcome to *The Remix Manual*! The aim of this book is to give you an overview of the world of the remixer and then guide you through the creative and technical processes of remixing—including sound design and mixing. When you've reached the end of the book you will have all the information and tools you need to forge ahead with your career as a remixer!

So without any delay, let's get started by looking at some remix history. We'll then move on to how that relates to your story…

Today the term *remix* is almost as familiar to the consumer as it is to the industry insider, since it has become such an important part of modern music. Most people, even those not actively involved in the industry, could give a pretty good definition of what a remix is. Their definition would probably be a little too narrow, however. A remix is, in general terms, simply an alternative version of a song (or more recently, a video). Most people associate the word *remix* with dance music—an area of the industry where remixing has developed the most. By taking a look at the history and culture of remixing I hope to present some of its broader aspects and, as such, give you a greater appreciation of what is possible.

In its most basic form a remix could simply be a version of a song where the balance of instruments is different, or perhaps instruments have been added or removed from the original version. Even creating a 5.1 surround mix of a stereo recording is, in essence, "remixing."

Prior to the introduction of magnetic tape recording in the 1930s, very little could be done to edit recordings. But as soon as the technology became available, people began to see the potential of creating alternative versions of songs by cutting and splicing the (mono) recording of the entire performance. The next step toward remixing came about after the advent of multitrack recording in the 1940s. The guitarist and inventor Les Paul developed the first 8-track recorder with designers from the Ampex Corporation as a means of enhancing band recordings with additional overdubs. This separation of instrumentation meant it was now entirely possible to change individual parts of a final recording.

This really did change the music industry in a major way, as it allowed the creation of far more complex and elaborate recordings and the ability to change them after the initial recording session. In essence, the remix was born. Of course, back then nobody considered this would become the specialized industry it is today. It would be another couple of decades before the true potential of this technique would come to light. And when it came, it was from what many would consider an unlikely source.

By the late 1960s multitrack recording was well-established and people were starting to see its creative potential over and above what it had been used for thus far. A new music movement in Jamaica called dancehall, which embodied elements of ska, rocksteady, reggae, and dub, was where remixing really began to develop. Some of the pioneering and legendary producers from that time such as King Tubby and Lee "Scratch" Perry, started to create alternate versions of tracks to suit the developing tastes of the audiences at the time. They started out by creating simple instrumental mixes of the records but this soon developed into a more complex format involving echo effects and repeated vocal "hooks." Later they began removing certain instruments at specific times and repeating other elements of the track to create new and extended arrangements of the songs.

Around the same time a similar revolution was happening in the disco scene in the U.S. DJs were creating extended versions of hit disco records of the time to keep people on the dancefloor longer. They used simple tape editing and looping to repeat certain sections of the songs. This was the start of the modern club remix. The pioneer of this genre was Tom Moulton, and what he did happened almost by mistake. In the late 1960s, at the beginning of his career, he was simply making mix tapes for a disco; the tapes became very popular and garnered the attention of record labels in New York. Soon these labels were employing him during the production stages of the records to advise on what would work well in the discos. The respect for his knowledge grew and he became very in-demand for creating versions of tracks mixed specifically for the clubs. This marked the first time that record labels would create alternative versions of songs specifically for the disco and club markets, but it wouldn't be the last. In fact, many of the basic techniques that we take for granted nowadays are direct descendants of Tom Moulton's work. A quick listen to some of his better known remixes can bring chills to those of us familiar with the remixes of today. To this day Tom Moulton is a legend. In his 30-year career he has worked on over 4000 remixes and is considered by many to be one of the most important people in the history of dance music.

The first commercially available 12" single was not the work of Tom Moulton. That honor goes instead to Walter Gibbons with his extended "re-edit" of "Ten Percent" by Double Exposure, in 1976. Prior to this, 12" singles had only ever been available to club DJs. With this release, label Salsoul Records took a chance because they believed the format had commercial potential. Ultimately they were proved right and, through the late 1970s, the 1980s, and into the 1990s, the 12″ single was a great commercial success. During those years Salsoul Records would also be the home of many remix luminaries, including the legendary Shep Pettibone, whose work lay the foundations for the remixers of today.

Indeed, Shep Pettibone, having found his feet in disco, went on to do remix work for The Pet Shop Boys and Level 42 in the mid 1980s before working with Madonna. His first achievement with the material girl was his "Color Mix" of "True Blue" in 1986 and this led to further well-known remixes of "Causing a Commotion," "Like a Prayer," and "Express Yourself." Soon the remixer became the songwriter and producer when he collaborated with Madonna on her single "Vogue" and her album *Erotica*. This was really a major step because, prior to this, remixers had been just that, remixers. Pettibone's collaboration with Madonna on "Vogue" really made an important statement that the world of remixing had developed and achieved a level of sophistication equal to that of actual production. Of course, remixers are now usually perceived as being equals in terms of skills and technical knowledge, but prior to "Vogue" in 1990, this hadn't been the case.

The third major development in remixing also happened at around the time of the Salsoul Records remixes, and was born of a fusion of the Jamaican dancehall cultures and U.S. disco sounds following an influx of Jamaican immigrants to New York. Iconic DJ Grandmaster Flash was one of the first people to use "cutting" and "scratching" techniques to create musical collages in real time. While these techniques would later be incorporated into remixing in a more planned way, this was a valuable addition to the remixer's toolbox and it also became one of the trademarks of hip hop music.

During the 1980s most remixes were based on rearrangements of the existing musical parts and, as such, they didn't particularly change the genre or market of the songs. However, with the birth of house music in the early 1980s and its phenomenal growth throughout mid and late 1980s, a new form of remix emerged that also took advantage of improvements in music technology during that time. Throughout the 1970s and 1980s artists such as Kraftwerk, Jean Michel Jarre, and Vangelis had been exploring these new synthesizers and sequencers, and this laid the groundwork for house music, which in turn developed into the numerous forms of dance music we have today. Many actually consider some of the hi-NRG forms of disco to be the direct ancestors of house music because of the increasing use of drum machines and sequenced bass lines. While the actual sound of Donna Summer's "I Feel Love" with its now legendary Giorgio Moroder sequenced bass line might be a far cry from some of the early Chicago house music, there are many similarities, not least of which is the rock-solid timing and largely repetitive nature.

Following the explosive growth of dance music during the 1980s, remixes started to take a slightly different direction and evolve more toward what we generally associate with remixes today. The remixers of the late 1980s and early 1990s were using less and less of the original instrumentation in their remixes and using more newly created musical parts. The rapid advances in sequencing and MIDI-based music made this easier and, as the technology evolved, so did the creativity of the remixers. Perhaps one of the earliest examples of the "modern" remix came in 1989 with Steve "Silk" Hurley's remix of Roberta Flack's "Uh Oh Look Out." What was novel about this remix was that *only* the vocal from the original track was used, with *all* of the musical parts being newly created. This method of remixing is standard fare these days but, at the time, it was revolutionary and really was a landmark in the evolution of remixes.

Arguably one of the most famous dance records of all time was M.A.R.R.S' "Pump Up the Volume." Shortly after its release in 1987 it went to number 1 in the U.K. charts; it also received a Grammy nomination in 1989. Although not technically a remix, the song used many popular remix techniques of the time and was composed almost entirely of samples of other records. It spawned a whole "sampling" subculture, which, during the early 1990s, would lead to a wave of tracks that used sampling to great effect.

As sequencing technology moved away from dedicated hardware units and became more computer-based and the sometimes prohibitively expensive and comparatively limited analog synthesizers and drum machines started to be replaced by more versatile and cheaper digital ones, it became possible to do more for less money. Also, around this time, more affordable samplers started to enter the marketplace. The arrival of this "triple threat" radically changed the dance music genre, and the rise of dance music to the top of the commercial charts began in earnest.

The rave scene in the U.K. in the early 1990s spawned a huge number of tracks that sampled other songs (or dialog from films or TV shows) and used these often sped-up samples as the basis for new songs. Although these songs were often marketed as "original," they were, to some extent, based around the same principles as the remixes that were happening at the time. As this was the dawn of the sampled music generation, many of these songs (including "Pump Up the Volume") were commercially released without legal action. Today it is extremely doubtful that a single one of them would be released without securing both sound recording and publishing clearance. At the time, however, these things were, if not ignored, at the very least not pursued so diligently as they are today. As a result of this, yet another new kind of "remix" evolved: the mashup.

Although a mashup isn't a remix in the sense thus far described, it is, to some extent at least, still within our initial definition of a remix: an alternative version of a song. A mashup is a mix or edit of two (or sometimes more) other songs to create a new one. Most often mashups contain little or no actual music added by whoever created them. In many ways they could have been (and often are) created by a simple DJ setup of a couple of decks and a mixer. Sampling and sequencing technology obviously makes this easier than trying to mix, beat-match, keep in sync, and recue records. Advanced technology facilitates more elaborate and clever mashups, and this style of remixing is still very popular. Creating a mashup can be technically difficult, since you are dealing with mixing together two full tracks whose musical elements don't always complement each other, or even match at all. On that basis, a DJ or producer often looks for an a cappella track (purely vocals) and mixes this over an instrumental track. This method reduces many of the technical problems associated with making mashups, but doesn't overcome concerns about legal usage.

As remixing culture grew throughout the 1990s, more and more mainstream artists embraced the idea of the remix. In some instances this was simply by speaking out in approval of the remixes of their work (both official remixes and "bootlegs" or mashups). Over the years, artists such as Nine Inch Nails and Erasure have made the original multitrack recordings of their work available publicly to allow fans to create their own remixes. Some of these remixes have even ended up being released commercially by an original artist's record label. This is quite a rare occurrence, however, and to this day there can be quite a lot of resistance to the (sometimes radical) changes a remix can bring to the original recording. Some artists have always been more open-minded, and some more protective of their work. I don't imagine this will change anytime soon.

Another way in which artists have shown their support and embraced the remix culture is through the recording of additional vocals specifically for a remixed version. This dates back to the very early 1990s when Mariah Carey became the first mainstream artist known to rerecord a vocal for a remixed version of one of her songs and, within a few years, nearly all of her songs had rerecorded vocals for use in the club remixes of her tracks. Many other artists have done similar things in terms of rerecording vocals specifically for a remixed version and some have actually written and recorded new material for the remix. This scenario is more prevalent in rap and hip-hop music where remixes with "guest" appearances by other artists/rappers have become the norm. As remixes grew in popularity, once again, so did the possibilities that existed with rerecorded vocals, new material, and guest appearances.

Over time, the unstoppable rise of the remix continued, fueled by the growth and increasing diversity of dance music itself. Slowly and surely remixes became pretty much indispensable and began to be seen as an integral part of the promotional plan for many commercial releases. So much so that in 2007 Def Jam Records commissioned remixes of every track from Rihanna's massively successful *Good Girl Gone Bad* album, subsequently releasing a "limited edition" of the album that included the remix CD. There can be no denying that remixing has become an important aspect of modern music culture. John Von Seggern, from the University of California's Ethnomusicology Department, states that a remix is

> a major conceptual leap: making music on a meta-structural level, drawing together and making sense of a much larger body of information by threading together a continuous narrative through it. This is what begins to emerge very early in the hip-hop tradition in

> works such as Grandmaster Flash's pioneering recording "Adventures on the Wheels of Steel." The importance of this cannot be overstated: in an era of information overload, the art of remixing and sampling as practiced by hip-hop DJs and producers points to ways of working with information on higher levels of organization, pulling together the efforts of others into a multi-layered, multi-referential whole which is much more than the sum of its parts.

This is a pretty telling sign of the extent to which remixes have become a part of the modern music industry and something that shows no sign of slowing down. With the increasing fragmentation and diversification of genres within dance music there are increasing numbers of bases that record labels are keen to cover and, in order to cover those bases, they turn to us, the remixers. The future looks bright indeed!

THE FUTURE OF REMIXING

In the last decade there have been such massive and swift advances in computer technology that it has been almost dizzying to try to keep up with them. The tools available to the producer, and hence the remixer, these days are quite remarkable. They offer a bewildering variety of ways to create new sounds, and more and more time and money is being spent on developing tools that will, by design or otherwise, allow the remixer more creative freedom than ever.

Software tools allow you to adjust the timing and pitch of a vocal (or other monophonic) recording as freely as if it were MIDI data, and with pretty decent results as well. But even more impressive, the latest version of the software Melodyne allows you to take a polyphonic audio recording (mixed backing vocals with harmonies, for example) and manipulate the individual melody lines. The implications for this are immense and will be covered in much more detail later.

There is also other software in development that claims to be able to take a final mix of a track and isolate and extract individual elements within that mix. If the results are good, this software, in conjunction with the other tools available for manipulating audio files, will blow the world of remixing wide open and give the remixer virtually full control over every aspect of a remix even if the content provided is less than ideal.

Perhaps the most interesting development in remixing in recent times is the concept of the video remix. While this is not technically included within the scope of this book it is worth mentioning. In a later chapter when we cover the issues of promoting yourself as a remixer, we will see that now more than ever you need to embrace the rapid developments in multimedia and be fully aware of the importance of video and sites such as YouTube. Music videos have, since the 1970s, grown in importance to the point where no commercial release expected to do well is released without an accompanying video. In recent years there has been a growing trend to have the videos "reedited" to the remixed version of a song. This is not a simple process, and generally happens with whichever remix is considered the "lead" remix. Because the video is shot with the original track in mind, there isn't normally enough footage for an extended (often more than six-minute) club remix of a track; some creative editing is required to fill the extended time with the existing footage without being too repetitious.

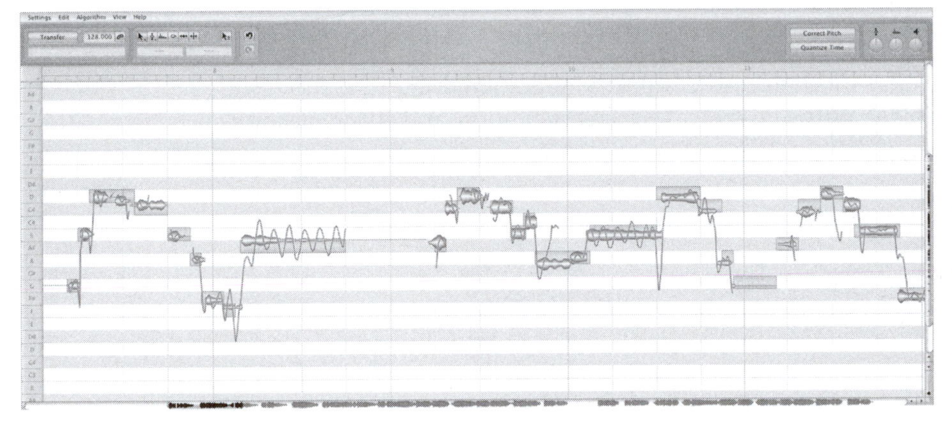

FIGURE 1.1 Celemony's Melodyne software allows the manipulation of audio data in ways previously only possible with MIDI data. It represents a real breakthrough for remixers and producers alike.

There is a growing number of people offering their services as video remixers, and while this is still very much a new industry, it is one that will grow as time goes on. If you feel you have any ability in that area, or any interest in learning, it might be worth delving into. It can only help promote your remixing/production skills by having a promotional video of the tracks that you are mixing edited to your remix. Along with blogs and tweets you now have to consider websites such as YouTube and other video content providers as valuable (and free) marketing channels.

With video remixing there is, of course, the issue of whether or not it is actually legal. As always, if in doubt you should speak to the record label that commissioned your remix, to seek their permission or approval before reediting the video. Some will probably turn you down, but others might give permission for a video remix as long as it is for your own promotional use and you do not intend to freely distribute or sell the remixed video. There are specific clauses under copyright law (often referred to as "fair use" clauses) that deal with this issue and allow remixers certain rights to a work in which they do not actually hold a copyright interest. Again, we will look at this more in a later chapter.

In the last three decades, the art of remixing has turned from a very underground phenomenon into a very mainstream one, and the fees commanded by today's top remixers can be, well, very respectable to say the least! It *can* be a lucrative business, but it certainly isn't an easy road. Modern technology does a lot to help the creative remixer, but its relatively affordable nature and easy availability means that there is *a lot* of competition out there. If you are still interested in taking that road—and I'm guessing that you are because you are reading this book—then please join me as we continue our journey into the nitty-gritty of remixing.

In subsequent chapters we will consider the artistic and creative side of the discipline, the technicalities, and the legal and business side; we'll "walk through" a typical remix showing some of the issues that might come up, with tips and advice on ways to deal with them. Let the journey begin!

CHAPTER 2

Being a Remixer

Seeing as you are reading this book you are obviously interested in being a remixer. But what does that actually involve and how is it different from producing your own material? Well, in essence, the technical process is very similar for the most part. The technical skills you would use in working on a track of your own are all very much the same for working on a remix. Only a couple of tasks you would encounter on a remix are things that probably wouldn't come up if you were working on your own material, but that's about it. The biggest difference lies not in the technical side of things, but in the *creative process*.

After much thought, the following is my summary of what it's like being a remixer. I think this describes the discipline quite well:

Think of every song as a story, a collection of words that conveys an emotion, or a journey; something unique and something personal. That story has its own language. That "language" might be pop, it might be R&B, or it might be rock. Our job, as remixers, is essentially that of a translator. We have to take that story and translate it into a different language, our language (which might be house, it might be trance, hip-hop, or dubstep), while keeping at least the meaning of the story intact. It doesn't have to be a literal translation, it doesn't have to follow every nuance and subtlety of the original story, but it has to still retain the meaning and the "message." If we can retell the story in our own language and still make it so that people understand it, then we have done our job well.

So, what do we need to be able to do our job properly as remixers? And how is this different from what we might do if we were writing and producing a song of our own? As I have already said, for the most part, at least, the technical skills are the same. You still need to program sounds, you still need to program beats, you still need to work out chords and melodies, and you still need to apply effects and mix the music. But it's probably true to say that, if you are working on your own song, you wouldn't ever need to time-stretch vocals (because the vocal recording would normally come after you had written the music). There may, on very rare occasions, be times when you feel that the whole track needs to be a tiny bit faster or slower (it has happened to me more than once) so you end up time-stretching the vocals, and any other audio

The Remix Manual.

parts, by a couple of beats per minute. But if the change is that small then a simple time-stretch will probably suffice and you won't need to resort to any of the trickery that we will go into later.

Closely related to this is the key of the track. Any chord sequence in any key can be transposed to a different key: it's simply a matter of shifting every note up or down by a certain number of semitones. If you were working on an original track and things were sounding good but then, in the chorus, you went to a pattern with different chords and one or more of the bass notes now felt just that little bit too low, you could change the key of the entire song to shift everything up into a better key. Conversely, if some of the bass notes, for example, sounded too high and didn't really have any "weight" to them you could do the opposite and shift the key of the song down until the lowest bass notes no longer felt too thin.

In a remix the key is already defined. Yes you can change the key from a major key to a relative minor key or vice versa (more on this in a later chapter), and it is even possible in some circumstances and on some tracks to shift to harmonically related keys. C major and G major, for example, are very closely related to each other with the only difference being that C major uses an F whereas G major uses an F#. The following table illustrates this more clearly:

C major	C	D	E	F	G	A	B	C
G major	C	D	E	F#	G	A	B	C

With remixes, however, in a vast majority of cases you *will* have to time-stretch the vocals. Sometimes this may only be by a small amount and, again, in those cases you shouldn't have too much difficulty. But as more and more pop music ends up being remixed you could find yourself having to deal with some really problematic vocals. Being able to find a solution in those kinds of situations, and being able to do it well, is possibly one of the biggest assets you can have as a remixer (from a technical standpoint at least). This is why there is a whole chapter dedicated to time-stretching vocals later in the book.

But for now let's put aside the intricacies of the technical process and think a little more about the creative process of remixing. We can compare it to the process of writing your own material. From more or less the very beginning of the process (depending on exactly how you work) things are different. If, for example, your music theory knowledge isn't great, it might not necessarily be that much of a problem when working on your own material because you wouldn't have to know about keys and intervals and inversions when working out the chords of the song. You could simply mess around on a keyboard or a guitar until you came up with something that *sounded* good, even if you didn't really know or couldn't explain (in music theory terms) *why* it sounded good. But as a remixer you don't have the option to just transpose the music and backing track up or down how ever you please. Yes there might be options of alternate keys other than the one you are in, but you don't have the flexibility that you would on your own compositions.

With a remix you have to always fit the music to the vocals, and if the song has complex melodies and harmonies that could be quite a challenge. The more music theory knowledge you have, the easier this part of the process is. But even if your music theory is well up to scratch and you don't have to think too hard to work out which chords might work with the vocal melodies, your ultimate choices will be restricted to a greater or lesser extent, and certainly much more than they would be if you had a "clean slate" (musically speaking) to start from.

Once you have a basic chord pattern, you might start work on building up a groove. If you're working on your own original material, that groove can be whatever you want it to be. With a remix you don't have quite as much flexibility. If the vocals have a really distinct "bounce" to them (meaning quite a lot of swing on the groove) you would have real problems trying to fit those over a more "straight" and regular groove. The opposite is also true.

That said, adapting a straight vocal to a swinging groove, or a bouncy vocal to a straight groove isn't totally impossible—very few things are. However, it would mean quite a lot of work entailing regrooving (effectively *quantizing*) the instrumental parts. Recent software such as Apple's Logic 9, Avid's Pro Tools, and Ableton Live have made this much easier (with their "flex time," "elastic audio," and "beatmapping" technologies, respectively), but it is still an extra stage of possibly quite intense work; work that could introduce unwanted artifacts into the audio that you might not be able to disguise later.

Then there is the issue of the actual choice of sounds. With your own productions you can choose more or less whatever sounds you feel would work best, bearing in mind, of course, any vocal track that will be added in later. Choosing sounds that are too overbearing for the vocal (even if they sound fine without it) is one of the easiest mistakes to make when writing your own material. In remixing, you have to consider what the vocal is doing at any given part of the song, as well as taking into account the actual tone of the singer's voice. If she has a very deep and warm voice then you will probably need thinner and more cutting sounds to complement her tone. If, on the other hand, the singer has quite a high-pitched and thin voice, deeper and warmer sounds may be a better match. And all the while you still have to keep one ear on the overall style you are working toward; some sounds are just not acceptable in certain genres.

In the arrangement of the track we see yet more differences. With a song of your own you can work with the chord movement you have, looping it and building it dynamically until you feel the time is right to go to the bridge, chorus, or go back to the verse. In a typical remix you will have a very clearly defined song structure and that will, for the most part, dictate where things go and how long sections last for. You might feel that the verse needs to go on longer to build some tension but you won't be able to do that without inserting potentially troublesome gaps in the vocal or repeating or delaying certain lines of the vocal. These are things I do regularly and aren't horrendous to deal with. The biggest problem area is when you feel that the lyrical verse goes on too long and that the music behind it needs to change or move to the chorus. What do you do with all of those extra words? Earlier in this chapter, you'll recall I said that even though the "story" didn't need to be retold exactly word for word, you still want to get the same message across. We've all heard "hack jobs," where a remixer has removed words, lines, or even half of a verse just so the vocal fits in with the musical arrangement he or she likes. We will look at arrangement in more detail in Chapter 6 but, as a general rule, you should try to keep the vocal structure as "intact" as possible if you have been asked to submit a "vocal" mix. If there is a particularly drastic edit that you would like to do to the vocal then you should always consult the client before doing so as this can be a "deal breaker" for some vocalists/artists.

One more thing worth mentioning is the overall "feel" of the track. With your own original material you have the freedom to deliver whatever emotion you want; if someone else is writing the lyrics and melody to the song he or she will be working to your brief and perhaps to a working title as well. With a remix much of that mood is already defined, in the musical progressions and in the melodic and actual lyrical nature. With any remix I am working on, the *very* first thing I do is time-stretch the original track and then listen to it a few times at its new tempo. On the first listen I am just listening with the track on in the background; on the second I will be thinking ahead about chord movements, melodies, and harmonies. On the third time through I listen carefully to the lyrics themselves to identify any lyrical cues I can use in my remix. Then I would probably listen to the whole song all over again, with the music on quietly in the background and possibly even on a different set of monitors so I can gain some fresh insight into the original song.

As you can see, there are several differences between the approach to a remix and the approach to an original track. You can think of each aspect of doing a remix (working out chords, bass line, groove, etc.) as stepping stones, each representing a step on the path toward the finished version. The actual path you take and the decisions you make as to which of the stones to stand on and in what order is what identifies one remix from another. Some of the metaphorical "stones" may actually disappear as you get more into the remix. For example, you may want to change certain things that the original song (mainly vocals) won't allow because of musical conflicts. It might be that the option to work in a totally different key is impossible and so that particular stepping stone disappears. Rest assured though, there is *always* a path to the other side. As remixers, we have to use our persistence to push ahead and find it, even when it doesn't seem too obvious at first.

CHAPTER 3
Choosing Your Style

There are so many different genres of music, and nowhere is this more true than in "dance" music. In recent years this ever-expanding genre has been spawning new subgenres by the month. For example, one style of dance music is "house." The following is a list of all the subgenres of house that I could find at the time of writing.

Acid house, ambient house, Chicago house, deep house, disco house, dream house, electro house, fidget house, French house, funky house, glitch house, handbag house, hard house, hip house, Italo house, Latin house, minimal house, New York house, progressive house, pumping house, tech house, tribal house, vocal house.

To further confuse matters there is often a great deal of debate as to what exactly forms the essence of each of these genres. What is electro house to one producer might be tech house to another; and if you played the same remix to 12 different DJs you would probably get at least three or four different descriptions of the genre. So, while it is undoubtedly good advice to keep up-to-date on all of these variations I would advise you not to feel too restricted by the rules of any particular genre or subgenre. That said, there are two parts to the decision of what style you wish to work within: one has a wider importance and the other is more relevant to the remix you are working on. Let's look at each of them in turn.

Be True To Yourself

The genre(s) you choose to work in will largely be dictated by your personal tastes and preferences. Many producers and remixers work exclusively in one genre. For example, they would consider themselves a "house" or "trance" producer. This normally isn't a conscious decision as much as a natural gravitation toward the kind of music you like to listen to and, most likely, make. There are, of course, many other producers who work in several different genres and sometimes they do this under different act names or with pseudonyms purely to avoid brand confusion. This makes a lot of sense because, ultimately, the hope is that your "name" as a remixer will become recognizable and in-demand. If you became known for making trance remixes and then, all of a sudden, you start putting out house remixes under the same name, it could cause confusion and ultimately cost you work! That isn't to say that it's impossible to move between genres under one pseudonym, but it is much more likely to work if it is a gradual evolution rather than a sudden change.

Taking that into account, it is clear that this wider genre you work within (at least under one pseudonym or act name) is pretty important and it has to be something you feel comfortable working in. Remixing is often about working to tight deadlines, so you wouldn't want to try something new when you are under pressure to get a remix done. Experimenting in other genres is best done under a new artist name, within which you can build up your confidence and then look for work. The obvious downside to this is that whatever profile you have built up under your original artist name will be lost when you start using a pseudonym within a new subgenre, that includes any brand recognition you had; this could make you less sellable or valuable to the record label commissioning the remix. That said, the dance music industry does allow a huge amount of flexibility to those working within it because of the possibility of working on multiple projects, under multiple names, and in multiple genres. If you have the ability and the desire, this can be a great way of working because it gives you creative outlets for pretty much whatever direction you want to go in creatively, while allowing you to focus each individual "brand" to its own market.

But Don't Limit Yourself

The second concern about your choice of style involves embracing flexibility. As mentioned previously, a genre such as house has wide variety of subgenres and while your production/remixing brand may be identified as house (or perhaps even more narrowed down to something like tech house), because of the large number of variations of style it is likely that your remixes will, at times, cross some of the boundaries between the subgenres. You will, most likely, get away with this without surprising anybody. So being consistent in your style is good, but don't feel you should be *too* consistent. There is a degree of flexibility open to you when choosing how you want to approach each remix, and you would be well-advised to take advantage of this.

All of the greatest remixes have a number of factors that make them great. There's the technical quality of the remix itself (e.g., EQ, levels, choice of sounds, mixdown), the originality and creativity, and finally there's a certain "sympathetic" quality to the remix or the choices that were made and how well they sit with the track itself. Perhaps it was the lyrics of the track that inspired the remixer, perhaps it was the melodies, or perhaps it was the singer's tone of voice, but there is always the impression that these remixes "gel" better than others because they *feel* like they are right. As a remixer it is comparatively easy to come up with chords, sounds, grooves, and melodies that *do* fit with the vocals, but that doesn't mean they are necessarily the best chords or sounds or grooves that you could find to go with the track.

So, before we have even really gotten started, we come across the first of many situations where you have to make a judgment call based on the interests and expectations of others. You have your own genre and your own "sound" that the remix will fit with, but there will also be the requests of the record label and sometimes the artist. If you work in a particular genre you might be asked to remix a track in your style and you may feel that the track just won't work with your "sound." If that happens, you have to make a difficult decision between compromising what you normally do and doing something that doesn't really work. I have heard many remixes that have obviously been done simply to appeal to the latest market taste; most sound dreadful, usually because the vocal (lyrics, melody, voice tone) just doesn't seem to fit within the genre of the remix.

Sometimes you have to embark on a project before you realize it's working out okay for you. There have been times I have started on a remix feeling that I wouldn't be able to get the vocals to sit comfortably only to have a flash of inspiration halfway through the mix and it turn out to be one of my favorite remixes. But there may be times when you feel that you wouldn't do a good job on the remix and, as hard as it might be, you have to be able to recognize when that is true and not be afraid to turn work down. It might seem counter-intuitive given that so much of this industry is about raising profile and awareness, but sometimes turning work down is beneficial. Not only because a "bad" remix will damage your profile and reputation but also because it gives the impression that you are professional and care about what you do. There is also some truth to the idea that the less "available" you are, the more (to some extent) people will want you. DJ and producer Eric Prydz suffered from a fear of flying for many years that prevented him touring outside of Europe, but the more he turned down gigs, the more people were desperate to book him. The money he was offered to travel then went up and up! Of course a fear of flying is very different from turning down work by choice, but this example just illustrates how things *can* work in the opposite way to what you might expect.

The Freemasons have had a massively successful remixing career and they turn down far more work than they take on. There are many reasons for this but the fact that they typically do less than ten remixes per year gives their work a certain "exclusivity" it simply wouldn't have if they did everything they were offered. There are always alternative approaches and some people are very successful by doing the exact opposite and by working on as much as they can to raise awareness of their work by the sheer volume of it. Both approaches can be successful and it is certainly worth considering which you feel more comfortable adopting.

As you can see, there are many factors to take into account before deciding how to approach a remix in the broadest sense. Many early remixes were very similar in style to the tracks they originated from, but it didn't take long before people became more adventurous and started taking things much more out of context. How far *you* are prepared to take things really depends on your confidence and your creativity, and it will vary from track to track and remix to remix. But there are usually common factors to take into account. The lyrical content of the track being remixed is obviously a *big* factor because if the original vocal is very positive and uplifting then it most likely won't feel right in the context of a musically and sonically "dark" remix. It *does* happen and people *can* get away with it, but to me that is the exception more than the rule. The opposite also applies, of course: if the original track has a very somber and "down" vocal it just won't feel right in a happy, bouncy, "up" remix.

You also have to consider the melodic structure of the track. If the original song has very elaborate and complex chord movements, then a monotone-based genre probably won't suit the track as the vocal won't sit over a repetitive chord structure; a more complex chord structure wouldn't work in the genre. To some extent this can be worked around if the record label is happy for you to deliver a mix that doesn't use the whole vocal, but for the purposes of this discussion let's assume that you are required to deliver a vocal mix.

Finally, there is the tone of the singer. If he has a dark, brooding, and gravelly tone of voice then certain styles of production will suit that more than others, and vice versa. I have to make the point here, though, that what I am telling you are not "rules," merely suggestions and observations from my personal experience. If you feel that you can be successful in taking something way out of context then please give it a try!

Most of what we're discussing will soon become second nature to you and as time goes on and you gain experience you're bound to become more diverse within your genre. The music scene changes rapidly and, along with it, your tastes, creativity, and abilities. I often listen to work I did a few years ago and am actually shocked at how much my style has changed; but if I listen back to the remixes I have done in sequence I can hear the direction that things moved in.

As I mentioned before, if you feel that you want to branch out into other genres then you should definitely do so; there is no downside to it that I have yet seen, assuming, of course, that you take into account what I said earlier about "branding."

We can't really talk about choices of style without at least touching on the issue of inspiration versus imitation. With the sheer volume of dance music available these days and the increasingly narrow genre definitions it can be difficult, at times, to be original while not straying too far from what is popular at the moment. Not to mention that as a remixer you will generally be working to extremely tight schedules! There may even be times when you are given another track as a reference track for your remix. It could be one of your previous remixes or productions, or it could be a remix from somebody else, but if a record label or artist already has a direction in mind then that may possibly help with that initial creative spark. In general, though, if this happens, I recommend only listening to the reference track once or twice, just enough to get a feel for it but not long enough to get it ingrained in your mind, because at this point it is very easy to end up with a remix that sounds *too* close to the reference track, or worse still, copies it.

One final thing to take into account is that different people, and especially people from different pockets of the music industry, often use different language to convey the same musical ideas. Sometimes, even more confusingly, they will use the same language but mean something different! Over a period of time of dealing with people regularly you learn what exactly they mean when they try to convey an idea but, if you're working with a new client and you're having problems understanding each other, a simple reference track can go a long way! You'll also find that as you establish your remixing identity people will be more willing and happy to just leave you to do your thing, and let you decide what is best for the track. Don't assume that will happen 100% of the time, however; in my experience, it just doesn't.

Now that you have an idea of which direction to take the track, you can get to work starting on the remix. Depending on which software you use you might be able to set up a number of different templates to work from (I use Apple Logic Studio and this allows you to have multiple templates). As already mentioned, remixing (perhaps more than producing your own tracks) is more often than not on a very tight deadline so, over time, you will find ways to improve your workflow and generally speed things up. Saving a few minutes here and there in the setting up of things might not seem like a lot but it can soon add up. I will cover this in more detail later but, for now, if you have the ability to do so, I strongly recommend creating at least one template with some favorite synths preloaded into the channels with some "inspiration" sounds set up. Possibly a bass sound, a pad sound, a piano (always good for working out chord structure, I find), and a couple of other sounds relevant to the genre you are working in are good places to start. I also recommend having, at the very least, a kick drum (synthesized or sampled) available from the get-go, as this is often the very first thing you'll want to put into the track.

The fact that these "template" sounds are preloaded doesn't mean that you have to keep them in the final remix, but it will give you a means to quickly switch to another track to try out an idea. Sometimes ideas will pop into your head and, if you spend too long loading up a synth and trying to find that really nice bass sound, well, they might just pop on out of your head again! The reason I suggest having multiple templates is that you can set up a different set of these preloaded sounds for a few different variations within the genre you are working in (as well as having totally different ones if you work in different genres). Chances are that these templates will change as you go along—swapping sounds from time to time—but remember, they are only there as a basic starting point to help inspire and get things moving quickly.

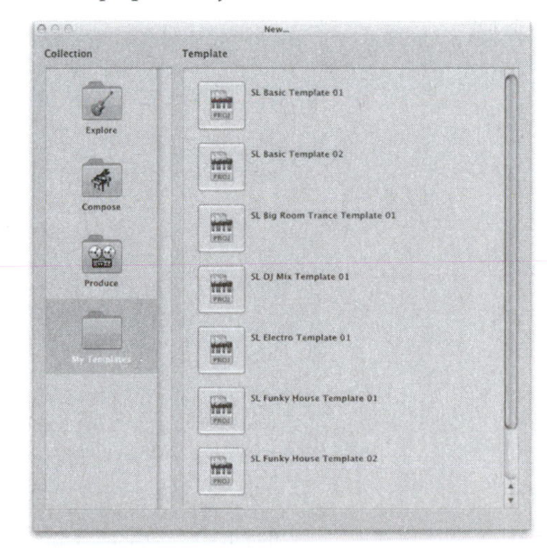

FIGURE 3.1 Apple Logic Studio allows the creation of several template files you can use to get started quickly by preloading certain favorite sounds and settings.

Each person works in a different way. Some prefer to get a groove happening with drums and percussion and then figure out a bass line or a bass groove, even if they subsequently have to adapt it slightly to fit the key or the chord changes of the remix. Personally I have found it more productive and inspiring to have at least some basic idea of chord structure in mind before I start working on the groove. I have found that you can sometimes hit upon a great groove that just doesn't work that well when you have to make it follow a particular chord sequence you put in later on. At that point it can be hard to find something else that you like because you have grown accustomed to that first groove, which no longer fits! Whichever way works best for you is what *is* best for you but, as I tend to work out basic chords as a bare minimum first, that is what I will be covering next.

CHAPTER 4

Keys, Chords, and Melodies

What you'll read in this chapter is in no way an exhaustive look at musical theory. Remixers span the gamut of those who have had some formal music training, those who have just picked things up as they go along, and those who have absolutely no musical theory knowledge whatsoever. So rather than get too deeply caught up in the intricacies of music theory and attempt to explain *why* everything works the way it does, I will instead focus on how we can use the musical "rules" that exist and what they mean to a remixer, explaining some theoretical things along the way as and when required. So let's take a look at what is probably one of the most useful pieces of information for a remixer: the key of the song.

A song's key is simply a way of describing which notes are "allowed" (and I use that term loosely) in the chords and melodies. Most, although far from all, songs have one key (perhaps with a key change to a different key later in the song) that defines which chords the song is made up from. There are two main groups of keys: major and minor. And within each group there are 12 different keys, one for each different note on a piano keyboard. These two types (major and minor) form the basis of the vast majority of modern western music and each is defined by a series of individual notes. Major and minor keys each have seven notes in them (they are sometimes listed with eight notes but the last of the eight is actually the same as the starting note) but they differ in the spacing between the different notes in the series.

Each key has an associated scale with it and the scale for any major key is formed in the following way:

Root note (first note in the series) > +2 semitones > +2 semitones > +1 semitone > +2 semitones > +2 semitones > +2 semitones (and ending with +1 semitone if describing the key with eight notes instead of seven)

Given that there are 12 different musical notes (C, C#, D, D#, E, F, F#, G, G#, A, A#, and B) you can now see how we get the 12 different major keys and associated scales. A minor key and scale is formed in a similar but slightly different way:

Root note (first note in the series) > +2 semitones > +1 semitone > +2 semitones > +2 semitones > +1 semitone > +2 semitones (and ending with +2 semitones if describing the key with eight notes instead of seven)

As with major keys and scales, there are 12 different versions, one starting on each of the 12 musical notes. It is very useful to have some kind of chart listing the different musical keys and the associated notes in the scales. To save you a little time and effort I have included two charts—one each for major and minor key.

Major Key	**Root**	**+2**	**+2**	**+1**	**+2**	**+2**	**+2**
C	C	D	E	F	G	A	B
C#	C#	D#	F	F#	G#	A#	C
D	D	E	F#	G	A	B	C#
D#	D#	F	G	G#	A#	C	D
E	E	F#	G#	A	B	C#	D#
F	F	G	A	A#	C	D	E
F#	F#	G#	A#	B	C#	D#	F
G	G	A	B	C	D	E	F#
G#	G#	A#	C	C#	D#	F	G
A	A	B	C#	D	E	F#	G#
A#	A#	C	D	D#	F	G	A
B	B	C#	D#	E	F#	G#	A#

Minor Key	**Root**	**+2**	**+1**	**+2**	**+2**	**+1**	**+2**
C	C	D	D#	F	G	G#	A#
C#	C#	D#	E	F#	G#	A	B
D	D	E	F	G	A	A#	C
D#	D#	F	F#	G#	A#	B	C#
E	E	F#	G	A	B	C	D
F	F	G	G#	A#	C	C#	D#
F#	F#	G#	A	B	C#	D	E
G	G	A	A#	C	D	D#	F
G#	G#	A#	B	C#	D#	E	F#
A	A	B	C	D	E	F	G
A#	A#	C	C#	D#	F	F#	G#
B	B	C#	D	E	F#	G	A

These charts will most likely prove very useful to you until you start to become familiar with certain keys and scales, some of which are much easier to remember than others. For example, C major has the notes C, D, E, F, G, A, and B in sequence, which, fortunately for us, are all of the white keys on a piano keyboard. So C major is a very easy key (and scale) to remember. Some of you might have also noticed that those same notes (albeit starting on A this time) appear in the scale for A minor as well: A, B, C, D, E, F, G. In fact, each key/scale has a matching counterpart in the other chart. Each major key has what is known as a "relative minor" key and each minor key has a "relative major" key. This one fact alone can be very useful to you as a remixer because in many circumstances switching between the two allows you to change the underlying emotion of the backing track from happy to more melancholy (or vice versa) with comparative ease.

The difference in feeling and emotion between a major and relative minor key might seem a little strange given that they both include exactly the same notes, but the way you put those notes together into chords and how you put those chords together into a pattern determines the overall mood of a song. A song in a minor key tends to have more minor chords, which will certainly give a more "down" feeling to a song than a major key that has chords that are more upbeat and uplifting. We will take a better look at chords in just a moment but, before we move on we should look at a couple of different techniques you can use to actually figure out what key a song or vocal (or other instrumental) melody is using.

Every melody is made up of a series of notes. Once you establish what notes are used in the vocal, you will have an idea (sometimes more definite than others depending on the complexity of the vocal melody) of what key the song is in. I do say "idea" rather than using some kind of "definite" because, if the vocal melody was quite static and didn't move around much melodically then you might find that it only used say four or five different notes. As a result, when listed out, that particular combination of notes could belong to a few different keys.

Let's take an example of an imaginary vocal melody that uses only the notes A, B, C, and E. Looking at the previous charts you can see that several options of keys contain those four notes: C major, G major, A minor, and E minor. If you have the original version of the track that you are remixing available to listen to then you can probably tell from listening to the track as to whether it was intended as a major or minor key track using the *major = happy* while *minor = sad* school of thought.

Not all tracks are that simple though, and even the example we gave (which is far from being out of the ordinary) could lead to a little bit of indecision if you're not careful. Obviously even though the vocal melody only uses four notes we may still have other instrumental or musical parts in the original that would help us figure out what's going on. In our example, the vocal melody notes are only A, B, C, and E, but if there were, for example, a lead synth sound that had a melody that included an F# in it somewhere, that would tell us the original track was in G major (or E minor), which can give us some peace of mind.

However, even if we are 100% sure that the key of our imaginary song was G major, that doesn't mean that it has to stay that way. Because of those more linear vocal melodies it means that, if we're not using anything other than the vocals for our remix, we have the option to change the feel from major to minor (or vice versa) and change the main key of the song from G major (assuming the synth melody that includes an F# as described before) to C major.

Step one, then, should probably be working out the vocal melodies and, as is often the case, there are many ways to do this. One of the easiest is to use the pitch analysis features of Celemony's truly excellent Melodyne software. Once you have your audio file loaded into the software (imported in the Editor version and "recorded" into the Plugin version) the analysis is automatic, and after a very brief wait you should see a series of "blobs" on the screen that look a little like the MIDI notes you see on the "piano roll" type MIDI editor on your digital audio workstation (DAW). From here you can simply scan through the full length of the vocal and note which of the 12 different musical notes are used (the octave is unimportant at this point). Once you are sure you have them all (and assuming there is no awkward key change part-way through the song), all you have to do is take that list of notes and refer to the charts shown previously. Once you find one (or more) keys/scales that contain all of the notes you "extracted" from the vocal melody then you have your key (or choice of possible keys!). So now that you know what key the song is in, what *exactly* can you do with that information?

One of the most useful things you can use this for is as a very simple way of working out possible chords that might fit with the vocal melody. Of course, if you are already at least partially musically trained then most of this will be second nature to you, but if, like many people, your musical knowledge isn't that extensive, it can sometimes be quite hit and miss trying to figure out chords. Earlier, when describing the difference between major and minor keys being mostly one of a different emotion I also mentioned that a song in a minor key is likely to contain a majority of minor chords whereas one in a major key is likely to contain a majority of major chords. Major and minor chords, like major and minor scales, are formed by using a similar pattern of note intervals to the actual scale itself but a much simpler version. Put simply, a major chord is made up of:

Root note > +4 semitones > +3 semitones.

Whereas a minor chord is made up of:

Root note > +3 semitones > +4 semitones.

Again there are 12 different versions of major and minor chords to match up with the 12 different musical notes, so I have included a similar chart that shows different major and minor chords in all of their variations.

Major Chord	Root	+4	+3
C	C	E	G
C#	C#	F	G#
D	D	F#	A
D#	D#	G	A#
E	E	G#	B
F	F	A	C
F#	F#	A#	C#
G	G	B	D
G#	G#	C	D#
A	A	C#	E
A#	A#	D	F
B	B	D#	F#

Minor Chord	Root	+3	+4
C	C	D#	G
C#	C#	E	G#
D	D	F	A
D#	D#	F#	A#
E	E	G	B
F	F	G#	C
F#	F#	A	C#
G	G	A#	D
G#	G#	B	D#
A	A	C	E
A#	A#	C#	F
B	B	D	F#

Once again I do have to point out that there are, in fact, many other types of chords (including minor 7th, major 7th, suspended 4th, diminished, augmented, 9th, 11th, and others for which I have included similar charts on the companion website to this book) that have similarly simple mathematical construction formulae to the major and minor chords we are currently looking at. Some styles of music make quite frequent use of some of the more complex chord types but, as a general rule you can always start from major and minor and then work out the more complicated variations further down the line.

Working out which chords will work in any given key is incredibly easy. All you have to do is list out the notes that appear in the key/scale and then have a look at the previous chord charts and see which of the chords listed can be made up using only notes that appear in the scale. Any of the chords that meet these criteria will work in the key you are looking at. From this point you can start trying different sequences of chords together to create something that starts to flow. It might help to take another look at the actual melody of the vocal, as the notes used in the vocal at any given point will help you figure out which possible chords you could use at that point in time.

For example, if your song is in the key of E minor then you would have figured out that a D major chord will work within the key of E minor but, if the vocal at a particular point is singing a B and you try to play a D major chord underneath that, it won't necessarily sound that great. Technically it isn't *wrong*, but it might not be the effect you had hoped for. In this particular case the result of layering a D major chord (made up of D, F#, and A) with a B is a combined chord of B minor 7th, which, as I said, isn't *wrong* but will possibly not be the "feeling" you were expecting to create with a major chord at that point. If that *is* what you were hoping for then a much better choice would be a G major chord as this already contains the B note of the vocal within the basic notes comprising the chord (G, B, and D).

As you start experimenting with different chord sequences you will soon start to discover certain patterns and combinations of chords that work really well together and sometimes you will find yourself naturally inclined toward using those again. This isn't necessarily a major problem; after all there are only 24 different "basic" chords available to us, so if we were to take a 4-chord pattern there would only be 331,776 possible different chord sequences available. That is an absolute maximum but, if we take into account that in any particular major or minor key there are only six possible major or minor chords we can use, that number decreases to just 1296! The numbers increase if we start including more complex chords, of course, but still, you shouldn't feel too bad if you find yourself using a chord progression that either you have used before or have heard on another record.

It would be virtually impossible to create a chord progression that hasn't already been used (and probably used many times) and, after all, the actual chords you choose are only a part of the overall sound of the remix you will be creating. You will be adding sounds and applying an overall production style over the top of those chords, and this can make a surprising amount of difference to a basic chord progression.

Obviously if you can find different chord progressions that work well on any particular vocal melody (and more often than not there are at least a couple of different progressions that could work) then it can keep things more interesting if you change things around from one remix to the next rather than reusing the same progressions for a few remixes in a row.

I should perhaps mention here the issue of plagiarism. This is basically where you take someone else's creation (music in this case) and then, essentially, copy it in whole or in part. Now that is clearly wrong (both morally and legally), but there is a massive gray area that surrounds this whole topic because people who listen to music for any amount of time will have melodies and chord patterns, or perhaps drumbeats and sounds, that stick in their head long after they last hear the song. At least you would hope that they would, as this is what I, as a producer and songwriter as well as a remixer, hope for every time I work on a track. If you have these other songs stuck in your head then it is quite possible that you are subconsciously influenced by the elements you remember without ever consciously humming the melody to yourself or ever deliberately browsing through presets on a synth specifically *looking* for a particular sound from another record. This is acceptable and, in fact, unavoidable.

It really comes to a question of *imitation versus inspiration*. Most songwriters, producers, or remixers would be flattered that someone thought enough of a track that they have worked on to want to do something similar, but would draw the line if they found out that this same person had taken whole chunks of what they had done and was passing it off as his or her own work. That can be a very difficult line to define let alone avoid crossing and I, like probably most of the other people actively working in the music industry today, have almost certainly crossed the line into "imitation" on at least one occasion but never with the intention of making money at someone else's expense. Most indiscretions of this nature can be forgiven and even forgotten as long as you don't start making a regular habit of it and trying to benefit financially from someone else's work. It does happen, sadly, but not that often.

To finish up this chapter we will take a look at how everything we have seen so far helps us when it comes to working out melodies and harmonies. Once we have worked out the key of the song, we have given ourselves a limited number of notes to work with (excluding, for now, the issue of so-called "blue notes," which are notes that don't technically fit into the key of the song but which, if used as a transition between notes, and only held briefly, can sometimes work) so coming up with melodies becomes that much easier. The most obvious melody in most songs (those with vocals anyway) is the main vocal melody, but that doesn't mean that you, as a remixer, will never need to come up with melodies yourself.

Certain genres of music are more dependent on groove while others are more focused on chords and melodies. If you happen to be working in a genre where it is relevant, creating a memorable melody line can be the make or break of your remix. That melody could take the form of a smooth and flowing "lead" melody or it could be a more arpeggiated melody. In fact, even some bass lines could be considered melody lines in their own right. So clearly knowing where you stand in terms of notes you can use is a good thing.

The same applies in this situation as I mentioned earlier when we were discussing working out chords from a vocal melody (albeit in a situation that was the opposite way round). There may come a time when you have a melody line you're creating that naturally leads you to a note that is in the key correctly but happens to not be in the chord that is underlying at the time. As stated earlier this isn't *wrong* but it can lead to an implied change of the chord that may or may not be acceptable to you. In the end you will probably make a decision more on what feels right rather than whether your choice of melody note turns your innocuous major chord into a "questioning" suspended 4th, a "triumphant" major 7th, or a "soothing" 9th.

Sadly, I cannot give you any advice on writing good melodies as is purely down to inspiration, a good sense of "feel," and, in some cases, moments of pure inspiration. Actually, let's not forget that sometimes it can be bad luck that gives you a great melody. On more than one occasion I have been trying out different melody ideas on my keyboard and have tried to hit a particular note but missed it and hit one I didn't mean to and yet that turned out to be a far better choice. Sometimes it just happens that way. And sometimes it happens that way with working out chords as well. But either way, the more remixes (or songs of your own) that you work on and the more familiar you get with different keys and scales, the more familiar you will be with the chords that go with them.

Eventually you will probably get to a point, even if you aren't a classically trained musician, where you somehow just know what chord progressions will work with a vocal melody and you find yourself moving to the next chord in the pattern without really thinking about it too much on a conscious level. Those are the best times, when things just fall into place like that. But even if it does happen like that, it is certainly worthwhile recording that in as one option and then trying some other chords out instead—perhaps ones where it is more about a conscious thought process rather than a subconscious one. After all (assuming that you make sure you save the file once you recorded your first idea in, which of course you did… right?), what do you have to lose?

Harmony is another area where this small amount of musical theory can help us a great deal. This is, perhaps, less useful to you as a remixer than it would be if you were writing your own material, but there may be occasions when it could be useful, so I will just run through it quickly here. The simplest way to work out potential harmony lines that would work against a melody line is to look at the note the main melody is on, look at the chord underlying it, and then choose a different note from within the chord. This will work, but the risk here is that you might end up with a harmony part that isn't especially fluid.

To me, and this might be purely my own personal opinion, the crucial test for any harmony part I create is that it has to work and feel right as a melody in its own right. What I mean by that is if I were to mute the main lead melody and play only the harmony line, would that harmony "melody" sound fluid and natural or disjointed and uncomfortable? If it did sound uncomfortable then I would unmute the main melody, try moving a couple of the notes of the harmony part around, within the confines of the notes that are in key, and then try muting the lead melody again until I found something that worked better on its own. Of course, you don't have to do this. "Block harmonies," which simply build upon the main lead melody notes and build them into the full chords of the underlying music, can certainly work, although sometimes they can be a bit overwhelming. Still, you have the choice.

Hopefully what we have covered here will make the process of working out the chords for a remix a little easier and certainly a little less hit and miss. Like so many things in the world of the remixer, anything that will save you time will only help you. If you are so inclined, there is much more to learn about musical theory. While it isn't strictly necessary to have a more in-depth knowledge it certainly won't do you any harm. For example, it will make it easier to deal with songs that are a little less obvious in terms of alternative chord progressions, ones that don't seem to have one clearly apparent key, or contain lots of notes that seem to contradict one another in terms of which key the track might be in.

The other alternative would be to find someone to collaborate with who is musically trained (even to some extent), or perhaps even a session musician who simply works out some chords and perhaps a bass line for you and then hands the raw MIDI data over to you to then produce. This might be a quicker and easier alternative, but I honestly don't think you get the same sense of achievement this way. There is certainly something to be said for the buzz you get when you have worked out a really good chord progression for a particularly difficult vocal melody.

CHAPTER 5

Tempo, Groove, and Feel

To some extent the tempo of your remix will already have been at least partially determined by its genre. Most genres of dance music have a limited tempo range considered appropriate and if you intend to step outside of that range you run the risk of the remix not being accepted in that "scene." There is nothing to stop you, for example, doing a remix that has all of the production values of a drum & bass track but is only 120 bpm instead of the more usual 150 bpm or more. If you do that, it probably won't be considered drum & bass anymore.

With that in mind, there is normally at least a 10 bpm range for most genres so you still have some degree of choice as to the final tempo. One of the biggest influencing factors in this decision is, as always, the original song. For any genre (and tempo range) there is what I think of as a problem zone in which the tempo of the original track could lie. If, for example, you were working on a house remix then you would probably be looking at anywhere between 120 and 130 bpm as the final tempo. If you took 125 bpm as a target tempo then the problem zone would be anywhere around 93.75 bpm. Tracks with a tempo in this area are problematic because you would have to make a decision whether to speed the vocal up from 93.75 bpm to 125 bpm or slow it down to 62.5 bpm, neither of which is a particularly compelling choice because that would represent a time-stretch of about 33% in one direction or another. What often happens when you are dealing with time-stretching vocals by that amount is that, in one case (speeding the vocal up), you get a hugely exaggerated vibrato—"the chipmunk effect"—or, in the other case (slowing the vocal down) you could well be left with vibrato that sounds very inaccurate and lazy or a vocal that just sounds slurred.

We will look at the specifics of how to deal with this particular issue in a later chapter but, for now, let's look at intelligibility. Even if we can fix the vibrato issues, we still have to deal with the intelligibility of the words. If the original track is quite bouncy, even though it is at a tempo substantially lower than your target tempo, you could run into problems if you try to speed the vocal up because there might be too many words in too short a space of time and a listener won't be able to understand what's being said. In most cases this won't be acceptable to the original artist so you should try to avoid it. Songs like these that feel like they have quite a lot of pace even though their tempo might be slower are better candidates for being slowed down instead. You run the risk of losing a lot of the feel and energy of the original track, but at least it won't sound languid and slurred.

There will be some tracks that are much slower than your target tempo and very slow melodically and lyrically at their original tempo. Songs like these sound horrible if slowed down because the words become unintelligible where they take so long to be said. There is also the risk that the complete lack of energy in the vocal performance at a slower tempo would drain all the energy from the music and bring the whole track down. Obviously songs are better candidates for being sped up (technical and vibrato issues notwithstanding) and remaining intelligible. In fact, sometimes songs take on a whole new dimension when sped up in this way. In some cases on remixes I have worked on in the past, when the lyrics weren't completely at odds with the new feel, I have actually felt that the song sounded *better* once it had been sped up!

There was one remix I worked on where both myself and my colleague felt, after trying both options, that we simply couldn't use a slowed-down version of the vocal because it lacked any kind of feeling, passion, and energy, so we opted to use the sped-up vocal. That wasn't without its problems and it took us quite a while to get the vocal sounding anything even close to usable due to some heavy vibrato issues and dealing with what felt like too many words in a few places. But in the end we were happy enough with the sound and feel of the vocal. The record company, however, did not agree and the remix we had worked on, along with all the others that had been commissioned (which all chose to use a sped-up vocal), were rejected and a totally new set of remixes commissioned with a stipulation to the new remixers that they *must* use the slowed down version. Later I was able to compare the new remixes against some of the earlier ones that had used the sped-up vocal and they really sounded dreadful. There was nothing wrong with the production value of the new remixes, but the slowed-down vocal sucked the life out of the song. But those were the remixes the record label decided to go with. I chalked that one up to experience but to this day I still feel we made the right decision.

Songs that exist in this problem zone won't always present you with problems when it comes to choosing a specific tempo to work out. The reason for this is that, if you have an original track at, say, 98 bpm and you're going for a target tempo of 128 bpm, you would have to time-stretch the original vocal by 30 bpm, which is 30.6%. The vocal might feel rushed at this tempo and you may think that slowing down the target tempo a few beats per minute to 125 bpm might help, but in reality the vocal will still be being sped up by 27 bpm (a change of 27.5%). Although that isn't stretching it by as much, it's not like you are only stretching the vocal by 10% instead of 30%, which would make a more noticeable difference. That said, it might just about be enough to help you with a difficult vocal. Sometimes only the subtlest of changes are needed to cross the border from "not quite" to "just enough."

Not all tracks will have those tempo issues. Some will already be in the range you want as your target, and with these you have the most flexibility. The further you get away from your target tempo, the less flexibility you will have (up to a point). There really is no hard and fast rule though, and you really do have to judge each track on its own merit and not assume that just because the last track you remixed that had a similar tempo ended up sounding better sped up, that the next one will be the same.

The introduction of the Varispeed mode in Logic 9 made trying out different options a whole lot easier. Time-stretching isn't a particularly quick process, even on a powerful computer, so if you wanted to audition the vocal at 125, 126, 127, and 128 bpm it could take a while to time-stretch the original vocal four separate times (even longer if there are a few tracks of vocals, such as with backing vocals). The Varispeed mode in Logic adjusts the audio files more or less instantaneously, so you can quickly hear tempo changes and listen how they sound in your track. The tradeoff is the sound quality, but it is certainly good enough to give you an idea of what to expect. Using this mode allows you to try things and discover if, for example, you'll have problems with too many words in a short space of time or if the words are going to be way too drawn out to make sense. The audio can sound a little bit clunky or even a little "phasey" in places, so it's not a good function for listening to subtleties of the vocal performance. Nevertheless, the Varispeed feature can be a real time-saver for a quick and easy judgment of whether to speed up or slow down, and to determine which end of the tempo spectrum for your genre you should be aiming for.

Next we consider the groove—of both the original track and the production you are aiming for. Groove describes the feel, or timing. At one extreme you have what's called a "straight" groove, which is generally equivalent to a straight 16th-note quantization (or possibly even 8th-note), and as you travel along the spectrum there is gradually more and more a sense of "shuffle" or "swing," which corresponds to, progressively, 16A, 16B, 16C (and so on) note quantization.

In general, all of the elements in your track, particularly drums and percussion, need to be at least similar if not actually the same type of groove. For example, while rhythmic elements with grooves of 16, 16A, and 16B might work together, it is unlikely that those with grooves of 16, 16D, and 16F would work particularly well. The same holds true for all the other elements in the track that have a strong rhythmic element (in other words everything except sustained chords, pad sounds, or any sound with a long attack), in that they don't all need to be precisely the same or even need to be quantized at all (assuming that the playing was fairly tight in its timing), but there does need to be some similarity. Vocals definitely fall into the category of having a strong rhythmic element.

So what do you do if the vocal you're remixing has quite a strong swing feel to it but you want to make the music more of a straight groove? Well, there are a few options, some easier to implement than others and some more successful than others. All the options work on the same principle, which is, essentially, to *quantize* the vocal recordings. Two of the methods described here are manual and quite time-consuming, but give more predictable and arguably better results. A third method (which uses Logic 9's wonderful Flex Time feature set) *can* sometimes work well with little more than a few mouse clicks in your DAW but doesn't have quite the predictably or, in some cases, precision of the other two methods.

Let's start with the simplest method of the three, which involves using Flex Time in Logic 9 to analyze the vocal track(s) to try to determine where the transient points are in the vocal recordings. These transient points should, in theory, come at the beginning of most words as there should be a pause followed by a word starting. The difficulty is that sung words very often don't have a clear and sharp attack and very often singers tend to run words together so there might not be a clear gap (even a short one) between every pair of words. When this happens sometimes the transient markers are placed in the wrong place (perhaps a little too early or late) or perhaps they are missing completely. You can start adding markers in manually, of course. But first we'll describe the automated process.

Once the transient markers have been placed on the file you can then treat the audio in the same way as MIDI data in the sense that it can be quantized by using a menu command. This allows you to quickly change the feel of the vocal from 16 through 16F and all the way back again. The changes to the quantization happen more or less instantly, although it may take a few seconds to kick in depending on the power of your computer, the load on the CPU from plugins and audio instruments, and the length of the file you are trying to quantize. Whichever way you look at it, it is certainly much quicker than either of the manual methods we consider later. I often try using this method first because, since when it does work as you want it to it can take just two or three minutes as opposed to over an hour to do "manually."

FIGURE 5.1
Here you can see the power of Flex Time in Apple's Logic 9. The top audio file is an untreated drum loop and the bottom file has had Flex Time used to remove the "swing."

The second option again uses the Flex Time features and builds upon the automatic method described earlier. You can still start the same way by letting Logic place the transient markers, and then you can go in and move, add, or delete additional markers to help tidy up the quantizing process. One thing to consider is that many vocalists ease into certain words, so what may look from the waveform view to be the start of the word, might not be the part you want to be 100% on the groove. A little experimentation might be necessary to really get the markers in the right place. Once the markers *are* in the right place you have all the same options as you did on the automated process as described previously, but this time there is a better chance that the newly requantized vocal will be nice and tight on the timing.

The downside of these first two methods is that the process of moving the transient markers around causes the Flex Time algorithm to stretch or compress the sections of audio between them to make sure that everything fits and flows. While the quality of the algorithm is undoubtedly very good, and the amount of stretching or compressing is generally very small, it does still represent a further stage of processing for an audio file that may already have undergone at least one and possibly more stages of manipulation. Every time we process the original vocal files (through time-stretching or pitch processing/tuning) we are gradually degrading the sound. The lead vocal is such an important part of a remix, you want to aim for the least amount of processing and a minimum number of "passes." Even if you are going to have quite a heavily effected vocal sound in the end, it is still good practice to try to minimize processing.

To this end there is a third method that might be a better option but does take a little longer to work through. In fact, the basic principle is exactly the same in that what we aim to do is split the original file into "regions" at the points where we feel the words should be locked to a particular groove. Again, the point we should be quantizing might not be the exact beginning of a word. There could be a lead in to a word, perhaps a hint of an "S" sound, which needs to sit a fraction ahead of the beat so that actual point we need to quantize is just after the actual beginning. There is no hard and fast rule here and you might find that you need to make a few adjustments as you go along, but once you have made your best guess as to where to split the file you then need to quantize those vocals. The following describes how to do this using Apple's Logic 9. If you're using a different DAW, the process should be similar; consult your user manual or help files for specifics.

In Logic 9 there is an option to quantize audio files in the same way you would quantize MIDI information, although the method is slightly more convoluted. You first have to select all of the audio regions you want to apply the quantize to. Then open the Lists panel to the right of the main Arrange window and select the Event tab. You should see all of the audio regions you have selected showing up as individual events that are highlighted within the list. If you don't then you might need to go back to the Arrange window and make sure the audio track that contains the region you want to quantize is the currently selected one and that the specific regions you want are selected; then checking that those regions (and only those regions) are highlighted in the list. Once you have made sure that this is the case, then just above the top of that list you will see there are options for quantizing the events in much the same ways as you can for MIDI data. You simply select the groove you wish to apply and the effects are instant. You can then start playback to hear the effects of your choice. It makes sense to have a part of the track that contains the regions you are quantizing set to loop so that you can leave the track playing while you try different options.

At this point you might well hear some regions (words in this case) that are sitting in the groove nicely, but others might be a little too early or late because you didn't get the start point set quite correctly. If this is the case you can go back into the Arrange window and move the start point of the region a little earlier or later and then try reapplying the quantization to just that one region until it is right. Alternatively, you can zoom in and move the region around manually until it fits. In fact, sometimes, a very rigid quantization of vocal parts doesn't sound all that good. Even if you allow for the inaccuracies of setting your start points for the regions, that does, inevitably, give the vocal timing a slightly more relaxed feeling that a perfectly accurately quantized synth line would. You might still want to just nudge the timing of some of the vocal parts to push them ever so slightly before or after the exact beat.

At this point you may have a couple of issues with the vocal that need sorting out. If you have, indeed, cut the original vocals into regions at the beginning of every word (or every point in a word where there is an obvious change of note or other event that needs timing correction) then you could find your vocal now has a number of little gaps or overlaps. The Flex Time algorithm described earlier will kick in here and stretch (to fill gaps) or compress (to remove overlaps) the vocals.

Given that we want the minimum amount of processing possible we will look at alternatives. The easiest option is, of course, if there is a natural gap between words anyway because if we changed the length of that gap by a tiny amount (which it would be if we are talking about a change in the groove of a vocal) then the overall effect shouldn't be affected. If, however, the words originally ran into one another and there is now a tiny gap, then what I would normally do in this circumstance is drag out the end of the first region or drag back the beginning of the second region to cover that gap. Then I would zoom in as far as possible to make sure that the point at which one file ends and the other starts is at a *zero crossing point* for both regions. In order for this to happen it may be necessary to shift one or other of the regions

A zero crossing point is a point at which the waveform crosses the horizontal center line in the waveform display. More accurately it is the point at which the amplitude of the waveform is exactly 0dB or, in physical terms, the point in the waveform where the speaker would be at its "neutral" center position.

by a tiny amount. Given the small amount we would need to shift it by (only a few milliseconds usually) there shouldn't be any noticeable change in the groove. If you're lucky that's all you need to do. Now solo the vocal to check that the transition is smooth, apply a very short *crossfade* between the two regions if it needs a little help, and then you're are done.

A crossfade is a technique used to blend two separate audio files (or regions). Where two files overlap a fade out is applied to the preceeding file while simultaneously applying a fade in to the following file. This semi-automatic process is called crossfading.

Another scenario is where the words originally ran into each other but instead of there being a tiny gap now, there is a tiny overlap instead. In this case I would again recommend zooming in, moving either the end of the first region or the start of the second region until there are no overlaps, and then making sure that the changeover point is again at a zero crossing point for both regions (using a short crossfade again if necessary).

Because we are talking about very small differences in timing between a 16 and a 16A or 16B groove, we can probably get away with taking care of the potential gaps or overlaps in the way described previously. If you were requantizing from 16 to 16F, for example, then there might be one or two occasions when the gaps that need to be filled would be best served by a very small amount of time-stretch. If we do things this way, we can time-stretch as few regions as possible and leave the vast majority unstretched.

There are a great number of groove variations in music but most fall into a variation upon a straight 4/4 beat with a 16th note subgroove. If what you have to work with falls somewhere along that spectrum then the chances are you will be able to regroove or requantize the vocal quite successfully using one of the three methods described earlier. There is, however, one group of vocals that could cause immense problems: those that come from songs built around triplet grooves. You will know these instantly by feel if not in an actual technical way. If we think about a typical 16th-note–based groove and we emphasize the main kick drum beat we get:

One...two...three...four...*One*...two...three...four...etc.

Whereas with a triplet-based groove that would change to:

One...two...three...*One*...two...three...etc.

If we try to put a vocal that has a triplet feel over the top of an 8th- or 16th-note–based groove the results could be very uncomfortable sounding. I say "could be" because if the original song has a heavy triplet feel but the vocal rhythm and emphasis mainly follow the "one" beats with occasional words lying on the "two" or "three" then you might get away with it with a little bit of requantizing. But if the vocal has a lot of words in it, then chances are it will make much more use of the "twos" and "threes." If that happens, you have some difficult decisions and some serious work to do because a rhythm of threes does not sit comfortably over a rhythm of fours for any decent amount of time. For a couple of beats, maybe even a bar or two, you might be able to get away with it, but for a whole song, it would be unlistenable!

So what do you do in a situation like that? One option is, of course, to change your music to a triplet-based groove, if it works with your chosen genre. Very few dance music tracks are triplet based so you could be taking a risk if you decided to go that route. Plus it is actually quite difficult to change a groove from a straight 4/4 to a triplet groove because the rhythms you would normally use for your musical parts (bass lines, arpeggios, etc.) no longer work and literally everything has to change.

I have worked on over 250 remixes in my career and I think that only two have used triplet-based grooves in over 10 years! It's not that I didn't have more than two triplet-based tracks in that timeframe, but rather that the genres I was working in were not as open to such a radical change in the groove and I felt that the tracks would have been dismissed as being unplayable by club DJs. Remember that a DJ has to transition from one record to the next, so if your track has a triplet based groove and the DJ has to mix that with a track that has a regular 4/4 groove then he'll find himself in "train wreck" territory. No DJs want to sound like they have messed up the mix, so in that situation they might just avoid your track altogether.

You could always turn down the remix if you didn't want to do a remix of a triplet-based track. But if you like a challenge, read on…

In order for the triplet-based vocal to work in a 4/4-based club groove we need to do some *serious* requantizing. The techniques you can use are exactly the same as we described earlier but taken to a more extreme level and accompanied with some creative decisions along the way. The biggest of those creative decisions lies in you sometimes having to decide whether to push a note forward or pull it back in time. Unlike changing a 16A to a 16D where all of the notes remain in more or less the same position but with just a subtle change in swing, when you requantize a triplet-based vocal into a 16A groove there will be times when the natural timing of the original vocal sits somewhere between two 16th-note timing divisions.

To illustrate this let's consider a track running at 120 bpm. At this tempo each beat lasts exactly half a second because 120 beats per minute means two beats per second. This also means that each 16th note (1/4 of 1 beat) lasts 125 milliseconds. With a triplet-based groove, however, we will only have three subdivisions per beat so instead of each lasting 125 milliseconds it will last 166 milliseconds. Although the 16th-note subdivision at 125 milliseconds is closer in this case than the one at 250 milliseconds, it is still 41 milliseconds out and we can easily hear a delay of that amount.

The decision as to whether to move the vocal earlier in time to the 125 millisecond point or later in time to the 250 millisecond point will probably become much clearer when you actually try both options. Even though shifting it back to the 125 millisecond point is technically closer to where it originally was, the phrasing of the vocal and the length of the words before and after a particular word will have an effect on your decision here. You do have to make a decision though and, in many cases, whichever decision you make the vocal will still sound strange because of the complete change in groove. It doesn't take long for our ear and our memory to become accustomed to things, so after only a few listens through to the original vocal, the triplet nature of it should become ingrained in your memory and any large deviation from that will sound a little unnatural.

Because of this there is also a slight risk that the artist or record label who commissioned the remix could turn around and say that they don't like the remix because you have completely changed the groove and feel of the song. If this happens then you really do have to stick to your guns and tell them that you couldn't do it any other way, that this is your sound, and that if they aren't happy then there is very little you can do about it. Sometimes people do want the impossible or, at least, the *extremely-highly-improbable-given-the-time-constraints*, and in those situations you have to make sure you don't feel like you have somehow failed when you were presented with a track that was problematic from the start.

Sometimes I am asked if a track is remixable. Inevitably my answer is "yes." But then I have to ask the question back, "Why do you actually *want* a club mix of this track?" If you had a 96 bpm, triplet-based acoustic guitar and vocal, folk-style ballad and wanted a 128 bpm electro house mix I would have to ask the question of what possible good that house mix will do for that record and why the record label or artist even *wants* to promote that kind of record to this kind of market. But sometimes they just do. So if the opportunity comes along to do something like that, have a crack at it anyway. It will certainly push the limits of your creative and technical skills and you never know, you might just come up with something pretty special along the way.

To summarize, changes in tempo are almost inevitable when dealing with remixes. There will be occasions when the tempo of the track you are remixing is exactly the same as the tempo you would like to use for the remix, but that rarely happens. Changes of a few beats per minute in tempo normally pass without really being noticed, but when you are getting to changes of five or more beats per minute the effects will start to be noticeable. You have to respect what the label or artist wants, but at the same time, they need to understand what it is they are asking you to do. Even with a lot of work and using all the tools and tricks at my disposal there have been a couple of times when, upon receiving the finished remix, the client has said "We really like the remix but the vocal sounds a little, well, time-stretched." At this point, after I have rolled my eyes a little, I simply reply that it was always going to as you cannot take a 90-something beats-per-minute track and make a 120-something beats-per-minute version of it *without* it sounding time-stretched to some extent. Most labels and artists realize this but you do get the odd time that you will get pulled up on it. There is nothing you can do about it, so honesty is the best policy here. Just tell them that you did all you could with the vocal and there is nothing more you can do to make it sound any better.

Changes in groove might not necessarily have to happen but they can be a good way to make the remix sound a little different from the original even if the tempo is similar. They can also change the feel of the track slightly, so having the knowledge of how to do it is certainly handy and definitely gives you more options when thinking about the production style you want to use.

And finally, if a vocal is going to be hugely problematic then always remember that you *can* just say "no." While nobody likes to turn down work, turning down the odd job because you honestly don't feel you can do a good job with it will prevent you from putting out a remix that is below standard. In many ways your reputation as a remixer is only as good as your last remix, but in some ways it is also true that your reputation is only as good as your worst remix. If you release one that really isn't that good people will remember it for quite a while and you could find it damaging to your profile.

In that instance you have to question whether the fee you would get for taking on that difficult remix is worth the potential repercussions of you not doing a great job. And it is probably better to not take it on at all rather than take it on, give the client the assumption that you will be delivering the mix according to their impossibly tight deadlines, and then phone them up, a day or two before they are due to have the whole remix package complete, and tell them you won't be delivering a remix because you don't think you can do the track justice. Taking on board the points and issues raised in this chapter will, as you get more experience, put you in a better position to decide whether you will be able to do a good job just from having a few listens to the original track and will hopefully help you avoid that difficult decision between putting out a remix that isn't great or upsetting a potentially very valuable client by letting them down at very short notice.

CHAPTER 6

Arrangement

Any discussion about dance music arrangement will, inevitably, have to be quite general because arrangements can vary so much between genres, producers/remixers, and even different tracks by the same person. The process of arranging a remix involves building from nothing (at the beginning of the track) to a peak (where all, or certainly most, of the elements are playing) and then back to nothing again. Exactly how you achieve this will depend largely on what elements you are working with and how many of them you have available to build with. More minimal styles will probably use much slower and more drawn-out build-ups, with each stage of the build-up being a subtle change. Whereas a genre with a more dense production style would likely bring in elements more quickly and obviously, since there are more of them to work with.

Even with all this diversity there are still some things that are considered "standard practice" or common sense. We will take a look at those throughout the rest of this chapter and explain why and how you would put them into practice. First we need to consider the different kinds of alternate arrangements you might be asked to provide as a part of your remix. These are the club mix (sometimes called an extended mix), dub mix, radio edit, "mixshow" edit, TV edit, PA mix, and instrumental. Of these the ones you will nearly always be asked to provide are the club mix and the radio edit. We will look at these two first and in the most detail and then move on to describe the function and general form of the others after.

CLUB MIX

The club mix is traditionally where the main focus of a remix would lie but, in recent times, a growing number of remixes have been commissioned primarily to provide an "alternate" radio edit. If this is the case there is normally still a club mix provided anyway as a matter of course. The club mix is probably still the main purpose of a large majority of remixes. The basic framework of most club mixes is fairly straightforward. There aren't too many "rules" as such, but there are a number of things considered "normal" and which, if deviated from too much, may lead a club DJ to avoid playing the remix.

Here's an overview of the basic structure of a club mix of a "traditional" song, followed by a more detailed look at each of the parts.

- DJ-friendly intro
- Small "breakdown" before leading into first verse
- First verse/bridge/chorus
- Short instrumental section before second verse
- Second verse/bridge/chorus
- Main breakdown
- Last chorus (possibly double length)
- DJ-friendly outro

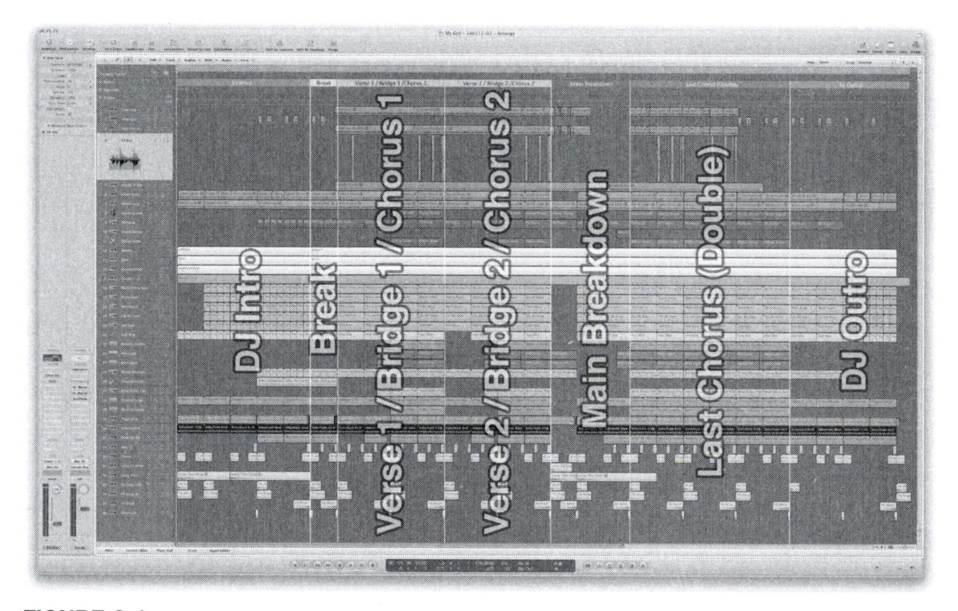

FIGURE 6.1
Here is the arrangement of a club mix I worked on recently. I have highlighted the different sections of the arrangement for you to see.

I realize that you are probably looking at that and thinking that it seems very formulaic and, to some extent, you would be right. But you have to remember that most songs are quite formulaic even before they arrive at the remixer. In fact, if you were to remove the first, second, fourth, sixth, and last parts of the previous list you basically have the "traditional" arrangement for most songs. Remixes are often quite formulaic in their arrangement as a direct consequence of the formulaic nature of the songs being remixed.

There is, however, still quite a lot of scope for variation within that basic framework. Of course, there is no reason why you *have* to stick to that structure; the song you are remixing might not have exactly that structure anyway. There may just be a simple vocal hook rather than a full verse/bridge/chorus. You might even be remixing an instrumental track, in which case I tend to view the main lead riff/melody as a vocal replacement and treat it as if it were a vocal chorus, placing it in the arrangement accordingly. Pretty much everything is open to variation and interpretation depending on the song you have been asked to remix. The two things that are pretty much a necessity are the DJ-friendly intro and outro. While you might have an idea for a great way to start the remix off or end it, if it isn't DJ-friendly (which normally means starting and ending with beats and very little else) then many DJs will shy away from playing it because it will be difficult or impossible to mix with other tracks. And if you're not getting support from DJs then a major reason for doing the remix has pretty much disappeared.

There is always the option, if you have the time and the inclination, to do an alternative club mix with a longer or more unusual intro that could perhaps be used for compilations or other situations where it wouldn't be mixed live by a DJ. There are a few examples of remixes I can think of with intros that wouldn't really be considered DJ-friendly. The most obvious one that springs to mind is the DJ Tiësto "In Search of Sunrise" remix of "Silence" by Delirium featuring Sarah McLachlan. This is very much an epic journey of a remix that starts off with little more than a sustained pad sound with no discernible beat for the DJ to use to mix in from the previous track. The beat comes in at around 1:30 into the song, so my guess is that some DJs who played that remix would simply start mixing the track in from that point. Others might allow the previous track to play out completely and then fade the intro pads of this remix in underneath the last eight bars or so of the previous track and then allow this one to play through from pretty much the beginning. The main danger of doing that as a DJ is you lose the energy levels of your set, which you don't want to do too often.

An important consideration here is the *8-bar rule*. In dance music nearly everything tends to work best when viewed in sections of 8 bars. Sometimes these will be doubled up into sections of 16 bars depending on the track you are working on and the genre. Radio edits are perhaps excluded from this rule in certain aspects but, largely, it even applies there. The reason for this is that dance music is about getting people dancing (obviously) so it is fair to say that you don't want to be challenging the people in the club to keep up with what you are doing. Repetition is a very important part of music as it helps us to learn. Certain aspects of a track get repeated and people become familiar with them. As they become familiar with them we can build further elements into the track until they then become accustomed to them. Getting people to feel comfortable with the track they are listening to is very important if they are going to settle into it enough to dance to it. Now I don't know the psychological reasons for it, if indeed there are any, but people don't tend to like odd numbers of bars. If you were to have a cycle that repeated every 7 bars, for example, it would probably throw an awful lot of people off their rhythm. While you might consider yourself *avant-garde* and musically adventurous and while that may gain you the respect of your peers, it would almost certainly consign your remixes to "back of the box" status.

So why 8 bars instead of 4 or 2 or 16? Well I guess that it just comes down to the fact that 2 bars, and even 4, aren't really a long enough period of time for our brains to become familiar with something. And along the same lines, 16 bars can be quite a long time for there to be no change in what is going on. Twelve bars could possibly work but I find that 12 bars just doesn't feel right. Like I said, that is my little "rule" that tends to apply to quite a lot of dance music so have a listen yourself and you will see what I mean. If you choose to apply that rule to your own music then great, but if you don't and you are happy to throw in some different lengths and you feel that it works then that's great too. Just don't go crazy with it!

I should mention, though, that, because of everything I have said, throwing in an occasional break of 1 or maybe 2 bars can, if done well and in the right place, bring out a heightened sense of awareness about a particular part of a track. I have heard a number of tracks and remixes that use this technique extremely effectively by including a 1-bar drop to nothing (barring the "tails" of reverb and delay effects) at the end of a breakdown. This total drop for just 1 bar actually increases the impact of the track when it slams back in a second or two later. But in general that is the kind of thing you can only really use once in a track. I wouldn't recommend throwing that effect in randomly. If you decide to use it, make it count and go for maximum impact with it.

The DJ-friendly intro (and outro) is important for a variety of reasons. Any club DJ is looking to physically mix the tracks into each other and choose the right ones to play. After all, DJing is part art and part science. The science of mixing songs, getting them beat-matched, and learning smooth ways to change from one track to another is in many respects the easier part of DJing. Many people, with the right equipment to practice on, and some time actually spent practicing, will easily be able to master this side of it. But the true skill of a DJ comes in a live club environment with a set that takes the crowd on a journey, picks up where the last DJ left off, and, through the course of two or more hours, increases the energy level and sense of excitement inside the club; all of this while reacting quickly and correctly to the likes and dislikes of the club patrons. With all of this to deal with, the last thing a DJ wants is to make his or her job more difficult by trying to mix into a track that will cause problems. Things that would be considered problematic in this context are tracks that don't have a clear timing reference at the beginning or start off in some other strange way; those that don't have a steady tempo during the intro or outro; those that bring everything in too quickly so that the DJ doesn't have time to mix out of the last track fully before the next one is fully into its stride; and pretty much anything that strays too far from the "norm."

One thing I struggle with from time to time when I am mapping out arrangements (for club mixes in particular) is that the intro isn't too long or boring or just lifeless. The thing to remember if you find yourself in this situation is that probably the only people who will hear your intro arrangement in the way you do in your studio are you and the DJ. For the most part the people in the club won't really be fully aware of your track until it is at least 45 seconds to a minute into it. The 8 bars you have at the very beginning of your remix with nothing but a kick drum aren't really going to be an issue in the club because nobody will actually hear it that way. While your 8-bar kick drum intro is playing they will still probably be focused on the outro of the previous track, that is, if they could even hear your kick drum at that point.

Most club mixes (or extended mixes) don't really get into their stride until at least one minute or possibly even longer and will normally have at least one minute of outro as well (again, possibly longer). These two one-minute periods are the sections of the track a DJ will overlap and use to *crossfade* between the two tracks: while the outro of one track is gradually dropping elements out as it drops down to nothing, the intro of the next track will be gradually bringing elements in. With these two processes happening simultaneously, the result to the listener in the club is a smooth transition from one track to the next with a fairly consistent dynamic level.

A fairly typical DJ-friendly intro will start with some kind of drum or percussion part. This is often just a kick drum because it provides a very clear and solid timing reference for the DJ to mix the track with. Something more offbeat, syncopated, or just less regular isn't as simple a *cue point* for the DJ to work to. This will probably run for 8 or maybe 16 bars and then perhaps an additional percussive element will join it, either by starting at an appropriate point or by being faded or filtered in over a period of time. At around 16 bars into the arrangement you would generally start introducing some musical elements as well as, perhaps, some additional percussion.

During the intro phase of most remixes, any musical elements that are brought in are introduced in either a filtered form or perhaps even a simplified form. For example, if the first musical element you wanted to introduce was a bass line then you could perhaps bring it in heavily filtered (either low-pass or high-pass filtered) and gradually open the filter up to bring in the full sound. Another option would be to bring in the full sound but with a simplified version of the main pattern, possibly with the same groove as the main bass line but just sticking to a single note rather than the pattern moving around to follow the chord progression in the song. It could also be a combination of both of these. The same thinking could easily be applied to whichever musical element you chose to bring in first. It isn't hugely important what you bring in first out of the musical elements. Bass lines and pads/chords are generally good choices because they can be blended in nicely and aren't always as readily identifiable as a lead riff; you don't want to give away the musical hook of the track as early as the intro.

Whichever element you have brought in first, this element will generally evolve—either through filtering, volume fades, or increasing complexity in the pattern used—over the next 16 bars, during which time you might start to bring in perhaps another one or two musical elements and more percussion. This will bring you up to the 32-bar mark, which is around 50 seconds to 1:10 into the track. By this time you could potentially start the main body of the track. In general, however, you would probably continue to build the intro for another 16 bars to take the track to around 1:15 to 1:45 (depending on tempo).

During this last part of the intro you might choose to introduce a few vocals as well (assuming, of course, that the song you are remixing has vocals), hopefully without giving away the main vocal hook too soon. You can often snip out individual lines or even parts of lines from later in the song and take them out of context by perhaps changing the timing position of the vocal so that it starts in a different place in the bar, or perhaps by repeating that one line over and over. Or you could think of other ways to avoid giving away the hook too early, for example, by giving the vocal a different sound with effects that create a more heavily processed sound, or by letting it sit further back in the mix. If you do this, you can almost make it sound like a completely new vocal line that only occurs in the intro (and quite possibly the outro as well) of the track.

This now brings us up to 48 bars into the track, which is, I think, a good place to bring in the main body of the song. Any longer than this and you risk things becoming a little boring. The DJ would almost certainly have finished her mix between the two tracks by this point and so anything after this would just be dragging things out for the listener. Of course, if you have an especially long intro, as I mentioned earlier, the DJ might simply choose to start his mix further into your track. So even if your intro is a little too long, the listeners will arrive at the critical point at the right time, that is, not too long after the DJ has finished mixing out the last song.

The next section is the small "breakdown" before, or leading into, the first verse. This can be quite genre-dependent, as some genres tend to use this technique quite regularly and others barely at all. It can serve a very useful purpose though, which becomes apparent a little later. With any club mix you need to build the dynamics of the track as it progresses. However, if you try to keep building and building and building from the very beginning of the track then you will quickly find you have used up everything you have available to you and have nowhere left to go before you're even halfway into the track! Obviously this isn't a good position to find yourself in, and this is why a small breakdown before the first verse can be good. By taking the dynamic level (or energy level) of the music down a little before going into the first verse you have given yourself more scope to build it again throughout the verse and bridge (if there is one) before going into the chorus. This has the effect of making the chorus sound bigger than it actually is. The human brain and ear combination adapts to what it hears very quickly so a drop in dynamic levels at the beginning of a verse will have all of these positive effects, even if the level starts to build again quite quickly.

This breakdown doesn't need to be as dramatic as the main breakdown later in the mix and could be as simple as just dropping back to nothing more than a kick drum and bass line, or perhaps keeping everything running as it was but removing the kick drum. It could feasibly be anything at all that drops the level of energy in this section of the track. From experience, I would say that the most effective things you can do are to remove the kick drum, the bass line, or both, or, strangely enough, to do exactly the opposite and remove everything *but* the kick drum, bass line, or both. Obviously you will be more familiar with the specifics of how these things work in the genre you're working in, but, hopefully, whatever those specifics are, you can see the reasons why you would possibly have a small breakdown at this point in the track.

Moving on to the first verse, bridge, and chorus sections is much simpler as, to some extent at least, the arrangement here is already taken care of for you. All you have to decide is exactly how much and how quickly to build things leading into the chorus. In some songs—and certainly the more commercial and "poppy" songs—the verse and the chorus have different chord progressions so this, in itself, will make the chorus feel like a "lift" from the verse. More recently, however, there has been an increase in the number of songs that use the same chord progression throughout. In these, the chorus is only really differentiated by a change in vocal melody and perhaps a different musical hook that only comes in the chorus, or the introduction of more complex percussion elements and other musical subtleties.

One possible area where a club mix can be a little more creative in the arrangement is in the introduction of an instrumental section, based on the chorus chord progression, either at the beginning or the end of the chorus. If you choose to have an instrumental section at the start of the chorus then that can help to provide a continued build throughout the chorus by first taking the changes in the chord progression as one "step-up" and then using the reintroduction of the vocal as a second step-up. The only risk with this technique is that the flow of the original song can be lost, especially in songs that have a vocal line that runs smoothly from the end of the bridge (or verse if there is no bridge) into the chorus. In cases like these it can be hard to find a way to end the line leading into the chorus properly and, similarly, it can be difficult to then lead back into the chorus vocal when it does return. If you wanted to do this, perhaps you could add a delay effect to the end of the last line of the verse/bridge, which would carry the vocal over into the instrumental chorus section and then possibly repeat the end of the last line of the verse/bridge as a lead-in to the chorus.

Conversely, if you choose to have an instrumental section after the chorus vocal it can serve as a way to step back down into the following section of the track without it being such a dramatic drop. Also, if you have a musical chorus "riff" then you could either leave this out completely during the vocal chorus or have it filtered down and running in the background. Then, once the chorus vocal has finished, repeat the whole chorus section but introduce this musical chorus hook in place of the vocal chorus in the repeat. These can be effective ways to increase the dynamic changes in a club mix but they are dependent on the length of the original track. Doing either of these will normally add at least 8 bars and maybe even 16 bars to the length of each chorus, which over the course of the whole remix could add anywhere from 30 seconds to 1:30 to the length of the whole mix and might, in that eventuality, make the whole mix just a little too long.

How ever you deal with the chorus, whether you keep it as a single vocal chorus or double it up to include an instrumental section, at the end of the chorus you will need to drop the energy levels back down again to go into the second verse. This usually happens during a short instrumental section following the first chorus. The aim of this section is exactly the same as we described earlier for the small breakdown before the first verse. By doing this we give ourselves more room to build into the second chorus. The main difference here is that this section needs to be a little different from the first. Remixers can approach this in a number of different ways, the most common of which is to simply not take the energy levels down as far as you did during the first small breakdown. If you keep a few more elements running during this section, and into the second verse, then you are at a slightly higher level and, if you build things throughout the second verse (and bridge) in a similar way then by the time you reach the second chorus the overall energy levels will be a little higher than at the same point in the first chorus. This is what we want to achieve: a gradual upward trend in power and energy throughout the whole track but with obvious peaks and troughs throughout this process.

Often the second verse, bridge, and chorus are dealt with in a similar way to the first verse, bridge, and chorus but, perhaps, with an additional element (musical, percussive, or even vocal) or two to differentiate it. The original song you are remixing might actually help you out in this respect because many songs will have something different in the second section to help raise the dynamics, so you may have extra harmonies or perhaps even a slightly shorter or longer verse or bridge. This helps to mark this section of the remix as being different but, even if that isn't there in the original track, there is no reason why you can't create that difference yourself. Overall though, this section of the remix is, perhaps, the one that needs the least thought. By this point in time people are already starting to become familiar with the vocal themes and melodies and, when the chorus hits for the second time, they will probably have started to remember it and will start to (inwardly if not outwardly) sing along. This certainly helps you as a remixer because you don't have to be quite so clever to hold people's attention so much during this part of the song.

If you did an instrumental section either before or after the vocal chorus first time around then you probably should include one again here. If you had one the first time around but not the second it might sound a little strange but, conversely, if you didn't have one the first time around then you can easily introduce one the second time around without it sounding odd. At the end of this second chorus you reach another of the crucial, and often "make or break," parts of your remix: the breakdown.

First let's ask the question: "Why do we even have a breakdown in the middle of a club mix?" There are actually two parts to this answer, both of which are pretty simple and make a lot of sense. The first part is that, at this point in the song, we are normally somewhere around the four-minute mark and so, assuming maybe two minutes or more at the end of the previous song and three to four minutes of this song, the crowd will have (hopefully at least!) been dancing like maniacs for six or more minutes and will probably appreciate a little bit of a rest. A breakdown gives the crowd a little time to relax, which, over the course of the whole night, probably means that they will be there for the duration rather than burning themselves out too soon or having to leave the dancefloor for a while to take a break. Our aim is to get people to the dancefloor and *keep* them there, so these little breaks give them just enough of a rest to mean that they don't need to actually take a bigger break at some point during the night. But, happily, they also serve *our* purposes because they provide a way for us to have a big build-up and an apparently huge lift into the last, and final, chorus.

The second part of the answer is as follows. We have, hopefully at least, been pretty successful in keeping the upward trend of energy in the track so far but, by now, people are quite familiar with everything we have going on so we have to give them one final push, one final "this is new" feeling, before we leave them to start enjoying the next song. And without quite a substantial drop in the dynamics and energy of the track this would be almost impossible to do. Over the duration of a normal breakdown we can drop the energy levels enough that, even if we just returned to exactly the level they were at in the second chorus, it would still feel even bigger this time around. By incorporating certain kinds of drum rolls and FX sounds we can add an element of tension that takes things even higher. Given that, sometimes at least, a remixer will introduce one final new musical element (possibly even something quite subtle), we have all the ingredients for that last big finale moment.

The specifics of a breakdown vary enormously across genres and remixers. There are, as always, many patterns that have emerged as being the most common and dominant, which means that a majority of breakdowns fall into one of three different main styles with one extra variation on the theme. The first is a really sudden drop to almost nothing with a slow and subtle build back up as the breakdown progresses. The second is almost exactly the opposite: elements are gradually removed until there is not much going on at all, followed by a quick build back up into the last chorus. The third is really a combination of both of these, with elements removing themselves gradually and then a reasonable length build back up into the last chorus. The final variation mentioned earlier, which crops up quite a lot, is really a variation on the first method but involves a little fake-out toward the end of the breakdown. Everything progresses as you would expect with a long build up, but instead of dropping back into the last chorus at the end of the breakdown, you get a further drop followed by another build-up (probably a much shorter one this time) that then leads back into the last chorus.

There may be a different chord progression in the breakdown and if the original track had a *middle 8* vocal section this can work very nicely as a vocal in a breakdown. If there is no middle 8 as such then a common thing to do is to create a variation on either the verse or chorus chord progression and use that instead. If there is no specific vocal to use here then another possibility is to do something similar to what we did for the intro section and actually create a little vocal hook by taking a little part of the chorus and effecting it heavily and using it in a more repetitive way.

Musically the breakdown will nearly always involve removing the kick drum and the bass line because those two elements coming back in at the end of the breakdown will add an immense amount of power compared to when they were not playing. This will really hit home the transition from breakdown to last chorus. Many of the other sounds that were in use during the other parts of the remix can also be used here but will most likely be either filtered or volume faded in or out as the breakdown progresses. FX sounds can also appear here in abundance with long *risers* being very popular as a way of building tension and anticipation leading into the last chorus. We want to use the end of the breakdown to lift people to a high state of excitement and then, at the end of the breakdown, drop them back down again so that it almost feels like they fall just a tiny little bit, even though they are actually at higher place than before. Try listening to a track or remix that has a really great breakdown and then try to visualize that image as you listen to the transition from the breakdown back into the chorus.

A riser is an effect sound, often non-chromatically tuned, which is quite long in duration and which creates a sense of "build up" by rising in volume, getting "bigger" and "brighter" in sound and perhaps rising in pitch (or possibly any combination of the three). These kinds of effects are often based on "white noise" type sounds and are, in a way, the dance music equivalent of a cymbal crescendo in orchestral music.

Lifting people up that high is surprisingly easy by combining *riser FX* with drum fills, filtering, and volume fades, but the difficulty can be that sometimes you can make the end of the breakdown *so* big that anything that follows feels like an anticlimax. With some careful forethought and judicious use of automation (for example, by automating the volume levels of all of the sounds downward by a couple of dB over the last few bars of the breakdown so that even though the sound feels like it's getting bigger sonically it is actually dropping in level almost imperceptibly), we can make sure that when everything crashes back in for the last chorus it really *does* crash back in a big way.

One final thing to consider with breakdowns is how long to make them. There is a temptation to make them too long because, with longer breakdowns, the effect of such a slow build-up to such a high peak can be mesmerizing. The trouble with this is that it is a balancing act. Yes, we want to give the crowd a break, but at the same time, we don't want them to lose their enthusiasm and liveliness completely. Because of this you should definitely avoid making your breakdowns too long. There are, of course, exceptions to every rule and if you have a chance to check out the breakdown of the truly epic Size 9 "I Am Ready" (club mix) you will hear a breakdown that, in total, lasts for close to 4 minutes and, during that time, goes from a 120-ish bpm house track into a 70-ish bpm breakbeat track and back again. To be fair, this is the only dance music track I have ever heard that goes to this extreme, but for some strange reason, it gets away with it. The TR-909–type snare drum roll that runs through the last minute of the breakdown builds so much tension that, by the end of it, you feel like you are ready to explode.

I think that perhaps the best way to appreciate breakdowns is to go to a few clubs that play the genre of music you will be working in and actually *see* how the crowd reacts to different breakdowns. Listening to them in a studio will give you a technical appreciation and possible insight into *how* they are done but they won't really help to explain to you *why* they work. Once you have seen it for yourself, you will have a better understanding of how to create your own effective breakdowns.

Moving on to the final chorus of the club mix, we are pretty much dealing with the same scenario as the first two. The principles are the same but, once again, you are helped along by the fact that, by now, there is an even greater sense of familiarity with the song and, given that you have only just come out of this big build-up section, people will hopefully be feeling excited enough to forgive you for not actually giving them anything (or much) that is sonically new. You may decide to up the ante a little by adding in another new musical element here but you really shouldn't feel that you absolutely must do this. Sometimes it is better to leave things out rather than risk making your production sound crowded with unnecessary sounds.

Once again though, the issue of the instrumental chorus section raises its head and my position remains the same. If you have included one in the first two choruses, or even just in the second, then you probably should include one in the final chorus as well. If you haven't included one so far then perhaps now is a good place to do so. But again, consider the effect this will have on the overall length of your remix.

At the end of this last chorus all you have left to do is give the DJ a nice smooth and easy way out of your remix and into the next track. Rather than go into huge detail over this I will just refer you back to what I wrote about the DJ-friendly intro but, instead of applying those principles in the order they were given, a good starting point is to apply them in reverse by taking things out in the opposite order you brought them in. This is a rather simplistic view of things and I would never advise you to literally make your ending a mirror image of your intro, but many of the same (well … opposite) principles apply. The DJ will still want a nice solid timing reference until very near the end of (if not the actual end of) the track, so a kick drum is a very useful thing to keep running. It would also be advisable to start removing the bass line, either by filtering or with a volume fade, a little way from the end of the track so you are giving the bass line of the new track some space to come in without clashing. Beyond that, do whatever you feel works best for the track.

I would certainly try to make things sound smooth in your outro because you want the transition into the next track to be gentle rather than a big, jerky change in dynamic level, which might catch the DJ out. I would also say that you don't always need to go all the way down to nothing on the outro. I think it is quite acceptable to keep a few more percussion elements going in the outro than you might ordinarily have starting off in the intro. On the whole though, if you apply the "DJ-friendly" principle, the outro needn't be a huge headache.

That's pretty much it for a very basic overview of a club mix arrangement. As I said at the beginning of this section, what is presented here might not make complete sense to you depending on the genre you work in. Some genres have slower and more gradual builds through the track without the ups and downs of the method I've described. In some way those kinds of arrangements are much harder to work on because you have to keep that build going continuously for a longer period of time. They often have more of a minimal set of musical elements to do it with so they can be a challenge in a different way. Even if this is the case though, much of why you do certain things remains the same.

Just as a final bit of food for thought before we move on to radio edit arrangements, Figure 6.2 shows a screenshot of the club mix arrangement from Figure 6.1, only this time all the tracks are arranged so that the elements start with the first ones at the bottom and gradually pile up, as they come in, as if you were building a wall.

The view shown in Figure 6.2 gives us a pretty good idea of how the dynamics and energy levels of the track build, even without listening to the remix. While I would never rely on a purely visual method of arranging a track, this visual overview can certainly help us, but it does come at a price. It is very easy

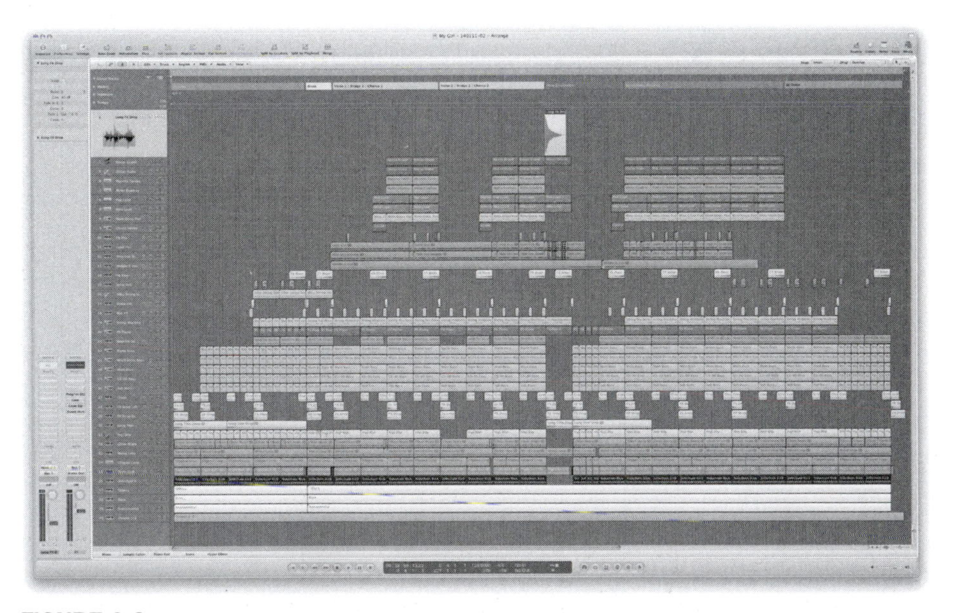

FIGURE 6.2
With the tracks arranged in a different order you can get a feel for how the track will build without even hearing it.

to start hearing things with your eyes in the sense that if you are looking at the screen as the track is playing and you can see that a certain sound or musical part will be coming in very soon, it prepares your brain for its arrival so much so that when it does come in it can feel like it has less of an impact because you had already anticipated it. I personally like to listen to tracks as I am working on them with my screen either turned off or, at the very least, with me facing away from it, so that I am hearing the sound of it objectively and in the same way a listener would. I'm trying to avoid the image giving me visual clues as to what is coming next.

RADIO EDIT

Having tackled the basic structure of a typical club mix we can now turn our attention to the radio edit. In some situations this can actually be more important to the client than the club mix. In any case, being able to deliver a good, clean, and punchy radio edit is a good skill to have. To some extent the arrangement of a radio edit can be laid out for you from the structure of the track you are remixing. Given that most radio edits have to be no more than 3:30 long, and sometimes you are asked to make them even shorter, there isn't always a great deal of scope for creative arrangements.

The original track (assuming you have been given a radio edit version to work from) will probably be around that length anyway so if you have sped the vocal up then you might have a little room to be creative because the running time of the original vocal will be shorter than it was before. If this is the case, you may be able to insert some extra space into the arrangement where you can. For example, you might have a short section with a newly created vocal part (like we described for the intro for a club mix in the last section) or even have a double chorus where there was only a single one before. Your options are still limited, but there still might be a bit of room for creativity.

If, however, you have had to slow the vocal down then you might run into problems fitting the whole song into the allotted time. Out of the two possible scenarios this one is by far the more difficult to deal with as a remixer, because it will most likely require you to cut something out of the original track. First, this isn't necessarily an easy thing to do while at the same time retaining the flow of the song, because it was written to follow a particular pattern and, by removing a part of that pattern you might end up with something that doesn't make much sense melodically or lyrically. A second, and bigger, problem is that you might find that, in doing this, you make the original artist unhappy because you have somehow "ruined" his song. You could argue that you really didn't have any choice but, understandably I think, most artists are, well, artists and as such are more likely to think about the situation creatively and emotionally rather than technically and logically.

One possible solution to this second scenario, and the one that is least likely to cause offence, is to shorten the choruses, or perhaps just the first one. Even this, however, is only an option if the chorus is very repetitive and longer than

8 bars to start with. If the chorus is only 8 bars long then you wouldn't really be able to reduce it in length at all as a 4-bar chorus (the next logical step down in length) is just simply unheard of. And if the chorus has a lyrical content that changes constantly throughout its length then, again, you would have difficulty reducing the length in any way. If you do find yourself in this situation then the best solution is simply to contact the client and explain to them, as clearly as you can, that you are having difficulties. If you tell them that as a result of you having to slow the vocal down to fit the tempo you are working at, the original song now runs for, say, four minutes and you explain that the only options you have are to edit something out of the original song or to have the song stay intact but at the cost of having a longer radio edit then you should receive some kind of guidance as to the best route to take. After all, no remixer, no matter how good or experienced, can do the impossible. At least if you ask your client for her input at the very early stages then you stand the best chance of coming up with a radio edit that is acceptable to everyone.

There are very few standard practice things about a radio edit, far less, in fact, than for a club mix, but the ones that do exist are equally important. They are:

- The first vocal in the track, whether a chorus or a verse, should come in after no more than 10–15 seconds.
- There should be no *dead air time*, which means that, unlike a club mix, a radio edit shouldn't contain periods in the track with nothing going on. This is most likely to be relevant in the sense of having no breaks between verses and choruses, between choruses and verses, and no extended breakdowns with long instrumental sections.
- It should ideally be between 3:00 and 3:30 in length (although U.S. record labels are usually happy with radio edits lasting up to 4:00 in length, but be sure to check that with each client).
- There should be a defined and clear end rather than a fade out. Fade-out endings were acceptable and often quite normal many years ago but today's market seems to demand a clearly marked ending to a radio edit.

A less clearly defined rule is the requirement to keep any breakdown short, although that tends to happen anyway given the length restrictions, since the breakdowns tend to not drop down dynamically anywhere near as much. In many cases they are simply the last 8 or 16 bars of the club mix breakdown edited to fit into the radio edit, perhaps with a few changes to make the transition into them from the previous chorus a little smoother.

Getting a radio edit started is generally quite easy because mostly it will either just start with a bang and come straight in or have some kind of drum fill, reversed cymbal, or some other kind of short FX sweep to lead into the track. Sometimes you would go straight into a chorus section (perhaps instrumental but usually with some kind of vocal included) that might only be 4 bars long, and would be the only time you could really get away with such a short chorus. Other times you might go into a 4- or 8-bar instrumental version of a verse. There are no hard-and-fast rules here and it depends largely on the song itself and the conventions within a particular genre and your own preferences as a remixer. You will figure out what works best for you. Keep in mind that 10–15 second time scale to get into the song properly whenever you are working on this initial intro.

From this point on most of the arranging will simply follow the progression and dynamics of the original arrangement of the song. More often than not you will go straight from the verse into the bridge (if there is one) and then straight into the chorus. From the end of the first chorus you will usually go straight back into the second verse but could, overall length permitting, include a very short instrumental break, often as little as two bars but certainly no more than four, before returning to the vocals. The same pattern will follow through the second verse, bridge, and chorus.

As for the breakdown I would really try to keep that down to 8 bars, possibly 12 although that might feel a little cumbersome. I would only ever use a 16-bar breakdown if there was a middle 8 vocal that actually lasted 16 bars and was only used in your breakdown section. Otherwise I think 16 bars is generally too long for a breakdown in a radio edit because you don't have the same reason for having it that you do in a club mix. In a club mix the breakdown is there to give the people in the club a chance to relax for a short while and to help you build the dynamics of the track back up again for the last chorus. People listening to a song on the radio in their car, or on their MP3 player on the train, or at home on a CD player in the kitchen while making dinner won't need that respite from dancing that a clubber would. Yes, it can still help to drop the dynamics just enough to keep that last chorus sounding interesting but, at this point, and in these listening environments, it isn't so crucial to have such a large range of movement in the dynamics and energy levels in the track.

Beyond the breakdown we really just have the last chorus, and quite often this is longer than the first two choruses. This is possibly one area where you might be able to lose a little time if your radio edit is running a little long. But if you can afford to keep the longer last chorus, or even make a longer or double-length chorus if there isn't one, then I do think it adds a little something to the end of the radio edit. It can provide a nice final farewell from the song or a triumphant exit using the most memorable part of the song, and it's a good way to really push home that vocal (and musical if there is one) hook.

Once this last chorus is done we have to end things. As I mentioned earlier, almost without exception, radio edits these days tend to end abruptly. This is mostly quite easy to achieve by adding some kind of small build-up—perhaps a drum fill or a rising FX sound—over the end of the last chorus and then finishing up with a crash cymbal or similar effect and possibly some kind of longer reverb or delay effect on a vocal line or other prominent musical part to carry things over a little and give a second or two of "mini" fade-out. Sometimes you might find it effective to have a bass note on this final downbeat as well. This could be in the form of a sustained bass note that fades out over a second or two or it could be a shorter, more staccato note that has some kind of delay effect on it to give it that same sense of fading out. Either way, sometimes a track feels like it needs some extra weight on that final downbeat so a bass note (or even a kick drum, or both) really marking out that final point can work quite well.

There used to be, and perhaps still are, people in the industry who primarily just made radio edits. I remember quite a few years ago being recommended someone to use to create a radio edit of a song I had written. Even then, when I had far less experience in the industry, it still seemed like a strange idea to get someone to work on a radio edit, but I took the recommendation on board and handed over the *stems* as requested. However, I was told that it would be a few days before this person could begin work so I decided to have a go at it myself. As I listened to the song it became quite apparent where I would make the edits and I didn't feel that there was any kind of magic to it. So I did my own edit and waited a few days to get the version from the other person. When I eventually got it back I was quite surprised, and quite disturbed. You see, this person had come with the highest recommendation and had quite an impressive résumé of previous clients but, when I listened to the edit, it was fundamentally the same as mine. There were a few places where things were a little smoother, where transitions had been edited a little better and a few spot effects had been put on to cover any slightly bumpy edits, but otherwise the arrangement was basically like mine.

Stems are audio files of groups of musically related sounds. A typical set of "stems" for a pop song could be made up of: drums, bass, guitars, keyboards, strings, backing vocals, lead vocals and effects. In the case of the guitar stem, for example, all of the different guitar parts for the song would be mixed together into just one "stem." In many ways it is a midway point between having every single track available and having only the final stereo master version.

This surprised me because I had, up until this point, assumed that there was something different that these remixers did but, and I say this with my hand on my heart, what they did then was nothing *that* special. In fact, I would say that my edit was probably 85% as good as theirs and, had I actually spent a little more time smoothing off the rough edges on mine, it would probably have come close to 95% as good as theirs. This troubled me because, at the time at least, some of these radio edit "specialists" were charging quite a lot of money and I really couldn't figure out how they could justify that. Since then I have done well over 250 radio edits on various original tracks and remixes and, I have to say, that in general most of them have been pretty easy on the creative side (as in, where I made the edits and which parts I edited). Some of them have been technically more difficult than others because of the need to lose a little time in order to get them under the 3:30 mark or perhaps because of a difficult musical transition from the end of the chorus into the beginning of a verse. But on the whole, I think that especially now that my technical skills have improved a great deal since those days, I am perfectly capable of doing my own radio edits.

Keep them as short as possible (within reason, of course), make sure everything is nice and smooth in the transitions between sections, there are no sections of more than five seconds or so within the track without at least some kind of vocal going on (assuming the track has vocals, and if it doesn't, replace "vocals" with "lead musical part"), and there is a well-defined ending. That has worked for me so far and I have had very few complaints about the radio edits I have submitted to clients, so I must be doing *something* right.

One final point I want to raise about club mixes and radio edits before we move on to other types of remix is that, in most cases, they are variations of each other. You could look at the club mix as being a radio edit with an intro, an ending, a longer breakdown, and a few instrumental sections thrown in. Alternatively, you could look at a radio edit as a club mix but with all of those parts taken out. As such, you would normally create one of these first and then adapt it into the other.

There is some degree of contention about the best way to do things. Some remixers will argue that you should start with the club mix and work your way back into a radio edit from there. They use the valid argument that you are creating club music so it has to work in the clubs first and foremost. That certainly used to be the main focus of remixes so I would be inclined to agree with them on one level. However, at the same time, if a client has come to you with the intention of getting a remix that could also serve as an alternate radio mix then they are clearly aiming for radio play as well. In those situations, you could argue that perhaps the radio edit was more important as this was the main focus of your remix. And again, you would be justified.

To some extent, however, I feel that arguing these points is quite unnecessary. Yes, your club mix has to work in a club environment and your radio edit has to work on the radio, but which one you do *first* doesn't really have any effect on the outcome of either unless you spend all of your time on one version and then rush the other one to meet a deadline. In general I tend to work on radio edits first simply for practical reasons. A lot of remixes, especially the more commercial ones, will be accepted or rejected by either the commissioning A&R person or the artists themselves. If they are very much in the pop world then there is a good chance they won't be into your particular brand of dance music and may not even understand the mechanics of the longer intro on a club mix. It would be very easy for them to get bored and you really don't want that to happen before they have even had a chance to listen to the best parts of your remix. I have lost count of the number of times I have been in A&R meetings with tracks of my own, have pressed "play" on the CD player, the club mix has started, and the very first thing the A&R person did was to skip through to about 1:00 or 1:30 into the track to hear the main body of it. Only then, if he or she liked what he or she heard, that person went back and listened to the full version.

So for me, a radio edit is very useful to send as a means of getting either just feedback or perhaps actual approval on your remix. After that is completed, you can then finish up the package by providing a club mix (and any others they might request), and you can do it in the quickest possible way. That method works for me and I hope that, having explained why at least, you can understand why I do it but if you want to work the other way around you are certainly not alone.

DUB MIX

Now we come to the other possible versions of a remix you might be asked to provide. Of these the dub mix is perhaps the one that requires the most work. Back in the very early days of remixing, and as I mentioned in the very first chapter, dub mixes were extended reworkings of records that often used a very stripped-back musical arrangement and made great use of delay and reverb effects on much more sparse vocal arrangements. Over the next 30 years or so the sound, and role, of dub mixes changed a little, but the fundamental principle remained in that they always had much less use of the vocals.

In the 1990s and even into the 2000s a dub mix would often be a different musical production. It would be based around the same musical elements but might have a few of them changed; for example, with different sounds, different percussion parts, and perhaps even simplified musical arrangements with less complex chord progressions. This, combined with a minimal use of vocals, would have been a typical dub mix of the time. More recently dub mixes have tended to be based around exactly the same production and sounds and percussion parts but with slightly simplified chord progressions and a slightly different arrangement. They still have minimal vocal use. To some extent practicality may come into play here. Schedules are ever tighter and budgets have certainly been squeezed over the last few years so the remixer may simply not have the time (or perhaps the inclination given the budget) to produce what is, at least partly, a *different* version of his remix. Some of the responsibility also has to lie with the DJs and the record labels themselves though. If they demanded dub mixes then the remixer would have no option but to provide them or not get paid. So if there has been a falling off in the level of effort that has been put in dub mixes then I think that all parties are equally responsible.

If you did have the time and the inclination to do a completely different dub mix, nobody's likely to complain but I am going to assume, for sake of argument, that your dub mix is fundamentally the same (musically) as your club mix and just focus on the differences in terms of vocal usage and arrangement.

What often happens with dub mixes is that there are just one or two lines of the original vocal, or perhaps a full chorus, used throughout the whole song. It could be that the chorus is used pretty much the same as it was in the club mix with a few chopped up and heavily effected vocal parts taking the place of the verses. Or it could be just one line from the song that is repeated, infrequently, throughout the entire song with the musical elements on their own being responsible for differentiating the different sections of the song. There really are no rules when it comes to how you use the vocals in a dub mix other than don't use too many of them and, often, make the actual sound of them a little less clean or clear.

In some cases this can be taken to quite an extreme level. If you listen to the original version of Roger Sanchez's "Lost" and then listen to the D. Ramirez Dub Mix you will hear that the *only* vocal used in the dub mix is the two words "I'm Lost" and even then it is mangled into borderline unintelligibility through what sounds like a vocoder. That one line is used quite sparingly throughout the track. At this point though, you have to ask if what you are actually supplying is a dub mix or a completely new track with little more than a cursory tip of the hat toward the original.

In terms of the music itself, because you have much less vocal usage to deal with you might want to simplify chord progressions either in the verse sections or throughout the mix. Many inspired dub mixes are actually quite linear in terms of their chord movement and don't change around that much. They build the energy of the track by introducing new layers to the production in place of the traditional chord progression changes that happen in the full vocal versions of the track. The overall effect of a dub mix should be one that captures the spirit of the club mix but presents it in a format that is less sing-along or anthemic and more about the groove and the music.

In terms of the arrangement, most of what you need to know has already been covered in the previous section on club mix arrangements (specifically the parts about the DJ-friendly intro and ending and the main breakdown) but, obviously, disregarding the parts specifically relevant to verse/bridge/chorus vocal structures. There is still a need to build and drop the dynamics and energy levels to keep the interest going for the full length of the mix, however. In some ways this is a little more difficult with a dub mix because you don't have that same pattern of builds and drops within the vocals to work against.

What you *do* have is the freedom to get a little more creative with the arrangement and the actual vocal sounds and effects. You could take one or two basic lines from the vocal and then run them on several different audio tracks with slightly (or wildly) different effects on each so that, even though the actual lyric and melody of the vocal were repeating, the sound of it was an ever-changing morph from one sound to another. Doing this would certainly help you maintain the interest levels throughout the full length of the track.

You might not even be asked to provide a dub mix as a part of the package for your client. If not but you still want to provide one for whatever reason, and can do it within the time scale you have been given, I am sure nobody will complain. But at the same time, if you end up rushing your club mix or radio edit, even slightly, because you want to supply a dub mix even though you haven't been asked to, then that's clearly not a wise use of your time. Believe me when I say that most record companies are not shy about telling you what they really want. And if they really *want* a dub mix they will ask for one! So if you haven't been asked it probably is a better use of your time to get the most out of the mixes that were originally commissioned.

INSTRUMENTAL MIX

This might sound like a stupidly simply job; after all, all you have to do is mute all of the vocal parts and record a pass of the track, right? Well possibly yes, but most likely no. As well as being important for putting across the message and meaning of the song, the vocals are contributing a musical role in your remix. If you had a live band with a drummer, bass player, a rhythm guitarist, lead guitarist, and a singer, and you were to remove any one of those elements it might sound odd. The same applies to removing the vocals from a remix. Unless the vocal contribution is purely a hook that comes in every now and then, the vocals are likely to play a substantial role in the structure of the song, so if you simply remove them you may find quite a few places in the track where things sound flat and boring.

This isn't a huge problem because you would never be expected to replace the vocals with an instrumental part, even though that would be the sensible thing to do if you wanted to maintain the same sense of dynamic variation of the track. But you will still need to do something to take into account the changes. Quite often this will be as simple as tightening up the arrangement a bit so that verses are a little shorter. Or perhaps the sound you have is filtered pretty much all the way down in the verses but is open in the choruses and could gradually open out throughout the length of the verse instead. A double chorus could probably easily become a single one if there is no vocal, and a breakdown might be similarly trimmed down.

However, beware, as I have been caught out with this before. I was working on a remix for a major U.S. record label and they requested an instrumental mix. So I followed my normal procedure and loaded up the club mix, muted all of the vocal parts and then started to make the kind of changes I described earlier. By the time I was done I was very happy with the result and felt it had a really nice flow to it. I sent it to the client and was told that, while it did sound good, what they actually wanted was basically the club mix with the vocals muted! So it was time spent that I didn't need to spend. You obviously don't want to come across as either insecure or amateur but I really don't think that asking those kinds of questions to clients will do you any harm. In fact, on some levels it shows professionalism because it shows that you want to make sure the client gets exactly what they require. Asking for a little clarification on certain points like that will only help you get it right for them. However the subtleties may go, an instrumental mix is generally pretty easy to figure out, so I won't insult you by going into more detail!

"MIXSHOW" EDIT

A "mixshow" edit is a bit of a strange beast and one that, so far at least, seems to be native to North America. You could look at it in one of two ways: either as a cut-down club mix or an extended radio edit. They are so named because they are used on mixshows on U.S. radio where basically what's needed is an extended version of the song, with all of the builds and drops of a normal club mix, but without the need for longer intros and outros. On these shows, radio DJs don't use the longer crossover sections, so outside of a club environment those transitional sections could seem a little unnecessary and boring.

The easiest way to create a mixshow edit is to take your club mix and reduce the length of your DJ-friendly intro and outro from the length it is to around the 30-second mark. That is usually long enough for the purposes of a mixshow edit. What this will do is remove between 1:30 and 2:00 from the length of your club mix and will probably bring down the overall length to around the 5-minute mark, possibly as much as 6 minutes.

Another way to look at this is that you take the essence of the remix, which in many ways is what you have in a radio edit, add on a 30-second or so intro and outro, and extend the radio edit breakdown so that it is a little more dramatic. Looking at it that way we would probably be adding between 1:00 and 1:30–3:30 or so of the radio edit to get us back up to that 5:00 or so length we are looking for. Whichever way you want to look at it, the process is fairly simple.

TV EDIT/PA MIX

In many ways these two versions of the remix are one and the same as they serve very much the same purpose. That purpose is to have a version of the track that is suitable for "live" performance by the singer. In terms of actual arrangement, a TV edit/PA mix doesn't differ a great deal from a radio edit. After all, for the most part this is the version most people will be familiar with so why rock the boat? There is, however, normally a request to have a slightly longer intro rather than just getting straight into the track. The reason for this is that during a live performance, be it a TV performance or a club PA (personal appearance), the singer might want a short period of time to interact with the crowd and get them involved, so having between 8 and 16 bars of intro before dropping into the radio edit arrangement gives them the time to do just that.

Beyond that, the main body of the track should probably follow the arrangement of the radio edit because this is the tightest version of the song that exists, and the one that gets the point across the most directly. There may, rarely, be a request to give a slightly longer breakdown to allow the singer to again interact with the crowd, but this doesn't happen that often. You may also get a request to double up the last chorus if the radio edit doesn't do this already to give the singer a chance to *ad-lib* over the end of the song. Finally, you may be asked to provide a short outro (4 or 8 bars or so) after the end of the last chorus to just give a small *playout* before the song ends.

Possibly the most important difference, however, is that both the TV edit and the PA mix will require the lead vocal to either be removed completely or dropped substantially in level. The reason behind this is simple. The singer will be singing the lead vocal line live so if the original lead vocal were in there and the singer were to phrase a line slightly differently it would sound very untidy. A differently phrased lead vocal line over a backing vocal probably wouldn't sound too bad though. As such, you are usually asked to take the lead vocal out altogether. However, some singers request to keep the lead vocal mixed in albeit at a lower level, often as much as 9–12 dB lower than it normally is. This is to help provide just a little backup to their live vocal performance and to add some additional power. As a matter of course I generally provide both versions, even if I have been asked for only one or the other, as it really doesn't take too much effort. A simple gainer-type plug-in inserted onto the lead vocal channel will be enough to provide the "lead down" version and then just mute the channel altogether to provide the "no lead" version.

One other thing to consider, although it doesn't come along that often, is in situations where your radio edit doesn't have any kind of clear beat running (either as a percussive part or some kind of musical part with a regular and clear rhythm). The reason it doesn't happen that often is because normally in a radio edit your breakdown wouldn't drop that low (dynamically speaking) but, on the off chance it does, you might need to amend it slightly for a TV edit/PA mix because the singer would certainly be helped out by having a clear timing reference through that section. Something as simple as a single percussive part (maybe a hi-hat or shaker that is already present the track) or a single musical part continuing through that section is probably all you need just to give the singer enough of a cue.

THE MIX PACKAGE

So now you have seen all of the possible variations (the ones I have come across so far at least) you might be asked to provide. Very rarely will you be asked to provide all of them for any single remix, but it can happen. For one recent remix I was asked to provide a club mix, a dub mix, an instrumental, a radio edit, a TV edit, *and* an a cappella (of the main club mix vocal) as well as stems (in this case there were actually 10 of them: kick drum, loops, FX, bass, pads, arpeggios, leads, others, lead vocals, and backing vocals) for *each* version! I was working with 24-bit, 44.1 KHz files and, as I always tend to do, I was providing mastered and unmastered versions of all of the mixes. So I had two versions of each of the six mixes and 50 individual stems, which, in total, came to about 5 GB of files to upload. Let's just say that it took a very long time to get everything uploaded and delivered to the client. As I said, it happens very rarely, but it does happen, so just be ready for it.

Hopefully this chapter has given you an insight into the basic structures of the different kinds of mixes and some idea why these different mixes are laid out the way they are. Dance music and remixes have been around for quite a while now so these conventions didn't just appear out of nowhere. They have been tried and tested on tens if not hundreds of thousands of tracks and remixes. I am not saying that you absolutely *must* abide by them, but at the same time remember that they exist for a reason.

CHAPTER 7
Anatomy of a Remix

In this chapter we look at two well-known remixes and go through the technical and creative aspects of them that, to me at least, make them standout remixes. I encourage you to listen to both the original and the remix versions as you study each section. (I have included YouTube links for all the tracks mentioned here but obviously I can't guarantee that those links will still be active when you are reading this. If not, you should be able to easily find the tracks elsewhere, if you don't already have them in your MP3 collection.)

I have chosen remixes that I think represent the different extremes remixing can achieve. The first is the Freemasons remix of "Déjà Vu" by Beyoncé (from 2006). I chose this because the remix still contains most, if not all, of the original vocals and still keeps a lot of the melodic and musical structure of the original version of the song but in the context of a very polished house production sound. By contrast, the second example is Armand Van Helden's remix of "Professional Widow" by Tori Amos. This remix is notable for taking literally a couple of lines of the original track and, with the help of some brilliant programming and a monster new bass line, turning it into pretty much a brand new song. There were literally hundreds of remixes that I could have chosen purely on the basis that I liked them, but I thought these two would be a good way to illustrate different approaches to remixing and to highlight what is technically possible with remixing.

BEYONCÉ "DÉJÀ VU" (FREEMASONS REMIX)

Original version: *www.youtube.com/watch?v=mp0jL38dDtg*

Remix version: *www.youtube.com/watch?v=epb6SLHnydM*

Tempo Change

The first thing to notice here is the difference in tempo. The original version of the song is at 106 bpm and the remix is at 126 bpm. While that isn't the largest tempo change I have ever heard in a remix, it certainly isn't trivial either. This is even more impressive when you hear how natural the vocal sounds in the remix version.

Groove Change

The original version of the track has a lovely "loose" feel, which gives the song its individual character. Aside from the tempo change, one of the most obvious things to me about the remix is that it has reduced the amount of swing or looseness in the groove. That's not to say that the remix is "mechanical," far from it in fact; it has an awesome groove, but it is definitely much more bouncy than the original version. This can potentially present real problems if the vocal is tied in too tightly with the original groove because having a really swingy element at the same time as a much straighter one can sometimes feel uncomfortable. The Freemasons may have done some manual retiming or groove correction on the vocals or perhaps the original vocals just happened to work over the slightly straighter groove that they used. Whatever the case, the results sit really well. In fact, listening to the original version, the vocals don't sound that "swingy" even though the groove does, so I'm guessing that the vocals were probably a straightforward stretch.

The remix groove itself is deliciously catchy and has elements of classic '70s disco as well as a bass line that picks the whole track up and carries it along on a relentless ride for the full length of the song. There isn't much going on in the groove but what is there really does count for a lot. Drums are taken care of with a pretty solid four-on-the-floor house beat with quite "live" sounding percussion (snare, hats, tambourines, and bongos) that have a nice swing to them. I would guess they used only a 16A or 16B groove, which is pretty consistent throughout the track. There are a few changes in percussion as the track progresses that accentuate certain sections of the song, but otherwise there isn't a great deal of trickery in the percussion. Given the style of the remix, this is a good thing. Remixes of this nature that are musically rich and have a very slick or glossy production shouldn't, in my opinion, be overly fussy with the drum programming because while the drums are important, this kind of remix is more about the music and, above all, the song. In contrast, more minimal styles of production tend to sound better with little changes and constantly evolving detail in the drum programming.

The bass line has a very "live" feel to it (I believe it was actually a live bass line as opposed to a programmed one), but again it isn't overloaded with detail and fancy fills. Instead we have a simple but detailed bass line that complements the drums and percussion and provides a solid base upon which the rest of the bass line is built.

Musical Change

Possibly the most significant difference between the original song and the Freemasons' remix is that the whole song now uses a single chord pattern (except in the intro and ending of the club mix, which is a simplified version of the same pattern). Unusually for a club mix this is actually more complex than the musical structure of the original. More often than not, club mixes have a simpler chord structure than the original song, but in this case the new chord pattern actually makes the song much more dynamic and interesting. The fact that the pattern repeats through most of the song isn't really a bad thing and it doesn't feel boring at any point because the dynamic changes in the song are taken care of by clever arrangement of the different musical parts to give subtle increases and decreases in energy throughout the track.

The remix retains some of the musical feel of the original by sticking with the original key; there is no shift from minor to relative major, which could have changed the mood of the song significantly. The choice of chords works especially well in this instance because they keep the feeling of the original song while being very suitable for the genre of the remix. Something I hear perhaps a little too often is an inappropriate choice of chords for the genre of music. As well as getting the sounds and the groove right, choosing the right chords to fit in with the melodies and key of the song as well as the genre itself is crucially important. An otherwise great remix can be spoiled by having "moody" chords in a bouncy and poppy production style; likewise, having really major and "up" chords in styles that are typically quite dark wouldn't work either. Here the balance is just about right.

Sounds and Rhythms

I mentioned earlier that there aren't a great number of musical parts in this remix. The following list identifies the individual parts of the song. Have a listen and see if you can identify each of them and how they are used.

- Drums
- Cymbals and FX sounds
- Bass line
- Thin background arpeggio (resonant sounding synth)
- Main piano chords layered with strings (with occasional piano runs and stabs)
- One-note string (held legato notes that last for the whole chord pattern cycle)
- String melody
- Short string stabs that appear briefly in the outro
- Sustained thin pad sound with sidechain compression (very subtle)
- Funky guitar chords
- Additional funky guitar (single notes rather than chords but similar pattern to the chords)
- Funky guitar riff that appears briefly in the outro
- Vocals

You can see that it really isn't that much. Having said that, in many ways the number of parts is similar to what you might find in a classic "live" disco track. In some respects that is what I think has made the Freemasons so successful in this funky or disco house genre. It is very easy to get carried away with putting lots of sounds into a production just because you can. But by keeping the sounds pretty much to the kind of things one would expect to hear in an older disco record, there is an air of authenticity here, which gives a pleasant and slightly retro feel to the whole thing.

Vocals

The vocals sound excellent, considering they have undergone something approaching a 20% change in speed. There are a few places where the vibrato on Beyoncé's voice sounds a little fast, but there are also places in the original song where the vibrato sounds quite strong, so it's inevitable that the sped-up version would also have these issues.

Later in the book we consider some techniques that can be used to minimize the effects of sped-up vibrato in remixes, but there are times when you can't actually do a great deal. I think this would be one of those situations because even the original tempo has a really quick vibrato. Ironically, because the remix is faster and more energetic, the vocal vibrato isn't as bothersome as in the original even though it is actually faster in the remix. This is one of those situations where it is all about context; you can get away with things in one situation that you perhaps couldn't in another.

Aside from the main lead vocal, which is pretty much as per the original song in terms of sound, there are also some really nice chopped-up vocals used throughout that have a more spacious and distant sound to them. I am a big fan of using vocals in this way because it allows you to hint at the song even during the intro when the listeners won't necessarily know what's coming. They also work well in the outro of the song because it carries the song on for a bit longer without getting too repetitive. Of course, you need to choose the parts carefully because you don't want to give too much away too soon. In this case a simple "Baby" is repeated, but it really does set things up nicely.

Arrangement

I think the most significant arrangement decision the Freemasons made in their remix was to not include the bridge vocals of the original version. Those vocals wouldn't have worked especially well (if at all) with their new chord pattern, and would have meant a change in the chord pattern for only 4 or 8 bars at a time, which would have really spoiled the flow of the remix.

Another interesting vocal issue in the original song is that the first verse is led off by Jay-Z's rap while the second verse is purely Beyoncé and then back to Jay-Z for an extended rap section in the third verse with no Beyoncé vocal in the verse. As much as I think this is interesting it's also quite an unusual arrangement for a vocal, so it's not really a surprise to me to hear that the remix vocals follow a more consistent pattern. Here is the difference between the original version and the Freemasons radio edit:

VOCAL ARRANGEMENT ORIGINAL

- Intro section
- Jay-Z ("I used to run bass"... "stand back")
- Beyoncé ("Baby seems like everywhere"... "compare nobody to you")
- Beyoncé bridge ("Boy I try to catch myself"... "I can't let it go")
- Beyoncé chorus ("You know that I can't get over you"... "swear it's déjà vu")
- Beyoncé ("I'm seeing things I know I can't be"... "it's like I'm losing it")
- Beyoncé bridge ("Boy I try to catch myself"... "I can't let it go")
- Beyoncé chorus ("You know that I can't get over you"... "swear it's déjà vu")
- Jay-Z ("Hova's flow so unusual"... "no déjà vu just me and my")
- Beyoncé ("Baby I can't go anywhere"... "that I'm having déjà vu")
- Beyoncé chorus ("You know that I can't get over you"... "swear it's déjà vu")

VOCAL ARRANGEMENT REMIX (RADIO EDIT)

- Beyoncé ("Baby seems like everywhere"... "compare nobody to you")
- Beyoncé chorus ("You know that I can't get over you"... "swear it's déjà vu")
- Jay-Z ("I used to run bass"... "stand back")
- Beyoncé ("Baby seems like everywhere"... "compare nobody to you")
- Beyoncé chorus ("You know that I can't get over you"... "swear it's déjà vu")
- Jay-Z ("Hova's flow so unusual"... "no déjà vu just me and my")
- Beyoncé ("I'm seeing things I know I can't be"... "it's like I'm losing it")
- Beyoncé chorus ("You know that I can't get over you"... "swear it's déjà vu")
- Beyoncé chorus ("You know that I can't get over you"... "swear it's déjà vu")

As you can see, the remix follows a more repetitive structure, which works well for the format of the remix (club track) because, in general, club music isn't about challenging the listener to keep up with the twists and turns. I know it might seem like a sweeping generalization but, in general, people in clubs listen to the lyrics less intently than people who are listening on the radio or on their MP3 player. Of course, there are vocal hooks they will sing along to in the club (inwardly or outwardly) but it's more the melodic nature of the vocal that pulls them in rather than the clever arrangement of vocals. And I realize that this might seem like I am advocating "dumbing down" the remixes, but I have to say I have had those exact words said to me by a record label. When I submitted a remix I was working on I was told "It's a really great remix and we love it... but is there any chance you can dumb it down a bit? It's a bit too clever at the moment."

Other than that, I really like the dynamic changes in the arrangement. The build-ups are generally quite subtle, which gives the song itself a lot of room to be in control, which is the way it should be with this kind of remix. I think that one of the best lessons I have learned over the years is humility. It might be different if you are dealing with underground remixes of underground tracks, but in the commercial world the song really is king and even if you have been employed to bring the song into a totally new market, and even if you are absolutely at the top of your game and in demand, I think it is still good to remember that the remix should be about the song you are remixing, not just about your production and remixing chops. Not everyone will agree with that, but remember, even if you do take a back seat, you will still benefit greatly from the marketing budgets spent on promoting the record, so there is something there beyond just the remix fee to make it worth your while.

Overall

To me this mix is a shining example of how to take a pop record and turn it into a club record that sits perfectly on that boundary between a big club track and a radio-friendly club track with one foot firmly planted in each. The musical content works perfectly and actually brings out the best in the vocal to the point where I think the vocals sound better over the remix chords than the original chords because they sound less repetitive and more emotional. The sounds and production overall are very well-balanced and, if anything, are slightly on the light side. Because nothing is heavily compressed, you are left with the impression, even at very high volumes, of something that you could listen to for hours on end without it becoming fatiguing.

Listening to it quite a bit, as I have done, has made me realize just what a great remix this is on pretty much every level. It is definitely a great choice to illustrate what can be done when the remixing brief is to keep as much of the original song intact as possible and aim for a commercial, radio-friendly yet dancefloor-friendly remix. That is almost the exact brief that I get for about 75% of the remixes I work on and it isn't an easy thing to do. Of course, you may not have much of an interest in these kinds of mixes; they might be far too commercial for your tastes, but I think that, no matter what your tastes are and what kind of music you make, it is still important that you can listen to other genres and be able to appreciate their technical qualities even if you're not particularly into their stylistic ones.

TORI AMOS "PROFESSIONAL WIDOW" (ARMAND VAN HELDEN REMIX)

Original version: *www.youtube.com/watch?v=o8NGWHZHZao&feature=related*

Remix version: *www.youtube.com/watch?v=BroNPS0uuNM*

Tempo Change

This time around the tempo change is even more significant—from 138 bpm to 256 bpm to give a "double time" feel at the remix tempo of 128 bpm—but I don't think it matters too much. This remix was made in 1996 so the quality of the time-stretching available at the time was far inferior to what we have today. The vocals used in this remix are minimal and are more of a "feature" of the track rather than the core of it, so any time-stretching artifacts are less important and less intrusive.

Groove Change

This is, perhaps, even more of a radical change than the Beyoncé remix we looked at. While the Beyoncé mix featured a change from funky to disco, this remix goes from a half-tempo, almost dirge-like feel to a driving, bouncy, and relentless club groove. This is, of course, made easier (perhaps even made possible) by the fact that very little of the original vocal is used in the remix. Had there been more use of the vocal, the more strident feel of the remix might not have worked that well with the more laid-back feel of the original vocal performance. One thing common to both the original version and the remix, though, is the very repetitive nature of the groove (with the exception of the breakdown of both versions). A lot of the magic of the remix version comes from the repetition of the groove and the bass line. Had there been a lot of changes happening in the musical bed of the remix version then I don't think the remix would have been anywhere near as good.

Listening to the actual sounds and feel of the groove today, it is surprising how undated it feels. Of course, production styles have moved on since then but you have to remember that this remix was done in 1996. When you compare it to a lot of other remixes from nearly 15 years ago this one has aged especially well. In fact, I am pretty sure that if you were a DJ and you were to play this now (assuming, of course, that it wasn't totally out of context for the music you play) the crowd would go crazy. A big part of that is probably its "classic" status but I think that in some cases there is more to it than that. There is something fundamentally catchy about this groove with its outrageously funky bass line and bouncy drums and percussion. In fact, I think this would have been a hit with pretty much any vocal placed over the top. The fact that it used a relatively unknown track goes some way toward proving that. Then again, the cheeky and exceptionally near-the-mark vocals are probably contributing factors as well. Sex sells, as we all know, and the innuendo in this track probably meant that there were many suggestive dances going on all over the world at the time of its release.

One final thing to note is that the drums and bass (which make up a good majority of the remix) have very much a "live" feel to them. I imagine that the bass line was a sample or a recording as it still has the feel (with all of the intricacy, dynamics, and detail) of something a live drummer and percussionist might create.

It's very important to take into account the finer points of the track you are remixing when considering stylistic choices. Obviously your own personal production/remixing style might not allow for a "live" feel, but there are still ways of making something feel more organic if the original track had that feel to it. If, on the other hand, the original track was more synthetic then it makes sense for the remix to echo that. Of course, you don't need to factor the details like this in when you are remixing but I honestly believe the small details make the difference between a good remixer and a great remixer.

sound. Aside from obvious FX sounds and cymbals that is pretty much it. In addition, there are very few fills/drops in the drums. Once they are in they pretty much stay there for the duration of the track, other than in the main breakdown. This is, perhaps, a little unusual for those of us more used to the club music of today where there tends to be a little more variation (or a lot more in some cases) in the drums and percussion throughout the length of the track. But, like so many other aspects of this remix, further complication simply isn't necessary!

Vocals

The original Tori Amos lyrics were, as I am sure you have heard for yourself if you have listened to the track, somewhat eye-opening in nature! In addition, they would almost certainly have not worked in a club record because of the nature of what they were. But what Armand Van Helden did was choose just a couple of lines from the original track and repeat them throughout. In doing so he also changed the emphasis and implied meaning of the track. The original song is, in my interpretation, about wanting, and being prepared to do anything to get, power and a rock-and-roll lifestyle. So in that sense it is about a kind of pseudomanipulation. It's quite deep in many ways. However, the remix changes all that and gives the song a huge sexual innuendo, which, I am almost certain, helped to sell the remix version into the clubs. It's no secret (or surprise) to anybody involved in this industry that (the right) music can have a very sexual edge to it. The repetitive beats (tribal-esque), the dancing, the flashing lights, it's all very primal in a way and, especially if those in the club are dressed sensually and provocatively, it can induce feelings of sexual tension. If you then add to that already powerful mixture a lyric that is extremely sexual in nature it will, of course, go down well in the clubs!

Aside from the "Honey bring it close to my lips" and the "It's gotta be big" lyrics, Armand Van Helden also used a different section of the vocal in the breakdown of his version. I can only assume that he wanted a more flowing section of vocal to use over the breakdown and therefore chose that part because, in all honesty, I can't see any lyrical relevance in that part of the track to the rest of his remix. But the effect is just what is needed: something more floaty and melodic. This serves as a great counterpoint to the rest of the vocals and provides a brief moment of calm before launching back into the incessant groove of the rest of the remix.

Arrangement

For the Freemasons remix we looked at the arrangement of a radio edit but the Armand Van Helden link I included is to the full club mix, so this will serve as an interesting comparison. In the last chapter we looked at the arrangement of a typical club mix and, in many ways, this Armand Van Helden remix conforms to that basic idea. The first 1:15 of the remix is purely drums and percussion, which works nicely for the DJ who would use that part of the remix to mix in from the previous song. At this point the bass line comes in and the pace starts to build up. The next main *hitpoint* of any vocal remix is the introduction of the vocal and, in this case, that comes at around 1:45, which is, again, roughly around the time where we would expect it. When the vocal does come in there is a drop to just the bass line, which serves as a small breakdown, as we might expect.

From there until around 3 minutes in, things continue fairly smoothly, with a few different variations on the vocal arrangement to keep the interest up. At 3 minutes, however, there is the introduction of another musical part in the form of a quite spacey and airy melody line. In addition, the vocal drops out at this point so, in effect, the new melody line takes over from the vocal in providing melodic interest. The next thing we might expect, given how far into the remix we are, is the main breakdown, and that is exactly what we get at around the 3:45 mark. The breakdown in this case is actually a pretty drastic one with a sudden stop, where everything except a held FX type sound drops out completely. This continues for almost 15 seconds at which point the different vocal section is heard ("Rest your shoulders...").

Toward the end of this breakdown vocal, the first drum loop we heard returns for a few bars before all of the musical elements return. What is perhaps unusual to me in this breakdown is the lack of any kind of dramatic build-up before the return to the main groove. The breakdown starts off with a big drop before the vocal comes in. That is all pretty normal but, at the end of the vocal, I would ordinarily have expected a bigger build-up with a more obvious cue as to when the rest of the music would kick in, but we don't get that at all. We do get, as I mentioned, the return of one of the drum loops, but that simply plays for a little while and then, with no warning at all, the rest of the drums and the bass line return. That is really the only part of this remix that doesn't sit that well with me.

The total length of the breakdown is around 1 minute, which, again, fits in with the rough outline I gave in the previous chapter. When things do kick back in we get just the drums and the bass line for a little while, and then the vocal returns in much the same format as it did at the beginning of the record. After the vocals have been running for about 1 minute we see the return of the alternate melody sound that we had in the first part of the track. If we were looking at a full vocal mix (or a remix of a song that had a more traditional verse/bridge/chorus–type arrangement, then this wouldn't really make much sense. This is because we have had the main breakdown and then things build up again over a period of time, rather than the more expected big chorus after the main breakdown and then a gradual reduction of energy and dynamics toward the end of the track. This does make some sense, though, in light of the fact that there is a very small amount of vocal to be used. And this variation in arrangement is highlighted even more by the inclusion of a second quite dramatic breakdown at around the 7-minute mark. This time, everything except the alternate melody line is removed, but, as you might expect, for a much shorter length than the main breakdown.

After everything kicks back in, things become more conventional again. The main bass line and all of the drums and percussion return briefly and then the bass line is removed, leaving only the drums as an outro to the remix, which will allow the DJ to mix into the next track easily.

Overall

The approach to this remix is radically different from the approach taken by the Freemasons in their remix of Beyoncé's "Déjà Vu." The Freemasons aimed to preserve a lot of the structure, meaning, and flow of the original song but simply shift it from one genre into another, whereas Armand Van Helden's remix creates something that is, to a large extent, a completely new song.

It is, nevertheless, a truly groundbreaking remix. I think it probably helped that Armand Van Helden was given complete freedom with this and was, in fact, pretty much told to do something "out there" and unlike anything else around at the time. Although it might sound a little dated today, it was very much ahead of its time when it first came out. Many of the remixes you hear today owe a lot to this pioneering remix. So, technically it might not be the most interesting remix, it might have very little going on, and it might be only loosely related to the original, but if you want proof of how good it is, play it on the dancefloor of a club in 2011, some 15 years after it was made, and just see the reaction it gets!

Just as a quick diversion, and to demonstrate another remix that is only loosely based on the original song, there is another remix I think you should take a listen to. It is a track by Roger Sanchez called "Lost," and the accompanying Ramirez remix.

ROGER SANCHEZ "LOST" (RAMIREZ "LOST IN RAVE" REMIX)

Original version: *www.youtube.com/watch?v=jtjBXwwT2xE*

Remix version: *www.youtube.com/watch?v=CVIJ38lEa_8*

In this case the remix version uses just two words ("I'm lost") from the original song and, even then, they really aren't recognizable. If the two tracks were played back to back, and if you didn't know that one was a remix of the other, I am pretty sure that most people wouldn't even realize that one was derived from the other, in spite of the "I'm lost" phrase. It just goes to show how far you can sometimes go from the original track when you are remixing.

In cases like this, you may be well-advised to talk to the client before doing anything so extreme because it is entirely possible that he would be of the opinion that your remix isn't anything to do with the original track. Labels and artists these days are more open-minded about things like this, but it's always worth checking, especially if you are working to a tight deadline.

CHAPTER 8
A Remixer's Insight

To give you a bigger picture of what it is like to be a remixer, I asked a series of short questions to a number of remixers, both past and present, who span a variety of different genres. I hope that by reading their different responses you will get a feel for the industry as a whole and see certain things that seem to be commonly agreed on and accepted by many of them and other things that they have very different opinions about. In addition to these, there are more interviews on this book's companion website.

How and when did you first get into remixing?

Glen Nichols: The first remix I ever did was possibly on a 4-track tape recorder when I was 15. It was of something I made and sang on but replaced the music and spun in the vocals live from a second cassette player. I got the bug of turning a track into something different back then after being influenced by remixes by William Orbit and Vince Clarke.

Adam White: I got involved in remixing, back in 2003, because it is part of the logical progression of a dance music producer. It's the opportunity for you as an artist to put your stamp on somebody else's record. It's a fundamental part of the career of a DJ/Producer.

Phil Harding: I started remixing very soon after becoming an engineer at The Marquee Studios in London, around 1978/1979. In those early days for me, disco projects were coming into the studio from all over the world, but I did quite a bit for French clients. Mainly the mixes were radio or album length, some may have been extended versions, but not what we would really now call a club mix or 12″ mix.

StoneBridge: I started out 1987 with re-edits of big tunes using reel-to-reel tape. The next step was to add samples to the mixes and finally getting the Atari running Notator and a 24-track tape machine around 1989.

Max Graham: I started producing back in 1998 but was really not very good until late '99 and early 2000 when I made Airtight and Bar None. Then I started to get interest as those records circulated and really casual remix offers started to come in.

Ben Liebrand: Just a pure passion for music and the electronics and audio equipment that come with it.

Justin Strauss: In 1983. I had been DJing at some great, influential New York City clubs, Mudd Club, The Ritz, Area, Tunnel, et cetera, and developed a sound and style to my DJing. I wanted to take that further and try to capture that sound on records. I always loved getting great remixes to play as a DJ and thought that I had something to offer. I would hear records and think about what I would do to make them work better for me as a DJ. I had some studio experience having recorded an album for Island Records with a band I was in as lead singer and some solo recordings I had made. My dad was also an audiophile and I grew up with tons of audio equipment in my house.

Mark Saunders: In 1984 I went from playing around in my home studio housed in an old cowshed on my Dad's farm in Hampshire, U.K. to working as an assistant to the production team of Clive Langer and Alan Winstanley (producers of all the Madness hits and Dexy's Midnight Runners' big hit "Come on Eileen") in their brand new top-class SSL studio in West London. This was a stroke of good fortune and being in the right place at the right time. I got to watch great acts and producers at work as I learned the ropes and made the tea. Alan and Clive were constantly throwing me in at the deep end before I felt like I really knew what I was doing, but it was great because it really sped-up my transition from assistant engineer to engineer. It was a year and one day after I started working with Alan and Clive that I got to work on "Dancing in the Street" with David Bowie and Mick Jagger and actually got an engineering credit on it—one of my first. I was pretty happy with that!

Vince Clarke: I was asked by the manager of Happy Mondays to do a remix of "Wrote for Luck" around 1988.

Bill Hamel: I was working at a local record store in Orlando with legendary DJ Chris Fortier. I saw him remix a few big tunes and wanted to give it a shot too. Back then I was DJing a ton and my favorite records always seemed to be remixes. So it seemed like I had found my calling.

Ian Curnow: Working at PWL, Pete Waterman was very focused on the dance market and myself and Phil (Harding) were ideally placed to do that.

Ian Masterson: The first mix I did was Mark Snow's "The X-Files" theme. It was originally a dance cover version I did with a mate, but my manager found out that Warner was releasing the theme as a single anyway, so we offered our version as a remix, which they accepted.

Alex Sowyrda: Probably around 1999.

What was your first "official" remix and how did you get the gig?

Glen Nichols: My first official remix was probably for a singer from the Netherlands called Sue Chaloner. A friend of mine produced and cowrote with her and I bugged him to give me the vocals, so I did my own thing by sampling each phrase into my Kurzweil K2000. That was in 1997 (I think!).

Adam White: The first official remix I did was while working for Euphoria. I was compiling one of the albums, and one of the tracks we wanted to include didn't have the right sound. I was sent the parts and worked with another engineer to create a remix that would be suitable for the album.

Phil Harding: Hard to say for sure, but it may have been some of the extended remixes I did for Stiff Records or even the Strawberry Switchblade remix I did for Warner Bros Records. For sure, the first true 12″ club remix I did was "I Wanna Sing" by the JALN Band for Magnet Records with Pete Waterman as the A&R man in the studio with me, guiding me through the process of extending a basic three-minute (version) and mixing it section by section onto ½″ analog and editing it together as we went along in a club mix format. That was 1982 I think.

StoneBridge: It was a very serious jazz artist in Sweden and the mix was more or less a re-edit with added effects and beats synced up from records. I can't believe how primitive things were back then, but the label loved it. The label had heard my re-edits and wanted something similar for their artist.

Max Graham: I believe the first real one was Bullitt "Cried to Dream." I had just finished Airtight and they were really into it. I did one mix that they hated and I went back to the studio and basically made it in that Airtight style and they loved it. I still get a lot of people mentioning it even today.

Ben Liebrand: The Limit "She's So Divine." Rob and Bernard of The Limit were customers at the club where I was spinning. They heard my edits I was playing to the audience and thought I'd be the man to edit their 12″. This was in 1982.

Justin Strauss: My first remix to be released was "Wild Rain" by a band called Velveteen. The bass player in the band, Sal Maida, was the bass player in the band I was in years before and when his new band got signed to Atlantic Records he asked me if I wanted to get involved in some of the recording they were doing. I jumped at the chance to get in the studio and work with them. I coproduced one song called "Niteline" and did my first remix with my good friend Ivan Ivan who was also a great New York DJ. We did a remix of one of the other tracks on the record called "Wild Rain." The remix version was called "Get Wild" and was an uptempo dance rock kind of song which we stripped down and dubbed out, using bits of the vocals with effects on them and extending the instrumental parts to give it more of a club vibe. It was exciting and I couldn't wait to get back into the studio for more remix work. Sometime after that I teamed up with another New York DJ friend of mine, Murray Elias. We formed a production company called Popstand Productions in 1984. We got our first remix from Elektra Records, to remix the song "Reunited" by Greg Kihn. He was just coming off a big hit "Jeopardy" and this was the follow-up. We were super excited… basically learning as we went along. We hired the engineer Jay Mark for this, who worked at Sigma Sound in Philly, and also worked on the "Jeopardy" remix. I learned so much about mixing and remixing from the great engineers I was able to work with over the years: Jay Mark, Andy Wallace, Hugo Dwyer, Daniel Abraham, and Frank Heller. I have so much respect for these guys.

Mark Saunders: I was still an assistant at Clive and Alan's West Side Studios when I got my first remixing gig. The head of Warner in the U.K., Rob Dickins, who was a friend of Clive Langer's, came to the studio and told Clive that he'd had an argument with Robert Plant about a remix of one of Robert's solo album songs. Apparently Rob hated it, Robert liked it, and the argument ended with Plant challenging Dickins to do his own remix if he thought it could be better. My boss, Clive, suggested that he let me have a go at a remix of the song, which was called "Little By Little." So, I worked on the remix, which I was thrilled about. Rob Dickins came to the studio, played pool and occasionally popped his head into the control room to take a listen, but mostly left me to get on with it. It was accepted by Plant and released—"Remixed by Rob Dickins, remix engineer Mark Saunders." I don't think I got paid for it, now I come to think of it!

Vince Clarke: Ditto (Happy Mondays, "Wrote for Luck").

Bill Hamel: I did a remix for Stress Records/DMC of Secret Life's "She Holds the Key."

Ian Curnow: I really can't remember. Pete Waterman probably got the gig for us, and it may well have been something like "Instinctual" by Imagination or some Mel and Kim singles, which we mashed with Chic's "Le Freak."

Ian Masterson: It was the Mark Snow track I mentioned.

Alex Sowyrda: The first track we (Koishii & Hush) remixed was Fuzzbox "International Rescue."

What would you consider to be your "desert-island gear?" (Name a few pieces of kit that you couldn't live without and a brief description of why.) and how important do you think equipment is as a producer/remixer?

Glen Nichols: Obviously first and foremost the computer with Logic, Ableton, and Pro Tools (and now also Studio One!); my Focusrite Red 3 compressor which is audio cream on vocals and bass!; Reaktor (a must for mangling sounds); and my Crane Song STC-8 for that super fat compression.

Adam White: Firstly my Mac. It is the mother ship of the studio and the heart of everything that happens. Secondly Logic Studio; this is my sequencing software and in my opinion is the best piece of studio software in the marketplace, though it is all down to personal preference. Thirdly are my Mackie monitors. Fellow producer Andy Moor turned me onto these bad boys and I really cannot live without them. They are as important as my computer. Lastly I'd need to take a coffee machine and case of peanut butter and bread. These are essential fuels in the long dark lonely hours of music production.

Phil Harding: Back in the 1980s it was:

a) AMS delay machine/sampler—great for replacing drum sounds and grabbing short samples.

b) Publison Inferno MIDI delay and stereo sampler—great for playing vocal sample locks on a MIDI keyboard and retaining the tempo of the sample.

c) Linn 9000—was great for quick drum programming and quick MIDI programming for bass and synth sequencer lines.

d) SRC SMPTE sync machine—great for getting you in sync with any code and allowing you to sequence your overdubs tightly.

e) Roland 3000 stereo delay machine—best and most functional delay machine ever made.

Now it is simply just Pro Tools. Equipment and the personal choices behind them are crucial to producers and remixers and these are our daily tools for the job, so there has to be a human/machine-love/work relationship.

StoneBridge: These days it's quite simple as you really only need a MacBook Pro and Logic to do 90% of everything. I guess a bunch of soft synths and plug-ins to go with it. In the studio I can't live without the MacBook Pro, Logic with

TDM hardware, my Tube EQ, and the trusted old (Roland) JV-1080.

Max Graham: Well I think everyone has gear they love and could not live without but if you take someone really talented and give them anything, they will end up making music from it. It's not the gear at all it's the person behind it. That being said, the gear really allows you to create those crazy ideas you have. Right now I love Ableton. The workflow is really clean and allows you to get ideas down quickly. Sylenth as a synth has been great find lately. Also my Simplon FabFilter, Massive synth, and probably the Spectrasonics Omnisphere.

Ben Liebrand: In former days, that would have been my Akai MPC60 and Akai samplers. Now it is a Mac Pro and Ableton Live. The third, fourth, and fifth pieces of equipment would be my amp and two speakers.

Justin Strauss: My favorite pieces of gear back in the day were my (E-mu) SP-12 drum machine, which I used on most of my mixes till we started doing more programming on computer. It was like the first sampling drum machine and had this quirky feel to it. A unique sound and vibe. I eventually would hook it up to my Akai sampler for the sounds, but loved the feel of the SP-12 buttons. I had great samples from all the Roland drum machines and other samples I collected and was able to program them on the SP-12. My two Juno keyboards, the 106 and the 60, which I still use and love till this day. Great bass sounds and really nice analog synth sounds. And I loved doing mixes on the SSL board. The board had "total recall," which was amazing when doing mixing, as you could just build your mix and keep adding to it. I have seen so many changes in the way records are made. I am not that hung up on equipment. I have heard some really awful records made with a lot of great equipment and some amazing records made with next to nothing. Without sounding corny, I believe the best pieces of equipment a remixer or producer has is their creativity and imagination and inspiration.

Mark Saunders:

1) Logic Pro—I've been using this since version 1.0 came out in 1994 and I love it.
2) Roland Juno-106. I love this old analog synth. It's hard to get it to sound bad!
3) impOSCar—GForce's software version of the OSCar analog synth. This little puppy has character.
4) Battery Drums by Native Instruments. Along with Logic's Ultrabeat, it's my main source of drum sounds.

5) Manley Variable Mu compressor limiter. This beauty makes (in conjunction with my NTI EQ3 stereo EQ unit) mixes gel in a way that's really hard to do with software equivalents.

Vince Clarke: For remixing it would be a Mac with Logic Audio.

Bill Hamel: I'd take my iMac i7 27″, my Mackie speakers including the subwoofer, a MIDI controller, my vintage Roland Jupiter 8, and my Lemur touchscreen controller.

Ian Curnow: Cubase (Mac), Nexus, Halion Sonic, Waves plug-ins, Neumann U87i mic. That's all I need really… the rest is icing on the cake. Equipment is only 50% of it, the rest is ears. If you have the ears you can make pretty much anything sound good, within reason.

Ian Masterson: Mac G5 plus all the software I have on it and the Avid Pro Tools kit I use, Mackie HR824 monitors, AKG C12 mic, Focusrite ISA430 II mic pre. I don't use much outboard equipment any more.

Alex Sowyrda: These days it's pretty much all "in the box." Mac Pro, Logic 9, Dynaudio BM12a monitors, and a nice selection of plug-in synths and effects covers most things. Favorite (plug-in) synths would be ReFX Nexus, Arturia Minimoog, and ARP-2600, and, lately, the G-Media impOSCar and Oddity. The combination of modern, cutting edge sound and the creaky and dusty old "virtual analog" stuff is key really. We (Koishii & Hush) are big fans of the Universal Audio UAD platform and have quite a few of their toys to play with as well. Other than that we have picked up a few old classic hardware synths lately including a Roland D-50, Yamaha DX-7, Ensoniq ESQ-1, and Roland JD-800. In terms of recording we mainly use a really nice Studio Electronics ribbon mic as it gives a beautiful warmth and presence which works perfectly with the kind of tracks we produce.

To be honest though the equipment (or lack of it) shouldn't be the deciding factor in how good or bad your work is. Sure, "better" equipment can help but there is nothing saying that you can't produce great music on pretty crappy equipment. Or, vice versa, pretty crappy music on great equipment!

What would you say is your trademark sound or style?

Glen Nichols: I'm best known for my Future Funk Squad breakbeat remixes so I guess that would be my style, although I have a few different aliases so my remix approach and style varies from project to project depending on what I'm asked to do!

Phil Harding: The PWL club and radio sound was my trademark in the 1980s. I don't really have one currently.

StoneBridge: I'm a song man at heart so I never completely abandon the original song when I remix. I usually follow the vibe, but with my trademark beats and bass lines. Sometimes you have to recreate the track completely and that can be an interesting challenge of course, but I'd say my style is based around songs.

Max Graham: I have a few things people tell me I do: my drums are never dry, I like a lot of delay and movement, I tend to overlay then strip down later, but it's led to some muddy mixdowns that I frown back upon! I use strings a lot and love big soaring chords.

Ben Liebrand: People find my mixes to be fluent and with a clear logic to them. My productions are inspired by those elements plus all that inspires me.

Justin Strauss: I am influenced by and love many different kinds of music, and have worked on many different kinds of records by a wide variety of artists. I have always tried to treat each remix on its own and do what I thought would suit the song. Doing dance mixes I always spent a lot of time working on the drum and rhythm tracks and people tell me they can recognize my mixes straight off after hearing the drums, so I guess that, and trying to make records that I would play as a DJ. Back in the day I would always do two mixes, one a bit closer in feel to the original and then my "just right" mixes where I would do very different interpretations of the songs, a lot of times infusing my love of house music into those mixes, which would take the song in a whole different direction.

Mark Saunders: Hmm… hard to say. People have told me that I have a "sound" but I don't hear it myself. I've worked on a lot of different types of material and I don't feel like I have a formula. I wish I did!

Vince Clarke: I don't think I have a trademark sound… it depends on the "toon."

Bill Hamel: Melodic U.K. progressive house.

Ian Curnow: No idea—kitchen sink production maybe?

Ian Masterson: It's changed a bit over the years but I guess driving, melancholic, epic pop dance.

Alex Sowyrda: Trance/club.

Remixer styles can tend to vary or evolve over time. Was there one turning point that you encountered during a remix that prompted a major style change?

Glen Nichols: Not really for me, I don't like to follow trends to be honest so I just do what I hear from the original track and interpret it exactly how I feel at that moment.

Adam White: A remix rarely ends up sounding like you intended it to when you started. 99% of the time, where you start and where you finish are two completely different ideas… that's a large part of the magic.

Phil Harding: Chicago house in 1986.

StoneBridge: I have been careful not to change anything too drastically, but I do follow the times and change bass sounds, drum sounds, et cetera, to make things sound current. It's important to me not to be too caught up in current trends as typical novelty based mixes tend to age badly.

Max Graham: I think after Crank came out I was on a totally different trip, more electro and farty sounds. But I'm back with my first love now, the energetic but emotional progressive sound that borders on trance.

Ben Liebrand: It actually evolved over time, where the gathering of knowledge and experience gradually kept raising the bar. As for megamixes and mashups, I believe that anyone still making mixes that clash tonally should be sent off to that desert island, preferably without any equipment.

Justin Strauss: The first few records I did (with Murray Elias as Popstand Productions), it was basically a learning experience: doing some additional production, extending the song, making it more dance floor friendly. When we got the job to do Debbie Harry's "In Love with Love," the original was pretty much a traditional rock song. At that time freestyle music was blowing up in the clubs in New York City. I thought it would be cool to remix the song in that style. I did the drum programming and it really suited the track and Debbie's voice. It took the song in a completely different direction and became a huge club hit, especially here in New York. Little Louie Vega pushed the record and it became an anthem at the Funhouse. He invited Murray and myself down one night to see the crowd's reaction when he played it and we were like "whoa." The place went nuts when Louie put it on. From that point on I knew we were on the right track. Not necessarily doing freestyle music, but with pushing the boundaries of what you can do with a mix.

Vince Clarke: I recently starting experimenting more with minimal techno grooves.

Bill Hamel: Every time I start a remix it usually ends up much different than how it began or I initially saw it in my head.

Ian Curnow: Soul II Soul ended the 4's (4/4 beat) kick drum monopoly and showed that you could make dance records at a slower tempo and with rare groove beats too.

Ian Masterson: Can't think of one, no. It's not very practical to go experimental when you've been commissioned to do a particular thing.

Alex Sowyrda: Not really. Koishii & Hush have always done various styles but have managed to stay within one main genre. There are other projects under a few other names if a major format shift is required.

Do you DJ yourself and would you consider a DJ perspective to be important as a remixer?

Glen Nichols: I do DJ and have been as long as I have been professionally producing. For my FFS project it was an integral part of the process as it enabled me to test early versions on the dancefloor and make a few tweaks here and there before completing the final product.

Adam White: Certainly! Being a DJ and dance music producer goes hand in hand. When I play a track that works in a club I make a mental note of what does (and doesn't) work. I include every aspect of what makes a dancefloor rock in my mixes.

Phil Harding: I don't and never have DJed, but I would now consider it a crucial part of a remix team.

StoneBridge: I DJ every weekend and feel I totally need it. If you do radio mixes only I suppose it's not as important, but for club mixes you need to play regularly to get the beats right. Also the arrangement, as non-DJing remixers tend to get a bit complex and not giving the mixes enough groovy sections.

Max Graham: I'm a DJ first and foremost and I think it's very inspiring to DJ other people's music then go and make your own. There are so many amazing producers out there and DJing their stuff really makes you want to go and produce. I don't think it's crucial but it's great to know how a dancefloor moves.

Ben Liebrand: Absolutely! It is exactly that understanding of what makes people move that gives you the edge over other producers. Yes, they can visit a club too, but they do not see the response to new tracks on the dancefloor each and every week.

Justin Strauss: I started DJing at the Mudd Club in New York City in 1980. I still DJ and consider it my best training ground as a remixer/producer. My favorite remixers and inspirations back in the day when I started, Larry Levan, Shep Pettibone, Francois K. were obviously amazing DJs. For me it is very important. To have that connection with the dancefloor and to see what people react to is a great thing. As a DJ, being constantly inspired by great records has kept me excited about remixing records now and then.

Mark Saunders: I'm a frustrated DJ. I've owned Native Instrument's Traktor for years but never really been brave enough to give DJing a go seriously. I love the idea though. I'm sure being a DJ helps but I don't really consider myself a "club" remixer. I don't really do full-on hardcore dance mixes. I feel like my remixes are more like alternative versions of the originals than club versions.

Vince Clarke: I don't DJ (usually in bed by 9 PM heh! heh!).

Bill Hamel: I started off DJing for two years before I got the urge to buy gear and learn how to produce and play music. Back in the early 2000s when I was DJing out every weekend people would always come up to me and say "You are a really good DJ for a producer. Most producers suck at DJing." Then I'd tell them that DJing was my first passion but I made a name for myself via my record label Sunkissed and my remixes. I've always embraced technology when I DJ. I was mixing records in key back in 1996 and I was the first person I ever saw burn CDs and carry a CDJ-100 to all of my gigs so I could spin my CDs of unfinished productions and whatnot. Now when I do spin out I use my laptop and my Lemur controller. I want to get back into playing out a bit more. I think it's a good balance for your productions. It allows you to feel the pulse of the dancefloor and see what's really working on the dancefloor at that point in time.

Ian Curnow: No, wouldn't know where to start. I did have a go but I was terrible! The perspective is important, but I don't think you need to actually DJ, just listen in clubs and absorb what seems to go down well there.

Ian Masterson: No I don't DJ. I spend enough time in the studio and in clubs already without trying to be good at something other people specialize in doing. You have to be really committed to be a DJ and I'm too comfortable in the studio. I think being a DJ can be of invaluable help if you produce in a certain way. I just make tracks that make me dance in the studio, and hope for the best.

Alex Sowyrda: No, but I do think it helps to have an ear for what is going down well in clubs.

How important do you think remixing has become to the music industry as a whole?

Glen Nichols: In dance music I think it has become pretty important but I kind of feel taking other forms and remixing in too drastic a style has cheapened the market to some extent and I also feel that now everyone can be a "remixer" the magic has gone a little.

Adam White: I think it has become extremely important. Dance mixes of tracks open them up to a whole new marketplace. The best example of this, in my opinion, is U2. So many great new bands and artists are having their music mixed to be played in clubs and this raises profile for both the artist and remixer. I am so pleased that dance music has started to be embraced this way in other genres of the music industry.

Phil Harding: Less and less so since the 1990s. It's still a thriving industry but fees have dropped dramatically.

StoneBridge: I think we may have gone too far and sometimes labels don't trust the project to be strong enough by itself and get way too many mixes done just to be on the safe side. This development is, of course, great for us remixers, but in the long run, it messes with artist identity.

Max Graham: Well producing as a whole has become crucial to your success, but just remixing won't get you as far as producing original material also. Remixing is really about helping songs that may not appeal to certain crowds cross over to those crowds. Like a techy mix of a trancey track that the techier dancefloors will like.

Ben Liebrand: It has always been important. Even with classic tracks way back in the '70s, many people do not even realize that what they consider to be the original version might already be the second or third attempt to mix down that song. Nowadays the dance market has become more divided into specific styles than back in the '80s when "dance" covered all genres within. So now people try to adapt their songs in several mixes to each different genre, often coming up with house or club mixes from downtempo "urban" tracks. Same goes for adaptations between house, club, and trance. This way one can optimize the chances of a hit by making a song available for several markets. The remix is also well-suited for rereleasing a track in the style today.

Justin Strauss: I have mixed feelings on this question. It has become very important and at the same time almost ridiculous. A tune nowadays can have so many remixes it just becomes silly. With the Internet and home recording studios, everyone and their mother has some kind of remix out there somewhere. With rare exceptions, most of it to me doesn't sound very good. Back in the day, you went to a remixer because you wanted a certain sound and you knew you would get a quality production. Not to say there isn't some great work being done out there because there is. But sometimes it just seems too much and who has the time to listen to 95 mixes of one song?

Mark Saunders: I think the remix thing has got a bit overblown these days, mostly since it became the trend for artists to be asked to remix other people's songs. There are a gazillion remixes out there but how many are actually any good and memorable in any way? Only a small percentage I think. There are tons of really average ones.

Vince Clarke: I think it's become more important because there are now so many web outlets (Internet radio, Beatport. etc.).

Bill Hamel: I think it's VERY IMPORTANT. How else will you get pop, rock, or hip hop into dance clubs? It's the perfect solution to get artists more exposure in a different area than they'd probably never get otherwise. There are some songs that shouldn't be though, nor lend themselves to a good remix though… lol.

Ian Curnow: Pretty crucial. By no means most, but quite a few, singles end up being cut up from the remix, and that's been the case for many years.

Ian Masterson: Well it's been central in helping artists get exposure across different dancefloors. Remixes tend to either be amazing, creative reinterpretations, or cynical exercises in remarketing music. Occasionally they are both.

Alex Sowyrda: I think remixes have been hugely important in getting tracks to a greater audience. With many remixers having their own fanbase, the original artist can benefit by having their track purchased by a fan of the remixer as well. I know I bought many a track having never heard it just because it was remixed by Matt Darey.

As an artist yourself, how does it feel when you get remixes done of your own tracks? Are you perhaps a little more forgiving, given that you know what it's like having to remix somebody else's song?

Glen Nichols: I love hearing other people remix my material although I only approach people I really respect in the first place, which helps. I think I have only ever turned down two remixes that have been commissioned by myself for my own projects.

Adam White: To be honest this is where good management and A&R step in. It is their job to source the best person for the mix, give them the correct brief, and let them know exactly what they want—after all they've been chosen to do the mix in the first place. This is a massive part of the remix process and sometimes gets overlooked.

Phil Harding: It's a wonderful feeling to have a great remix come in of something you have either performed or produced. A great remix can turn the record around and it can quickly become the main radio version in an edited form.

StoneBridge: I love remixes of my songs and get very impressed when people totally change them and bring out something

I didn't think of myself. Around the time I did "Put 'Em High," remixes came in handy to extend club play, but now with Beatport and other online stores, the remixes tend to split the focus and the more mixes you have, the less is the chance to get a top 10 record on there so it's worth thinking about right now.

Max Graham: I'm not sure what you mean by forgiving, but it's really cool to hear someone else's interpretation of your ideas. I love hearing remixes of my own stuff.

Ben Liebrand: I find it very difficult to have other people remix my songs. Even if they come up with a great approach, I still often am bothered by details that I would have done differently. Things like build-up of the song, or a brilliant sound or sequence which hasn't been utilized to the max.

Justin Strauss: I had mixes of my own stuff done and I love hearing how someone else reinterprets my work. I think that is the most exciting thing about remixes. I never understand why some people get so bent out of shape about remixes when they are very different to the originals. The original is always there for anyone who wants it and the remix is there as well. So it's the best of both worlds.

Mark Saunders: I think the first time that someone mixed a track that I'd coproduced and recorded was when Arthur Baker remixed Neneh Cherry's song "Buffalo Stance" in '87 (I think). That felt good!

Vince Clarke: If they are good, then great. I think I've become more forgiving now that I am more involved in remixing myself.

Bill Hamel: Totally not. In fact, I may go over it a bit more with a fine-toothed comb. If I hear anything that's not correct I usually work with the remixers to make sure it's up to par and to politely lend my advice and tricks of the trade that I've learned over the last 16 years. I want to see their remix turn out better than my original.

Ian Curnow: I love hearing my work remixed as it shows the tracks in a different light I would probably never have thought of. My only issue is with DJ mixers who don't understand chords. Too many remixes seem to be "harmonically challenging" and don't support the vocals at all. There's an art to finding the minimum amount of harmonic movement to keep a groove, but still support the song enough to make it work.

Ian Masterson: I like getting other people's interpretations of stuff I've done. I like it less when their remixes are better than my version.

Alex Sowyrda: If the mix sounds good we are always pleased. If it sounds like crap we are never as keen on it.

What is your favorite remix (by somebody else) and what is it about that mix that makes it your favorite?

Glen Nichols: Oh that is a really, really tough question! William Orbit's remix of Erasure's "Star" was possibly one of my earliest influences as a remix, just the way he added his blippy digital goodness to quite a live-sounding track! Blew me away!

Adam White: I think one of the best remixes I have ever heard is the Hardfloor remix of More Kante "Yeke Yeke." This is a real example of how a producer can take the essence of a "non-dance" record and turn it into something magical. It speaks for itself and it is 16 years old now and still sounds as fresh as it ever has done—I still play it!

StoneBridge: There are so many that it's hard to pick one, but let's say Armand Van Helden's mix of Tori Amos "Professional Widow" as he completely reproduced and rewrote the track and gave her her biggest hit. Amazing mix and groundbreaking in many ways.

Max Graham: I don't think I have any one favorite, but what I love in a remix is when the original is respected rather than just a tiny bit of it used…when the essence of the track is kept intact but made into the style of the remixer.

Ben Liebrand: There are too many to list them, but in general, my favorite would be a remix that takes the essentials of a track, leaves out all the boring stuff, and gives it a twist so it deserves to lead a life on its own.

Justin Strauss: This is the hardest question for me. It's like asking me about what is my favorite record because I really don't just have one favorite record or remix. I have been listening to music my whole life and DJing and producing records for a long time, and many records are special to me for many different reasons. But I'll say this. Shep Pettibone's remixes of New Order's "Bizarre Love Triangle" and "True Faith," Pet Shop Boys' "West End Girls," and "Love Comes Quickly," and Depeche Mode's "Behind The Wheel" will always be at the top of my list. Along with Larry Levan's Gwen Guthrie remixes among many others he did, and Francoise K.'s amazing remixing during his Prelude Record years, and Patrick Cowley's mega mix of Donna Summer's "I feel Love" all have influenced and inspired me and still do. Currently I love the work of DFA, The Glimmers, Prins Thomas, and Aeroplane, among others who are doing some amazing work.

Mark Saunders: Clear-cut winner for me is The Killers' "Mr. Brightside" remix by Jacques Lu Cont (Stuart Price). It's so incredibly musically rewarding to listen to. Stuart hooks the listener in right from the start with the synth riff and there's not a dull moment for in the whole 8:50. It's totally epic. When the bass comes in it makes the hair stand up on the back of my neck. I love the squelchy short delay on the snare too. There's no compromise with the vocal melody (which drives me nuts on a lot of remixes—where the vocal doesn't quite fit melodically with the new track). The track builds beautifully. It's sonically perfect too.

Vince Clarke: Changes from week to week but generally, the more electronic/unreal the better.

Bill Hamel: I have too many to name but a good example would be Rabbit in the Moon's remix of Sarah McLachlan's "Possession." The original is a song that I probably would have overlooked in general, but this remix was not only good on the dancefloor, it became a timeless dance classic.

Ian Curnow: Impossible to answer… too many.

Ian Masterson: This changes daily as I listen to new music, so I couldn't really identify one. As of September 2010 I'm currently hugely into Fear of Tigers, Rex the Dog, Aeroplane, and Anoraak.

Alex Sowyrda: The Paul van Dyk remix of Binary Finary "1998" is just epic. He managed to take a cool track and take it to another level. Also Matt Darey's remix of Gabrielle "Rise" is just phenomenal. The energy that the mix has blows me away each time I hear it.

There are so many social media websites today and signing up to them all seems to be a prerequisite for anybody in the music industry. Do you honestly feel that there is as much benefit to be had from them as is assumed, or do you think that the time spent updating profiles and "tweeting" could be better spent in other ways?

Glen Nichols: Yeah I simply think this time should be spent in the studio learning your craft and being the best at it! I think blogging and all that stuff should be left to a PR company in some ways! I understand as an artist you want people to feel connected to you but then again it never really hurt the careers of big pop stars before it was possible did it?

Adam White: Technology has changed the way you connect with people. At one time it was all about a great flyer, a great poster, and getting out onto the street to promote your product. Now in this digital age it's all about email, smart phones, and reaching people that way. It's made it possible for artists to connect with their fanbase and really hear what their fans are saying (in a positive and negative way).

Phil Harding: MySpace is still a must for most artists/producers/remixers, and I check MySpace at least two or three times per week—that's about it other than daily music news emails and regularly checking on websites like www.mpg.org.uk.

StoneBridge: It's great in the early stages of a new site. Like MySpace was a place you could find talent and careers were started. Then every single artist went on there followed by every producer, DJ, and remixer. When you get 500 newsletters a day you kind of ignore it and move to the next one and the whole scenario is repeated again. What we lack now is the power of word of mouth. People need to find music by themselves and talk about it with their friends and slowly build a following. This was possible with vinyl as there was no other way of hearing a tune but to go to a club and hear it drop.

Max Graham: Well how long does it take to tweet? Six seconds? If you ever spend your time in a recording studio you'll find there are plenty of small downtime moments. I may write a synth line then bounce it down, that bounce can take two minutes and that's *more* than enough time to pick up my iPhone, read my email, check my Facebook, and update my Twitter. I don't feel they are a "prerequisite" but they are an amazing way to directly interact with your fanbase, basically the people you are making the music for. I don't see anything negative with them at all. We can share music as it's made (Soundcloud), photos from our gigs (Facebook), and interact with fans all over the world while we sit on the toilet! I mean how else "could your time be better spent in other ways" while you're sitting on the toilet?

Ben Liebrand: It's not a question of believing, I know that there are better ways to spend your time than wasting it on social sites. Having said that it is simply frightening to see that loads of exposure appears to make up for a lack of quality or any actual substance. So my advice, have substance and make quality so the attention you might draw is actually based on something.

Justin Strauss: The Internet is a great and amazing thing. I use it all the time. Facebook for me has been great way to promote the work I've done and do, and have gotten in contact with many people who have loved what I've done over the years, and that has been very rewarding. However, I think the Internet and downloading is just abused. It should never, and will never for me, replace, listening to a record, reading the liner notes, the cover art, reading a book, a magazine, watching a movie. Everything is just disposable now and it's a shame.

Mark Saunders: I signed up with Twitter and then realized "I have nothing I want to Twitter about!" I don't really want to read other people's Twitters either. I think it's important to keep an updated discography online so anyone searching for me can see what I'm working on.

Vince Clarke: I think Facebook, Twitter, YouTube, et cetera, can really help an artist. Especially now that traditional record labels are on the decline.

Bill Hamel: Tweeting is if you have a large following. For someone like BT or Deadmau5 that has hundreds of thousands of followers, it's the perfect way to interact with your fans. It makes them feel closer to you and keeps them up-to-date on your releases and show dates.

Ian Curnow: I think people should stick to what they do best. A startup artist can spend so much time Tweeting and FaceSpaceing and stuff and lose sight of what they're actually trying to achieve, which is to make music and present themselves as an artist. There are simply not enough hours in the day to do it properly. PR companies have whole departments focused solely on doing the social networking part of a promotion for their artists.

Ian Masterson: I think it's important for artists and producers to stay in touch with the people buying their music, and to listen to what people are telling them, and sites like Twitter can facilitate that. But they undoubtedly fuel procrastination too. I guess, like anything on the Internet, social networking is only as helpful or harmful as you let it be.

Alex Sowyrda: I am torn on this. I see how these sites and services have their place but I miss the days of mystery. I do not need to know when someone has just arrived at a restaurant or watched TV. I think the openness of sharing everything about one's self has ruined some of the illusion and allure.

With the increasing power of computers and software and the relatively cheap cost of setting up your own "studio," do you believe that, to some extent, it has become too easy now and that as a result, some of the true "art" of music making is being lost? Do you think that more education (either in a college/private tuition setting or simply through books and online tutorials) can actually help to improve this situation at all?

Glen Nichols: In some ways I think the fact you can get hold of all of these plug-ins and DAWs for quite cheap that do produce great results is a good thing, but ultimately it still comes down to a definitive idea and that is where the "art" comes from not really the tools! They are simply "tools" and without the idea would be worthless.

Adam White: So many producers come and go and the scene moves a lot quicker than it did for this technological reason. A true artist with an understanding and knowledge of music will stand out from the masses. There are some amazing resources to aid in music production but there is no substitute for raw talent and natural ability.

Phil Harding: I believe that more education is crucial to the future of our industry and vital for future remixers/producers and artists. Dissemination of experience and opportunities for people like myself that can do that for students is an important platform that needs to be encouraged and grown throughout the world.

StoneBridge: It's pure democracy in a way actually. It used to be incredibly expensive and you had to invest massive amounts of money to get a tune recorded and put on vinyl so it was a very select group of people that could do it. Along the way you also had all these filters like A&R men, distributors, record shops, and major labels. Now you can do it all by yourself and this has led to 5000–15,000 new releases a week. How can you possibly find anything with that amount of music coming out? You can't and this is also why some records become hits one or two years after release as it takes time to rise above. Another sad thing is that whole groups of professionals have become obsolete. Studios have shut down, mastering engineers are not used, record shops gone… so yes, the art is somewhat lost, but good songs are still written and hopefully people will appreciate the value of good music again once we've reached the saturation point.

Max Graham: No, our ears know what is crap so even though there are more people making music now and thus more crap, it falls by the wayside very quickly. There are so many talented kids out there doing their own thing, Look at Arty. You're going to tell me because he didn't go to a college or school about sound design that his music is any less amazing? His remixes have been the biggest thing this year and he's producing, with no experience, on a PC with minimal software. I would say no, there's nothing wrong right now.

Ben Liebrand: The way of making music has evolved quickly over the last 30 years ever since the introduction of the first samplers. However, what it still boils down to is having great ideas. The technology nowadays only makes it easier for more people to realize their creativity and convert those ideas into actual music. So each time has its true art of making music, independent of the technology used. What would help without a doubt is have an understanding of music theory. Yes it is boring and feels like you're wasting your time, but the time invested in music theory will repay itself quickly in faster productions where arrangements take form fluently and the final result comes together quicker and better.

Justin Strauss: To learn studio techniques, how to master the programs available today, yeah sure. But like I said before the best tool a producer can possess is his ideas to make a record interesting and great. It's totally incredible what you can do with computers in regards to doing mixes and recording, but if the ideas aren't any good, what's the point?

Mark Saunders: I think that having access to cheaper gear and powerful computers has made music way more interesting. I think it's great when people get the gear and don't really have any idea what they're doing. It leads to a lot more quirky interesting material. The downside is that there's a huge amount of crap to wade through in order to find the good stuff. If everyone went to colleges to learn how to use the stuff there would be a lot less happy accidents, where people stumble across something new and different because they clicked on a wrong button.

Vince Clarke: There has always been good music and bad music. All that's changed is that today there is "more" good music and "more" bad.

Bill Hamel: I think it's awesome. I think it makes music better because affordable gear means more creative people can get their hands on the stuff. Music always needs variety and diversity. It makes music progress.

Ian Curnow: Yes it's easier, but it means that everybody and his dog are doing it, and not always with good results. Tuition can give the building blocks for someone to work with, but the talent has to be there to start with, and tuition simply accelerates the learning process. You need to have the ears first of all, and tuition can give the ears a helping hand (excuse all the body part references!).

Ian Masterson: Any equipment is just a tool. If you put rubbish into it, you're going to get rubbish out. It's no easier than it's ever been to be creative and love what you do. Any talk of "art" in music being restricted to people who can access studio equipment is, and always has been, complete elitist bollocks. Yes it does take time to learn practical skills to achieve the results you're after, but there are no hard-and-fast rules about that. I suppose education can help that, but I'd be wary of people picking up habits or ideas that send them in the wrong direction. Much better to get experience by trial and error and doing it yourself. The affordability of equipment now at least makes that possible. The only downside it that there is much more "noise" from the sheer amount of music being released.

Alex Sowyrda: 100% It is harder to separate the wheat from the chaff. Any courses and this book, for example, is a great tool to help someone with the ambition to gain the skills. Just because you can drive doesn't mean you are an F1 driver.

What would you say is the single biggest challenge of doing a remix?

Glen Nichols: I would say the biggest challenge for me is taking a track that is so well-known and doing something equally as strong, but yet for another market, whether that be genre or age group, as it is very hard to please people who already know and love the original!

Adam White: Staying original is the biggest challenge. With so many new sounds constantly coming through in the dance music scene you've got to always be thinking two months ahead when doing a remix.

Phil Harding: Finding a great groove (drums and bass) that will work on the dancefloors around the world.

StoneBridge: If the song is in a format that is completely different like a ballad or a folky kind of thing with weird time signature. I do take on things like this occasionally just to challenge myself.

Max Graham: Dealing with the A&R people trying to steer the mix in a certain direction. When you have a hit record and the remix offers start coming in, you inevitably get the "make it sound like your other record" comments when you really want to just do your thing.

Ben Liebrand: First of all, making a remix is a bit like making a custom car or hot rod for that matter. So it is of importance to start off with a song that is worth putting work into. But the biggest challenge is to make it into something which earns its own right to exist, next to the original. Sometimes this is achieved by honoring the original and giving the song the groove it needs to excel, other times it is taking a snippet and building a complete new track based on that original snippet.

Justin Strauss: For me, it's always making a record that I would play, and something that takes the record to another level. I've done all kinds of records by many different artists. There are many different people you need to try to please when doing a remix. The artist, the record company, but most important I think you have to make a record that you believe in yourself. The hardest records to do, I think, are the ones you love already. It's easier to take something that isn't so great and make it better. When you get a record that you already like, the challenge is to find something that makes it different, while keeping what you love about the record in the vibe.

Mark Saunders: To make it memorable and not just a four-on-the-floor kick drum, new bass line, and chopped-up vocals over the top. I'm looking to create a new track that the original vocal sits in seamlessly and gives the listener a totally different feeling than the original. But I want it to be hooky too—something that doesn't just sound good at high volume in a club—something that sounds interesting enough to give it a long shelf life.

Vince Clarke: Trying not to repeat yourself. Trying not to repeat yourself. Trying not to repeat yourself (see what I mean?).

Bill Hamel: Easy, polishing a turd.

Ian Curnow: Getting it to work harmonically within the stylistic framework you're looking for. Sometimes it can all work over one or two chords, but it depends on the original.

Ian Masterson: Pleasing myself.

Alex Sowyrda: Trusting in yourself that you have chosen the right direction to take it.

What was the most rewarding feedback you had on a mix, be it from a label, artist, or fan?

Glen Nichols: The most rewarding feedback I had was from Liam Howlett (The Prodigy) when I remixed "Spitfire" for him in 2005. He said he wasn't a huge fan of remixes up until then of his material, but thought my mix had an anarchistic approach and really loved it! (Maybe not those exact words, but words to that effect!) So coming from someone I grew up being massively influenced by, it was a huge compliment and self-achievement for me!

Adam White: The most rewarding feedback I've ever had was for a bootleg I made of Pearl Jam's "Better Man." I got an email from their manager saying that the band really loved what I'd done with their track. Sadly their management did not want them to appeal to a dance market so the mix was never released.

Phil Harding: Probably when the remix was so loved that the artist and label ditched the original record in favor of the remix version and then that going on to be a hit—like "Rise to the Occasion" by Climie Fisher on EMI Records in the 1980s.

StoneBridge: It's when radio picks your mix over the original or when fans request it on gigs. It's a great feeling.

Max Graham: I just had a vocalist ask me for my remix of her track so she could perform it live. Typically a vocalist would perform only the original so it's really flattering that they want to perform my remix.

Ben Liebrand: Foremost, being asked by the artist themselves and as feedback goes, one of the nicest ones was from Bill Withers, taking the time to call me and tell me how wonderful it was to see his kids boogie to his music.

Justin Strauss: I've gotten all kinds of feedback, luckily mostly great from the records I've done. Some artists have hated the mixes that, in the end, were some of my biggest mixes. But one of the most rewarding experiences happened when I was remixing Luther Vandross' "Never Too Much" in 1989. Obviously a classic, amazing record to begin with. After doing the overdubs, I went to Right Track Studios here in New York City to mix it. I didn't know then, but Luther was in the studio next to us and was working on Whitney Houston's record. He found out we were in there and came in to hear what we were doing. He said "I love this and would it be okay if I sang some new parts too?" WOW! So we set up his favorite microphone, an AGK 414, and he sang new parts and the intro and outro of the tune, which were incredible and took the mix to a whole other level.

Mark Saunders: It's been great since being on MySpace and Facebook and getting feedback directly from the public—it never really happened much before. My best feedback for a mix was for a track I worked on for a website dedicated to raising awareness to global warming called blacksmoke.org. They had a theme song that they sent out to loads of remixers and artists and I was one of them. Here's what they said:

> "Holy bat-shit, this is FANTASTIC. The frustrating downside in the development of a project with this scope and ambition is the inevitable stress that comes with securing anything truly innovative. The easy path is not exactly inspiring, although faster and less stressful. Typically we have avoided this like the plague… but the hard one can slightly tarnish the sense of brilliance, for us alone. The biggest payoff is the very rare occasion when something arrives that is completely beyond your expectation. This remix is a perfect and very rare example. Mark has clearly invested a huge amount of time and effort. The guitar sound hints at his Cure roots, but the complete sound is absolutely unique and very fresh. The vocal effects are flawlessly executed, with a sound that's subtle enough to feel completely intentional from the original recording. I'm actually stunned that Mark pulled that off so well. Full credit to him. It is an absolutely incredible remix, due in no small part to the completely fresh sound, new melody, and perfectly executed production. We are genuinely amazed. I'm so

pleased we invited Mark to participate in this unique initiative. The results of his work will help us in ways you can't imagine!"

Unfortunately, I don't think the project ever really got off the ground!

Vince Clarke: The best feedback I've gotten is from Andy Bell ('cos he's my mate!).

Bill Hamel: When I did Seal's "Get it Together," which ultimately got nominated for a Grammy, Seal and his legendary producer Trevor Horn called me up to say how much they loved the remix. They told me they were shooting a video to my mix and that when he played the song out from now on they'd be playing it at my tempo, which was 130 and the original was like 90 or something. I've always been a huge Seal fan.

Ian Curnow: First E17 single was a remix, and it launched the band and charted really high, top 3 maybe? Possibly number 1, I can't remember. Hearing your work on the radio and seeing it in the charts is the most rewarding thing, not because of the income it generates, but because it means that someone out there actually likes what you've done. It completes the performance—like the audience applause after a stage gig.

Ian Masterson: A guy contacted me and told me he met his girlfriend in Ibiza dancing to my remix of Karen Ramirez "Looking For Love." They got married to it 18 months later. That's pretty special.

Alex Sowyrda: Fan emails are always nice and appreciated. The guys from INXS really liked the mix we did of "Need You Tonight" and when we got to meet them at a show afterparty they raved about it. Danni Minogue also had nice things to say about the mix we did for her, so that was a big thrill.

What advice would you give to somebody wanting to get into the remixing industry?

Glen Nichols: Don't expect to make any money out of it, just do it simply for the pleasure and love of it, enjoy the whole process and try and pick tracks that you really think you can do something with!

Adam White: Stay original, listen to lots of different styles of dance music, and pick out the sounds and parts you like. Base your sound upon something that turns you on to start with and really work hard to develop your own sound and stick with it.

Phil Harding: Offer to do free remixes, stay focused and enthusiastic even when someone doesn't like what you've done.

StoneBridge:	Work the song first and foremost. It's more about the artist than it is about you. You might have great ideas that can take the track to a new level and this is what you need to focus on.
Max Graham:	Produce your own music and if those songs do well the remixes will naturally follow.
Ben Liebrand:	There is no set path to follow, so just pour all your energy and enthusiasm into your passion for music and hope it will get noticed. Knowledge of music theory is not a must, but is very useful if you have it.
Justin Strauss:	The whole remixing and music industry has changed so much since I first started. It's so hard to make money doing remixes anymore because everyone expects everything for free these days. But I still love doing remixes. My advice, do something you believe in and get it out there. It's so much easier for someone to get their music heard these days, via the Internet. The hard part is standing out from the crowd, with so much music out there. Have your own voice, your own style and sound.
Mark Saunders:	The way the biz is at the moment, you'd better win the lottery first to support your remixing habit! Or, you could try asking your local supermarket to speculatively submit some food to you and tell them that you'll decide, after you get home and taste it, if you want to pay for it or not.
Vince Clarke:	Do lots of remixes for free for your friends. Maybe enter a few remix competitions on the web?
Bill Hamel:	Just dive in… try and be a fly on the wall at other studios or around other remixes. Try and pick up on other people's techniques and tricks. Find a studio partner or someone that will help push you and pick up the slack when you get burnt out. Be eager and put your foot in as many doors as possible… be prepared for plenty of "no's," but never take "no" for an answer.
Ian Curnow:	Just do it. That's the only way you learn, push yourself but be humble and learn from what others do.
Ian Masterson:	I can never understand why people want to do remixes in themselves. Is there really a "remixing industry"? Surely they're a sideline to making and producing your own music? That's a lot of creative ideas you're expending on something someone else has written. And, quite frankly, there's not much money in remixing anymore unless you're in the top 0.1% people being commissioned. I wish I had spent less time doing remixes and more time writing original material during my early years of making music, so that's my advice.
Alex Sowyrda:	Please don't. There is enough competition already.

Can you talk us through one of your best known remixes and give us an insight into how that came about and let us in on the "process" of doing it a little?

Glen Nichols: Again, this would probably be my "Spitfire" remix. I actually did some preliminary work before getting the stems, so I took the stereo track from the album and chopped out all the bits that were more exposed and also did quite a bit of filtering of main bits and surgical EQ to extract pieces I wanted to use. Once the stems arrived I replaced some of them, although some bits sounded better taken from the stereo track as they just had a bit of character (kind of like sampling from vinyl!). I must admit, however, that the whole remix was done on my 15″ Mac laptop on my sofa as I was in-between studios at the time and had to do it pretty quickly!

Phil Harding: Read my book *PWL from The Factory Floor (Expanded Version)* which will contain many remix descriptions that weren't included in the first version of the book. Here is an example: the Harding/Curnow remix of Pet Shop Boys "It's a Sin."

Done a couple of months after the song had reached #1 in the UK, our mix was used as the main 12" mix for its release in the US, labelled "The Latin Vocal Mix" and the sleeve had a sticker on that said "Remixed in the UK for the USA". It went on to reach #9 in the US Billboard singles chart. Never receiving a release in the UK, it appeared labelled as "The Miami Mix" on an autographed Abbey Road limited edition (of just 20) CD called "The Pet Shop Boys Compiled" for competition winners of BBC Radio 1's "Pet Shop Boys Day" in September 1993. Ian and I weren't really aware that the mix was just for release in America but we threw ourselves into it with our usual gusto and we were quite pleased to be left alone to get on with it because Chris and Neil were known for wanting to be present during their remixes, as was to happen on the next remix we would do for them. We were blissfully unaware that this would not get a UK release.

The mix starts with the classic Miami high synth riff, which is not unlike the House style riffs we had been using for some time with Mel & Kim, which kind of added a PWL sound to it. Once the agogo percussion bell came in, followed by the typical bouncy Miami-style bassline, it's fairly obvious what type of mix we're building into. Fairly straight-ahead drumbeats are followed by our whip sound FX and some vocal triggers which are then followed by a female Spanish language spoken voice. It's not until 1'36" when the "It's A Sin" huge synth chords come in that the listener gets a clue that this is the Pet Shop Boys. I believe this was their first Miami sound mix, so it was all new ground for Chris and Neil and once we had the bass line and groove running under their song overdubs, and Neil's vocal, we were very happy that the mix was going to go down well with everyone. The Spanish female voice comes back in the middle breakdown. I'm not sure if we found that voice or whether it was already there on the tape. We then, rather strangely, got Ian to do a latin-style piano solo, as this had been a feature on quite few other Miami records. I think all of that combined with some of the thunderclaps and keyboards from the original record made for a classic mix. The breakdown under the last verse with a string arpeggio from Ian, which had featured throughout the mix, is also very effective, as well as some more rhythmic vocal samples introduced after that to keep the groove fresh and bubbling along.

StoneBridge: I suppose the Robin S. "Show Me Love" remix is my best-known work and the way it came about was very different to any other remix I've done. I had been a remixer for just a few years, mainly in Sweden, and wanted to get into the U.K. so I called a label I had done some licensing to (Champion) and asked them if they had something old in the basement I could remix, and they said they had this failed project with a great vocal. I got the tape and did three or four different mixes that they didn't like at all. In frustration, I decided to go in and do one final mix. I stripped the track to just the kick drum and vocal, changed the bass sound I had used for the latest mix and it happened to be an organ (preset 17 on the M1). I then found a snare drum from a record, but it had a kick in it, so it got this heavy attack that worked perfectly with the massive kick. I then put on two string chords in the chorus and put a distorted stab thing in the intro and it was done in a little bit over four hours, as I had a gig that night. On the following Monday I listened to it again and thought it was pretty bad, but my girlfriend convinced me to send it off, and the rest is history. The lesson here was that I just went in and did it rather than over-think the process. Believe me, I have tried to do that again, but this was a once-in-a-lifetime mix.

recording of this classic Cure song was nowhere to be found! So, I had to record the song again from scratch in order to have something to remix. I don't remember that much about recording it to be honest, but the band had played it a gazillion times live over the years and it was really quick to get it laid down. In those days we were still syncing computers up to the multitrack tape machines, which was always a bit a scary because there had to be SMPTE timecode recorded on the edge track of the tape (usually the last track, number 24). The scary part is that after hours of rewinding and playing the tape, the edge track could get worn down and it was not uncommon for the SMPTE code to drop in and out, completely buggering the sync between computer and tape machine, which could be disastrous and usually at 3 AM after you'd been working on the remix for a long time. I was probably only using an Atari ST computer and Cubase (not an audio version—they didn't exist yet I don't think). My only way of manipulating or moving audio was to sample it into my Akai S1000 sampler and trigger the samples from Cubase. It seems archaic now! I didn't have a lot of synth power back then. I probably used an Oberheim Matrix 1000 rack synth for most of the synth riffs.

I would create sections of the remix using the automation on the SSL board to mute and unmute tracks. When I was happy with a section I would lay it down to a two-track half-inch tape machine. I would then create a new section, lay it down, and cut and splice the tape to join the two sections together. There was no "undo" back then so if the edit didn't work then I'd have to peel the bit of sticky tape off from across the join in the tape and redo one the sections to make the edit work. It was a very time-consuming and fiddly process which I don't miss in the least! Most of the time it would take two days for a remix, and the second day would nearly always be a 24-hour day and I'd leave the studio at 10 or 11 in the morning when the studio's next client was knocking at the door trying to get me out so they could start.

Vince Clarke: The Happy Monday's mix was done with a Pro One, ARP-2600, and an Akai S900. We had one day, so there was no way it could be anything but "minimal." I think that kind of pressure often results in something urgent and exciting. And most importantly, the starting material was fantastic.

Bill Hamel: I don't really know what to say. I don't want to write a paragraph about time-stretching and composing music. That's boring… lol :)

Ian Curnow: VERY briefly:

- Get in sync with the original track (tracks used to speed up and slow down far more than they do now, which made life, errrm, interesting).
- Decide if the tempo will work and if not time-stretch to the tempo wanted.
- See how the harmonic structure works (work out the original chords then work at simplifying them, often taking a major key and putting it into the minor to sound a little tougher).
- Find a section that you can loop only one or two root notes and work up initial ideas going round that section.
- Work those elements into the rest of the song, and maybe inject some new sections that may have nothing to do with the original but give space for you to have some clear ideas.
- Give it to someone who can engineer to make my mess sound good!

Ian Masterson: I suppose the biggest early remix I did was Pet Shop Boys "A Red Letter Day." They commissioned the mix after we did a Trouser Enthusiasts mix of their track "Discoteca" on spec. To be honest it was so long ago I can't remember much about doing it, but I can tell you there was no lovely hard-disk recording. It would have been an Atari ST triggering Akai S5000/S3000 samplers, a couple of Roland Juno keyboards, a Yamaha TG500, and a Roland TR-909 over MIDI, using Cubase software, and mixed on a Mackie desk. I've still got the floppy disks it was saved onto. Now that's old school!

Alex Sowyrda: I guess it would have to be one of our most recent remixes "Twist Of Fate" by Bad Lieutenant. It was an interesting one to do for us because of the nature of the original song. Because it was more "rock" in nature, it presented a different set of challenges. It had some really strong melodies as well as a very strong identity, so we were very careful in trying to preserve as much of its personality as possible while shifting it into our genre. One thing we decided really early on was that it would have to be pretty energetic, so we started off by building up some really powerful and driving drum and percussion parts. From here it was a case of working on the bass line which would, in conjunction with the drums, drive most of the track. We used a few different layers for the bass sound in the end but the main bass sound comes from DiscoDSP Discovery which is an "unofficial" software recreation of a Nord Lead and is something that we have started using more lately. Once we had this basic groove laid down (using the original chords because they were just so epic!), we started listening through all the parts given to us and decided to use quite a few of them to really pull the original version into our remix. In the end we used most of the guitar parts and even a couple of the synth parts that we were given. It was such a good song to work with that we didn't mind incorporating so much of the original. Normally we try to take things a little further away but, in this instance, about 50% of what you here in the final mix is from the original track with our beats, bass line, and little stabs, pads and arpeggios backing that up.

Building on the beats and bass line, and original synths and guitars, was really just a case of trying out little riffs and chords and giving the overall track a more "synthy" feel but without dominating the original parts. It is very rare that we get to work on a track with great guitars and, to be honest, it is pretty rare that we get to hear great trance sounds mixed with great guitar! Trance and orchestral have long been known to work together but, for us at least, the trance/rock blend is just as good and also a little "refreshing."

Anyway, other synths that we used for the rest of the remix included Tone2 Gladiator, ReFX Nexus, and Korg Legacy Cell, along with drum sounds pulled from various sources including a couple of the Vengeance sample collections. They ROCK! Mixing and audio plug-ins were largely UAD stuff but, for the most part, it was just a little bit of EQ and compression here and there (with reverb and delay where used coming from Logic plug-ins), and final "mastering" was again UAD.

SECTION 2

The Science of Remixing

CHAPTER 9

Studio Equipment and Environment

The days of actually needing a professional studio in order to get professional results are long gone. What is achievable with a very modest set-up today would have been utterly unthinkable in anything but a pro studio ten years ago. With the increases in computing power and the corresponding increase in the quality of plug-ins, there are now more and more professional studios that rely quite heavily on plug-ins as well as on their collection of vintage outboard equipment. As a direct result of this it is possible for a small project studio to have much of the same equipment as a much larger professional studio. This dramatically levels the playing field in terms of what is achievable in smaller studios, but there are, of course, other factors than just what equipment you have that determine how good your results will be.

Obviously your skills and abilities will play a major part in the quality of the final result, as will the equipment you use, but the room you are working in also plays a major role. First, and most important, the one major advantage that many professional studios have over the average project studio is good acoustics. Most, if not all, professional studios have been specifically designed for that purpose. They have been shaped and acoustically treated to give the best possible sound, and are usually much larger and more isolated than most project studios, so that high volumes aren't an issue when they are required.

There are also the issues of aesthetics and ergonomics. If you, like many people, have a studio set-up in a room in your house, the chances are that it isn't all that big. It's unlikely to be the ideal shape or size, and the dimensions of the room probably force you into a layout for your equipment that isn't that great. Larger studios are laid out in a much more practical way and have more space to work in. They are usually air-conditioned (all that equipment running can make a studio *very* hot without it!) and have better lighting. All in all, a larger professional studio is a much more comfortable environment to work in for long periods of time than a home-based project studio. And trust me, as your career progresses, you will be spending *long* periods of time in the studio!

LOCATION, LOCATION, LOCATION

Having a studio at home has both benefits and considerable drawbacks. The obvious benefits are that you don't have (additional) rent to pay and, as a result, you don't end up rushing through things with one eye on the clock. But even this can have its own drawback. There is definitely something good to be said about having deadlines, because they force you to make decisions and, without that, it is all too easy to over-think things, over-work things, and end up ruining what you are doing. Sometimes it can be the best thing in the world to actually have to commit to something, make a decision, go with it, and, more importantly, stick with it. Procrastination is most definitely the enemy here!

Another benefit of having a studio at home is that you don't spend (or waste) time traveling to the studio every day. Even a half-hour trip each way can eat up a significant part of your day, and I am sure we could all happily accept an extra hour in our days. Lunchtimes become quicker (not to mention easier and cheaper) because you won't have to buy food from the nearest restaurant and take time out from working to physically go there to eat. The problem with this is, because you're not going out for lunch you might often find yourself grabbing something from your kitchen and taking it in to your studio to eat at your desk. Taking a break during the day, a *real* break, away from music and thinking about work, is a very good thing to do because it gives you a bit of space to clear your head and allows your ears to reset. Often when I have been working in more professional studios and have gone out for lunch, when I come back in the afternoon I hear things that I hadn't heard before because my ears had grown acclimatized to them and they had become almost subconscious. That little bit of time without listening to what I was working on allowed my ears and brain to focus on other things so that, when I came back, it was almost like listening to it from a totally objective point of view. Perhaps that doesn't last too long and your ears pretty soon get accustomed to things again, but it might be all you need to save yourself another day mixing because you don't have to wait for the "fresh ears first thing in the morning" effect.

Having a studio at home means that if your time is short you can spend half an hour or an hour working, whereas if you had a studio away from home you might not even be able to get there and back in that time. It also means that you could, if you needed to, break your working day up into a few blocks of hours and fit other things in-between. Again that would be totally impractical if each trip to the studio took an hour in travel time. It is also really handy if you have been working on something for an overseas client and then, at 9 PM, you get an email that reads "Sounds great, but any chance you could run off another version with the vocal up just a touch… and could you get it to us tonight as we are mastering first thing in the morning?" You can just go to your studio, fire up your computer, make a small automation change, and then bounce down a new version and email over the file—all in probably no more than half an hour. If you had a studio that was half an hour away, could you do it? Would you do it? The downside of this (depending on your point of view, of course) is that you can easily get drawn into the mindset of "well it won't take long… I will just do it quickly now, even though it's 11 o'clock at night" and, while your clients will undoubtedly appreciate your dedication, you do still need to have some boundaries. Where those are though is a very personal thing.

If those kinds of requests sound absurd then I think you need to brace yourself and prepare for exactly that kind of thing! You can, of course, choose to basically refuse to work or even respond to email outside of office hours but, as you will soon find out, that isn't really how most people in the music industry work. Personally, I do pretty much "switch off" once I have done my actual studio work for the day, and I don't really like working more than eight or so hours in any given day anyway as I find that my attention to detail and creativity tend to fall away quite quickly after that amount of time. I have, of course, done much longer sessions than that, though, but as a general rule I try to avoid them for practical reasons. Evenings and weekends are still, in some ways at least, work time for me, albeit the emails, phone calls, and admin kind of work.

Let me put it another way, I have been "working" as early as 6:30 in the morning with emails, paperwork, and phone calls, and I regularly get and respond to both phone calls and emails as late as midnight. (Midnight in the U.K. is only 4 PM in L.A. and while a response might not be expected from me at that time of night, in some ways I would rather get it over with there and then rather than get tied up with emails first thing in the morning.)

A final thing to consider about location is that it is sometimes very useful to actually leave the house and go to work because it can, and usually does, put you in a different frame of mind. The simple act of leaving your house and physically going somewhere else to work can make you more focused and productive. You are there for a purpose, you do what you have to do, and then you go home and have your life outside of work. Plus you don't run the risk of going into the house to make yourself a coffee and then, while you are waiting for that to be ready, thinking "I will just take a quick look at the TV" and then, before you know it, you are caught up in the woes and worries of the latest dysfunctional family appearing on a daytime talk show!

It is pretty obvious that there are benefits and disadvantages to both possibilities but, at the beginning at least, your choice might be made for you. Financial reasons may play a big part as well as practical ones. When you are starting out or even for some time after you have started, you may have another job you are doing at the same time to keep money coming in while you are building your profile as a producer/remixer. In that situation it makes no sense to be paying rent for a studio when you might only really be able to work a few hours a day or only certain days of the week.

When you are in a situation where your work is more regular and you are earning enough from remixing (and possibly producing and DJing) to give up the day job then you will have to make a decision about what works for you. The choice probably won't be between a small project studio at home and a full equipped mix room at Abbey Road or Ocean Way, but you may well have the option to get yourself a bigger and better room than you had when you were starting out, and a room that is somewhere away from home. What you choose to do at that point really depends on how you feel, how good your focus and concentration is, and perhaps a few other personal factors as well. But you should know that, assuming you had done at least some basic acoustic treatment in your project, or home, studio, there is no reason why you couldn't produce commercially acceptable (and hopefully successful) work in either environment.

As for me, I actually have *three* "studios" I work from! I have a set-up at home (in my third-floor apartment) that is just the most basic set-up of a Mac, monitor, USB keyboard, and small monitors and headphones. I only really use this if I want to throw a quick idea together or perhaps I need to do some preparation work: time-stretching remix parts, tuning vocals, sample editing, sound programming, and other things like that. The second studio I have is a small project studio, which is at the bottom of the garden of my parents' house. This is a pretty well-soundproofed room with half decent acoustic treatment and a little more space. Here I have another Mac (set up exactly the same as the one I have at home), monitor, a couple of synths and sound modules, bigger speakers, and my guitar and bass. This room is where I do the majority of my work and where I've mainly worked over the last three years. I have done remixes that have topped the U.K. and U.S. club charts that were created in their entirety in this small project studio.

The final studio I work out of is a studio owned and run by a friend of mine where I get preferential rates and sometimes free "downtime." I rarely feel the need to actually use his studio for mixing, even though he has a bigger and better room acoustically and bigger and better monitoring, but I do sometimes use the facilities and equipment he has there for recording purposes. Every now and then, if I am going there anyway for another reason, I will take some of my recent tracks or remixes to have a quick listen to in his control room and on his monitors, just to get a different perspective on how they sound. Over the last three years, I have grown so accustomed to the sound of my own monitors, headphones, and room that I really don't feel I am missing out in any way by not having a bigger and better studio to work in.

MAKING THE MOST OF WHAT YOU'VE GOT

In fact, great mixes can be made on pretty much any speakers or any headphones if the person using them is extremely familiar with both the sound of the speakers or headphones *and* how they sound in the room that will be used for mixing (obviously this second part isn't relevant for headphone mixing as the size and shape of the room isn't a factor with headphones). The same is also true of pretty much any other type of equipment you care to mention. Take hardware or software synthesizers for example. While there are certain undisputed "classics" (the Minimoog being one of the most ubiquitous), and while I would never criticize anybody for wanting to own such a thing, *not* having a Minimoog doesn't prevent you from making great sounds. They just won't be great *Minimoog* sounds.

Now if you think about plug-ins, a name that will almost certainly spring to mind to anybody who has been in the music business for even a short while is Waves. The plug-ins made by Waves have a thoroughly deserved reputation of being among the best available today. But they also have a price tag to match. Many people feel that if they were suddenly able to afford to buy these Waves plug-ins their mixes would magically improve. The truth, however, isn't quite that clear cut. Yes, the Waves plug-ins (and many other top end plug-ins) do sound amazing… *if* you know how to get the best out of them. In some situations, a better sounding or more flexible plug-in can actually sound *worse* if you don't know what you are doing, because the extra depth, clarity, and punch will be much less forgiving than a not-quite-so-good plug-in would be.

Going back to synths for a second, a good sound designer can get better sounds out of a very average analog synth than someone simply turning knobs and pushing buttons on a Minimoog without really knowing how to achieve what he or she wanted. In my opinion, it really does come down to taking the time to get to know whatever equipment you *do* have. Get to know its strengths and weaknesses, and all of its frustrating little intricacies. And take some time to actually learn a little bit of the theory about that piece of equipment—whether it's a synth, a compressor, or an EQ—because if you know those fundamental principles then when you do eventually move up to the better quality equipment you will already have a head start on knowing how to get the best out of it.

With all this in mind, one of the most frequent questions I get asked is "What's the best kind of…?" Most of the time people aren't satisfied with the pseudophilosophical answer outlined above and tend to get a little frustrated when I don't just say to them "Ah you want to buy a PhatTastic MegaSynth 2000! That's easily the best synth on the market." There are two reasons why I really can't give *any* examples of "what's best," the first of which is that, again, there really isn't a "best" as it depends on what you are used to and how well you know it. The second reason is that (taking as read that there isn't an absolute "best" anyway) what might work very well for one person's working methods and sound he or she wants to achieve might not work as well for the next person's unique needs and musical desires. There are certain "classics," as I already mentioned, but, depending on what genre of music you make and what your skills and abilities are, a Minimoog (or a Yamaha CS-80 or Roland Juno-106, etc.) might not be what you need to get the sounds that are in your head.

It is easy to fall into the trap of thinking that, just because something is considered "the best," that it will be the best for what you want to achieve. As an example, consider the question: "What's the best car?" Is it a Ferrari, a top-of-the-range Mercedes, a Volkswagen Golf, or a Ford pick-up truck? I am sure that you immediately made your mind up and you would, of course, be right! Personal taste is just that... personal. However, if I changed the question to "What's the best car for somebody who works on a farm and has to transport quite a lot of sometimes dirty equipment around from place to place?" then I think there would (hopefully at least) be a pretty general agreement on which car would be best.

So when you are asking "What is the best?" type questions you really need to be much more specific with your questioning in order to even get advice as to possible options and alternatives from which to make your decision. If you are asking about speakers, you need to think about qualifying your question a little more. What size is your room? Where is it located? Do you have acoustic treatment? Do you want active or passive speakers? Are they required to run really loud sometimes or would you always be monitoring quietly? What kind(s) of music will they be working on primarily? Have you had any other monitors in the past and, if so, for how long and were you in any way happy with the sound you got from them? Will these be your main monitoring source or do you have other sets you will be able to use for A/B comparisons? What about headphones? The list goes on. Once you have clarified exactly what you're looking for you should be able to get some better guidance as to a few possible alternatives. You should always check out and do some research for yourself wherever possible. And by "research" I don't just mean Googling a few specs, I mean getting out there and trying them out in the real world. In case you are curious about *my* thoughts and opinions, there is a *Buyer's Guide* section on this book's companion website where I have gone through each type of equipment and listed my own preferences and thoughts. It includes some reviews of particular favorite pieces of equipment of mine.

EQUIPMENT CHOICES

So what do you *really* need to get started? Well, as I mentioned earlier, my system at home is little more than a computer, a sound card, speakers, headphones, and a USB MIDI keyboard for playing in parts. In terms of software, a great deal can be done with just the plug-ins and sample libraries included in Logic. In Chapters 21 through 24 I guide you through the process of doing a remix, and everything is demonstrated using *only* Logic and its included "goodies."

Most will look at synths and audio plug-ins next, but that isn't always the best move. After all, if you're not hearing what you already have very well then "better" synths and plug-ins still won't give you the best results. My simple rule of thumb here is that you really need to increase all of the aspects of your system in stages. There is no point in getting better monitors if you are still using a very cheap sound card, as the sound going to the monitors won't be the best you could get. Conversely, having an amazing sound card won't give you amazing results if you are using average monitors. The same goes for plug-ins and synths (hardware or software) in that you won't get the full benefit of them if the rest of your system could use some improvement. And of course, through *all* of this, don't underestimate the importance of having a good room to work in.

The particular order in which you decide to update the components of your system is entirely up to you and often comes down to personal preferences and availability of funds. There may be other factors in your decision, but, on the whole, there are some things that will make a bigger difference than others for each step up the ladder you make. I think the biggest difference you can make to your result comes from improving the room you work in and your monitors, because *everything* else you upgrade will be dependent on these two things. A simple way of looking at it is this: Consider the order in which the sounds get from your imagination to your ears and then reverse it to get a good idea of the best order to upgrade things in. The order, or progress, is as follows:

Idea > input device (MIDI keyboard) > DAW > sound source > audio processing > sound card > monitors > room > ears

We can discount the first and last because money can't buy you improvements to *either* of those. But, apart from that, if we reverse the order then what we end up with is basically my advice for the order in which you should update your equipment:

Room > monitors > sound card > audio processing > sound source > DAW > input device

Even then I think that the last two will have a minimal effect on your end results, as the sound quality of pretty much any modern DAW is almost immaterial; and, unless you are a very good musician, the quality of the MIDI keyboard (or other MIDI equipped instrument) you play won't have a great deal of impact on the final results.

For many of us, most sound sources and audio processing take the form of plug-ins of one kind or another and these can range in price from absolutely nothing to many thousands of dollars. There is, perhaps, some correlation between price and quality but that doesn't mean you have to spend thousands to get characterful and original plug-ins. You just have to know what it is you want to achieve with them and then spend a little time researching and trying things out to figure out what works with your own workflow and production style.

An improved sound card can have a positive impact on your end results, but I have to be honest and say that this is perhaps the area where I think that you get the least bang for your buck once you are looking at devices costing more than a few hundred dollars. I personally use Apogee sound cards (at the cheaper end of the range) and, having had the chance to compare them to a few other (much more expensive) sound cards, I can honestly say that if there *were* any improvements they were subtle. There are a number of true high-end sound cards people swear by but, given the cost and the amount of improvement you would realistically see, I have to wonder whether the cost would be justified in most situations. Perhaps one day I will hear something that makes me change my mind, but for now I'm sticking with my Apogee.

However, when talking about monitors and your room itself, it's a totally different story. Here we consider an entirely plausible situation and examine a few options. Let's say that you have a basic system and you work from home, perhaps a spare room, and you have managed to save up $1500 to spend on upgrading your equipment. You decide that there are four ways you could spend it: on software (plug-ins and synths), on a sound card, on monitors, or on acoustic treatment.

1. **Software**—$1500 could certainly buy some really nice synths and even leave you with change left over for a few choice audio plug-ins. However, even if we take out of the equation the fact that you might not yet have the skills and experience to really get the best out of these plug-ins, will you really be able to hear all of the subtleties of what they are doing for your productions in your current room? I somehow doubt it. You will hear *some* change and improvement, of course, but will you be getting the best value for money? We will see.
2. **Sound card**—You could get quite a nice sound card for your money here, especially if you only need a small number of inputs and outputs. However, I feel that most people would connect the new sound card expecting to be blown away by the improvements to their sound only to be massively underwhelmed.
3. **Monitors**—$1500 will buy you quite a lot in monitoring terms. If you're not working in a big room then you really won't need huge amounts of power, so a system with smaller monitors and, perhaps, a matching subwoofer might be a better way to go. The extended range of the system will prove invaluable in working on your mixes. However, the increased bass response could actually *cause* as many problems as it solves. Which is why we need to look at…
4. **Acoustics**—Basic acoustic treatment for your room can make a *massive* difference, and it needn't be that expensive. Some well-placed acoustic tiles (and no I *don't* mean egg cartons) can clean up the sound you are hearing and bring things a lot more into focus. These can be had for as little as several hundred dollars. Beyond that, you could look at more sophisticated *bass trapping* to help sort out any problems with the low frequencies. There are a great number of different acoustic treatment options, in the same way as there are a great number of acoustic problems. Some of them can be quite expensive, but, for the most part, even the more basic ones can still really help the sound of your room.

Bass trapping is the technique of using acoustic devices called "bass traps" to control the low frequency response of a recording or mixing room. Bass traps can vary a lot in design but are, essentially, devices which absorb acoustic energy in a particular frequency band. Often bass frequencies are the most problematic in the studio environment so a great deal of time, effort and expense can go into making the bass response of a room more controlled and even.

My recommendation, based on all that information, on how to spend your $1500 would be to spend part of it on acoustic treatment and part of it on upgrading your monitors. If you don't feel that the $1500 budget is enough for both of these then I would recommend spending maybe $750 on acoustic treatment (and you will get a lot of it for that amount) and putting the other $750 aside until you have saved up some more, enough to get new monitors. There is no point in spending $750 on monitors if you really want $1500 ones, or if you already have $600 ones. There will be very little difference between $600 monitors and $750 ones, but there will probably be a noticeable difference between $600 monitors and $1500 ones.

Each time you manage to save up money to invest in your studio I would seriously consider at what level each part of it is and then look at ways in which you can bring it all to the same level.

I generally consider there to be four levels for studio equipment: amateur, semiprofessional, professional, and audiophile (otherwise known as "money to burn"). How you would classify any particular piece of equipment is open to discussion, but most people would be able to differentiate between amateur and professional; it's the differences between amateur and semiprofessional and semiprofessional and professional that seem to cause the biggest debate. And with each increase in level, we are subject to the *law of diminishing returns*: each time you double the value of the equipment you get less of an improvement. If a $160 sound card is twice as good as an $80 one, a $320 one won't be twice as good as a $160 one. Instead it might be only 70% better. If we then move from a $320 one to a $640 one, the improvement might only be 40%, and so on. Of course, it is incredibly hard, if not impossible, to actually quantify these things, but you get the picture.

Once you get into the realm of crossing the boundary from professional to audiophile, then the differences can be as little as a few percent and may even be inaudible to those with untrained ears. If, however, like most of us, you strive to have the absolute best, then you will, if the opportunity presents itself, probably ignore this fact and buy the new equipment anyway. And if you do, congratulations and good luck to you!

One thing I can definitely recommend, outside of the normal kinds of things you might want to buy, is some good back-up software. I really can't overstate the importance of being able to revert back to a fully working system quickly and easily. If, or rather when, you get software updates either for your plug-ins or your main DAW, or even an upgraded operating system, you can never be 100% sure that everything will go smoothly and work perfectly. In instances such as these, there is often an easy fix if things don't go according to plan, although there could be occasions when it takes quite a bit of time to get everything back to how it was before you upgraded. If you have the time to fix it, great, but what if you a sudden "rush job" come in that you have to turn down or delay the deadline on due to having to mess around getting your computer and software working again? Having back-up software won't prevent the problem from happening, but if you do regular back-ups you have a way of getting back up and running in as little as a few hours rather than (potentially) a few days.

One alternative to this is to actually *clone* your main system hard drive before doing any updates; that way you simply have to remove the hard drive that has the issues and replace it with the clone, and you can be back up and running in minutes rather than hours. The actual creation of the clone isn't always 100% foolproof and you may find yourself having to re-enter serial numbers for software and things like that, but once the clone drive is fully set up you would be hard-pushed to find a better back-up solution for your main hard drive, especially in an industry where time is money and time is often very tight.

SUMMARY

Hopefully all of this has given you a good overview of the various different ways in which you can spend your money and, on a more serious note, a bit of an insight into what might serve you best at different stages of your career.

There are no hard-and-fast rules as to what any one person will need, however. Primary factors are what experience level you are at, what systems you may have used in the past, what style of music you make, and, ultimately, how much money you have to spend. The good news is that the days of needing multimillion-dollar recording studios just to get a track that is of a high enough quality to be heard on the TV or radio are long gone. With time spent learning how to get the best out of whatever equipment you do have, learning some useful tips, tricks, and techniques (from this book hopefully), and a relatively small financial investment, you too could, and hopefully will, be producing tracks and remixes that are of a high enough technical quality to get on the radio. The rest of the magic is down to you!

CHAPTER 10
Sound Design: Introduction

The term *sound design* often conjures up images of technical alchemy and a secret society of geeky synth programmers in white coats lurking around a huge modular synth. But sound design in itself can range from a ground-up to a simpler top-down approach. You don't necessarily need to know everything about a particular form of synthesis to be able to do basic sound design. However, some understanding of the fundamental principles is certainly a great asset in being a good remixer.

Sound design can be traced back to the origins of music itself. Today the term is often used in the context of creating a single sound. But when we hear music we don't hear a single sound. We may be able to isolate a particular sound within a piece of music by careful listening, but what we are actually hearing is *one* sound that is made up of a collection of individual parts. So when the great composers of the past were creating their masterpieces, they were sound designers. When the orchestra was split into smaller bands, that too was sound design at work. And when synthesizers were developed, so too was the latest incarnation of sound design!

So how does all this relate specifically to remixing? Well, one of the key points of being a good remixer is a sense of *sympathy* for the track you are remixing, and your choice of sounds is an essential part of this. There are times when you will want to find sounds that blend in with elements taken from the original version of the track, and there are times you will want to find sounds that contrast or complement elements from the original track. There are also genre considerations to take into account. As good as it is to distinguish yourself from other remixers working in the same genre, there are often some expectations from record labels, DJs, and listeners themselves as to what works in any particular genre. These expectations are quite fluid and seem to change on a month-by-month basis, but it is still something to take into account.

In any case, being able to apply even some basic sound design principles will help you achieve good sound choices more effectively and quickly than running through the often huge collection of presets provided with modern synths (hardware and especially software-based). And, like so many aspects of the music business, time is often of the essence. So before we take a look at the basics of sound design and explore the different types of synthesis you are likely to meet, we should first consider the preset libraries that come with most modern synths and see what is good about them, and what is *not* so good!

PRESETS: GIFT OR BURDEN?

My opinion on this question is actually divided. In many ways having a large number of presets programmed by experienced programmers is great! It enables you to quickly audition some sounds and at least find something close to what you are looking for. In addition to this, just browsing through presets and auditioning them *in situ* can sometimes lead to you finding a sound that is totally different from what you were looking for but, somehow, works better. I have lost count of the number of times this has happened for me! But for many people, that is where the story ends. They may try to knock a preset into shape with a little EQ and other effects, or perhaps even try a little editing on the filter or amplitude envelopes, but beyond that they just make do. And this is where having hundreds or even thousands of presets can become a burden. Not only do you run the risk of having sounds that aren't as good as they could be for the purpose you want to use them for, but they can often be heard in other people's tracks and, as such, not have the uniqueness that distinguishes a really good track or remix.

Moreover, if you can learn some more in-depth sound design techniques then you are well on your way toward establishing your own *sonic identity*, which is invaluable if you want to start building a brand of your own. There are more than a few remixers who have "signature" sounds and, unless you are lucky enough to stumble across a preset sound that *hasn't* already been used to death and suits the genre you're working in, you will probably be well-advised to invest a little time in learning some of the techniques detailed later. This will benefit you more than you can imagine, but is often overlooked, as people want to get straight down to the process of actually making tracks.

SOUND DESIGN BASICS

It is certainly helpful to know a little about how the sounds we hear are made if we wish to get involved in making them ourselves. The theory behind it is, perhaps, not as complex as you might think. A vast majority of sounds, including some that might not *seem* that way to start with, are actually *pitched* sounds. That is, they have a discernible "note" to them. Some of these are obvious of course, like a piano or a guitar—instruments that can produce a variety of different notes. Some less obvious examples of pitched instruments include drums. Most people would classify tom-toms as pitched, but what about the kick drum, snare drum, and cymbals? The truth is that all of these are pitched sounds, albeit less obviously so. Any sound generated by an object that vibrates (the drum skin or the cymbal itself) is always a pitched sound. Finally, we have sounds that don't seem to have a "note" to them. Imagine the sound of a car door closing. Do you hear a "note?" Probably not. But there *is* still a pitch to the sound and, as such, it falls in line with the other sounds here.

The main difference between the types of sounds described previously comes from how obvious that note is. And that is determined by the harmonic content of the sound. Any sound you hear is made up of a collection of different frequencies of varying levels and durations. Think of an orchestra. When you listen to a piece of orchestral music you hear the overall sound. But that is made up of many different components. Sometimes you can easily pick out a flute or a cello, especially if they are playing distinct melodic phrases. But imagine every instrument in the orchestra playing the same note at the same time. Do you think it would be easy to pick out a particular instrument now? Probably not. When all the instruments are playing the same note in unison it becomes much harder to tell what is making up the sound. All of the individual instruments tend to merge into the sound as a whole. And this is what happens within every sound you hear.

Any pitched sound will have what is called a *fundamental frequency*, which determines the perceived pitch of the sound. Lower fundamental frequencies make lower pitched sounds while higher fundamental frequencies make higher pitched sounds. But there are (with the exception of one special case) always other frequencies being heard at the same time, and it is the balance between these *other* frequencies that gives the sound its character. These other frequencies fall into one of three categories: mathematically related frequencies, mathematically unrelated frequencies, and noise components.

Frequencies that are mathematically related are whole-number multiples of the fundamental frequency, called *harmonics*. Each harmonic is a multiple of the fundamental frequency. If we assume that a sound has a fundamental frequency of 50 Hz, then it will have harmonic frequencies of 100 Hz, 150 Hz, 200 Hz, 250 Hz, and so on (being multiples of 2, 3, 4, and 5 of the fundamental frequency in this case). In theory, these harmonics can be an infinite series, but, for all practical purposes, we are not concerned with harmonics that are out of the range of human hearing.

Frequencies that are mathematically unrelated (at least in terms of whole number multiples) are often called *inharmonic spectra*. These frequencies are often present in sounds we describe as "metallic." These kinds of sounds often sound "grating" and can be unpleasant. But *inharmonic spectra* are just as important as normal harmonics in creating some of the more interesting sounds you hear.

Finally, we have noise components. In theory, pure "white noise" is a sound that contains spectral (harmonic) content that is equally balanced across all frequencies. As such it doesn't have a dominant (fundamental) frequency. Because of this you might not see the use for it in sound design. But it is *exactly* this random nature of the sound that can make it useful in creating realistic percussive sounds and for adding realism to any sounds that attempt to recreate "blown" sound sources (such as a flute). The sound of human breath blowing would be almost impossible to recreate through harmonic methods, but a very good approximation can be made by using filtered (frequency limited) white noise. Also, as many cymbals have very complicated and dense high frequency components, it is often easier to use filtered noise as a source of this content rather than trying to recreate it using normal harmonic methods.

All of the aforementioned are sound *sources*. They relate to the *generation* of sound. The relationships between frequencies and individual harmonics can, and most often do, vary over time in acoustic instruments. And if we are trying to recreate acoustic instruments, creating this movement and change in the sound is one of the hardest things. Most synthesizers deal with static waveforms, in other words, they do not change over time. So in order to give life and realism to our sounds we need to create changes over time. That is where modulators and modifiers come in.

A modifier is anything that modifies the sound source. The most common of these is a filter—a fundamental component in sound design. Modifiers also include waveshapers (which modify the harmonic content of a waveform) and other effects that can alter the harmonic content—either adding, in the case of distortion effects, or subtracting or shifting. All of the aforementioned modifiers give *static* changes to the sound source. In order to truly introduce movement and life to a sound you need modulators.

Modulators most commonly take the form of LFOs (low-frequency oscillators) and envelopes. An LFO is simply a device that generates a steadily fluctuating output, normally cyclic in nature, that can be used to change certain aspects of the sound in a fluid and regular way and repeats indefinitely. Envelopes, on the other hand, output a signal that can be used to give a predictable and controllable change in a sound design parameter over a fixed period of time. Some envelope generators do include the ability to *loop* the envelope, giving this particular modulator the characteristics of both a traditional envelope and an LFO, but most envelope generators (EGs) are simply linear; they do what you ask them to do and then stop. It is by combining sound sources with modifiers and modulators that you get the building blocks of pretty much all modern synthesizers. The type, variety, and number of these vary enormously from synth to synth, but you can usually identify each of them with a little bit of digging.

Now that we know what makes up any sound we hear, we can turn our attention to the different ways in which we, as sound designers, can recreate these sounds or, indeed, create new ones that couldn't exist in the real world. For the most part, synthesis will use one of (or a combination of) four main types of synthesis: *subtractive*, *additive*, *FM* (frequency modulation), and *modeling*. There is also sampling to consider, but as that uses recordings of actual sounds as the basis of its sound generation, it is, in some ways, outside of what we are discussing here. We will nevertheless take a look at it briefly, as it is often incorporated into a subtractive synthesis model. So, let's take a look at each of these synthesis types in more detail.

SUBTRACTIVE SYNTHESIS

Subtractive synthesis, as the name implies, works by subtracting harmonic information from a complex waveform to get the desired sound. An analogy is the way a sculptor can take a block of stone and chip away at it to form the final sculpture. In sound, this is done with the use of filters to remove harmonic content in a static, cyclic, or dynamic way. For this to work, the waveforms feeding the filters have to be rich in harmonic content to start with. So at the start of the process for subtractive synthesis are *oscillators*. The oscillator is one of the key components in a subtractive synthesizer (in fact in *any* synthesizer). In a subtractive synth, the oscillators will generally have a selection of different waveforms to choose from. Numerous waveforms have rich harmonic content. Most real instruments have very complex harmonic structures, but since these structures change over time they don't work well as a basis for synthesis simply because of the harmonic variations already embedded into the sound. Usually single-cycle waveforms are used (see Figures 10.1 through 10.5). In the early days of synthesis these waveforms were produced in real time by the oscillators on the synth, and so the options were limited. Early pioneers of synthesis chose a limited selection of waveforms from which to build their sounds, and that selection has become the backbone of analog synthesis to this day.

The waveforms most often seen in subtractive synthesis are:

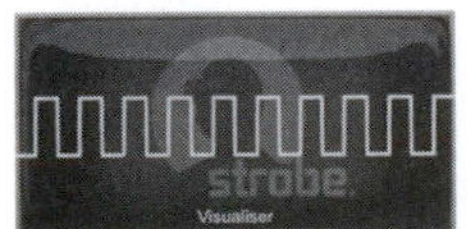

FIGURE 10.1
Square wave.

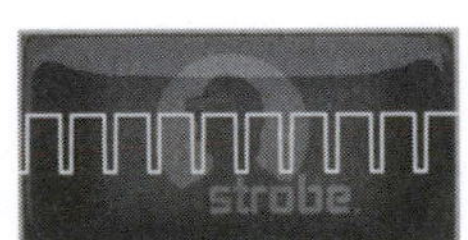

FIGURE 10.2
Pulse wave.

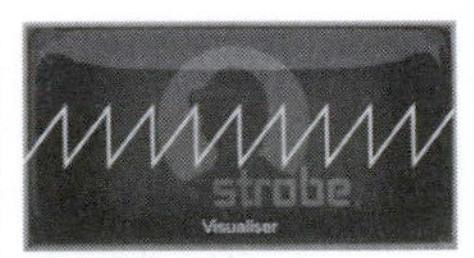

FIGURE 10.3
Sawtooth wave.

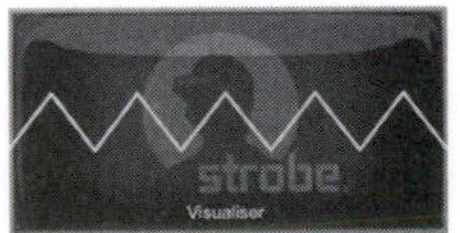

FIGURE 10.4
Triangle wave.

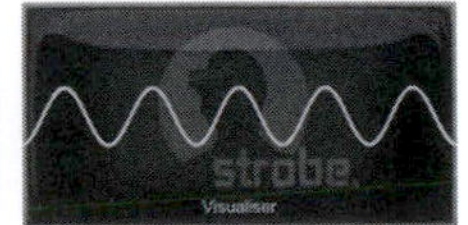

FIGURE 10.5
Sine wave.

As you can see, a square wave and a pulse wave do *look* similar, and as you might expect they can sound similar. On the other hand, a sawtooth wave looks very different and, not surprisingly, *sounds* different! These three waves form the complex waves from which most subtractive sounds are produced. A triangle wave is not as complex as any of the previous three waveforms. A sine wave represents the purest waveform: a single frequency with no additional harmonic information. Given what was said earlier about subtractive synthesis being all about subtracting harmonics from a complex signal, you might be wondering what the purpose of a waveform with *no* harmonics is in this process. And you would be right to wonder. Right, that is, if we were talking about a synthesizer with only one sound source. But as most synthesizers contain at least two sound sources, there is a use for harmonically simpler waveforms, and this will become clear later.

If you listen to a *raw* square wave (with no filtering) and a raw sawtooth wave you will hear the sonic differences very clearly. And, when you get used to programming your own sounds, you will probably be able to listen to sounds in other tracks and determine whether they are square-wave or sawtooth-wave–based simply by the dominant tonal characteristics. The reason for the difference between the two waveforms is down to what harmonic information they contain. If we refer back to the earlier section about harmonics, we can explain the differences between the characters of these waveforms.

- A square wave *only* contains the fundamental frequency and any *odd numbered* harmonics (third, fifth, seventh, etc.). The greater the number of harmonics included, the closer to a *real* square (or sawtooth) wave. Square-wave derived sounds are often described as *hollow*, which in many ways is a fitting description as, compared to a sawtooth wave, there is information missing. So if a sound you hear sounds hollow there is a good chance it's derived from a square-wave source.
- A sawtooth wave contains the fundamental frequency and *all* of the harmonics. As a result, it sounds *fuller* than a square wave. Sawtooth waves are often used as the basis for classic pad sounds because of the warmth they have. They can sit nicely in the background of a track without sounding harsh.
- A pulse wave differs from a square wave in that the shape of it can change from being equally spaced *positive* and *negative* sections to mostly *positive* or mostly *negative*. This ratio is known as the *duty cycle* and greatly changes the tonality of the sound produced. As the duty cycle moves further from the center point (50%, which gives a square wave) the sound produced will become *thinner* and less rich. Often this duty cycle, or *pulse width*, is modulated to give the sound more life and movement.
- A triangle wave is, in fact, related to a square wave. It has a similar tonality, but the relative levels of the harmonics decrease as the harmonics increase, meaning there is less high frequency content in a triangle wave compared to a square wave. The resulting sound is less bright.
- And finally we have a sine wave. A sine wave contains no additional frequencies other than the fundamental frequency. In effect, it is a *pure* tone. I cannot think of a single example of a sine wave that occurs in nature. There are, however, sounds that come close to approximating a sine wave, such as whistling, running your finger around the rim of a crystal glass, and the sound made by a tuning fork, but even these have additional harmonic information. As such its uses in subtractive synthesis are limited, but it does have a place. The sine wave comes into its own in other forms of synthesis.

All but the most basic of synths have multiple oscillators, and the outputs of these oscillators are normally variable in level so that the two (or more) waveforms can be balanced against each other. Sometimes this is a static balance, but in more complicated synths you can modulate the balance of the oscillators individually to allow for more complex and evolving sounds. The signal from the oscillators is fed to a filter section, which is then used to shape the sound. Some more complex synths have multiple filters you can assign to individual oscillators or to the combination of them, in series or in parallel. Whatever the routing of the signals, the purpose of a filter is always the same: to remove something, in this case certain frequencies.

As synth designers have more power to develop new and exciting products, especially with today's seemingly endless powerful computers, they are coming up with new and exotic filter designs. But, for the most part, there are three different kinds that will be most useful to you: low-pass, high-pass, and band-pass filters.

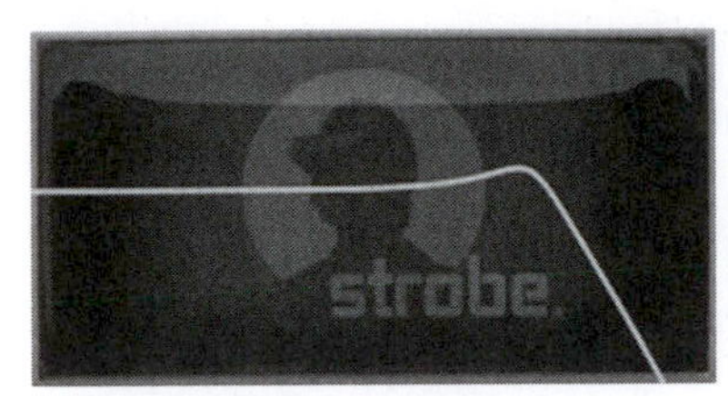

FIGURE 10.6
A typical frequency response of a low-pass filter.

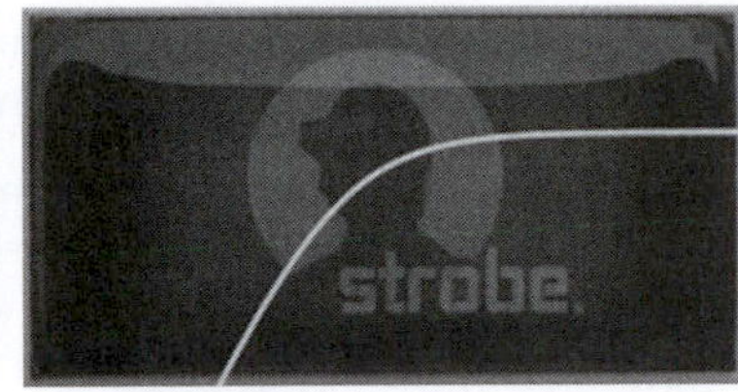

FIGURE 10.7
A typical frequency response of a high-pass filter.

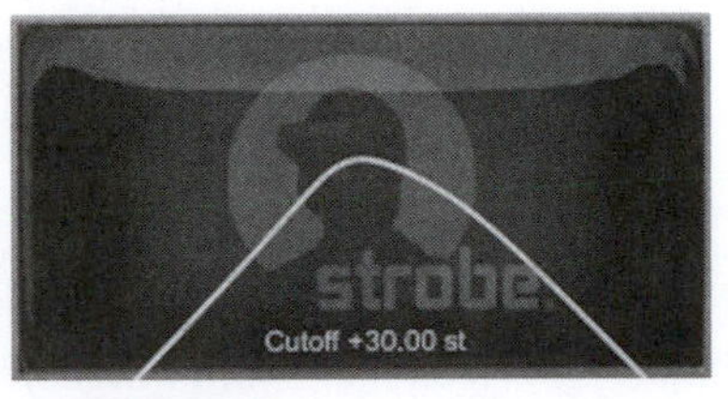

FIGURE 10.8
A typical frequency response of a band-pass filter.

- Low-pass filters are used to remove high-frequency content from the source waveform(s). The name is derived from the fact that they allow *low* frequencies to *pass* through the filter unaffected. There is a control for the point above which you want to start removing frequencies. This is called the *cutoff frequency*.
- High-pass filters work in completely the opposite way to low-pass filters, and are used to remove low-frequency content. As you can probably guess, the name is derived from the fact that they allow *high* frequencies to *pass* through unaffected. In a high-pass filter the *cutoff frequency* is the point below which you want to start removing frequencies.
- Band-pass filters are basically a combination of a low-pass and a high-pass filter. If you were to place a low-pass and a high-pass filter in series, the low-pass filter would remove some of the higher frequencies before passing that signal on to the high-pass filter, which would then remove some of the lower frequencies. What you would be left with is a *band* (or range) of frequencies that *pass* through. Normally with a band-pass filter you don't have separate controls for the low and high elements, but instead have a *center* frequency that controls the cutoff frequencies of *both* filters. Sometimes you can offset the low and high-pass *cutoff frequencies* from this center point to allow for a wider *band* to pass through.

Aside from the cutoff frequency the other main parameters you would find on a normal filter are *resonance* and *filter slope*. The resonance control can be used to emphasize the frequencies around the cutoff point. This can actually introduce a lot of character to a sound if used correctly, and some filters can even *self-oscillate* so they can, in a roundabout way, function as a sound source in their own right. The other control, the *filter slope*, controls the rate at which frequencies above (in a low-pass filter) are removed. It is normally listed as a measurement of decibels (dB). Normally the options are 6 dB, 12 dB, 18 dB, or 24 dB, although some filters can go higher than this. Basically, the higher the decibel value, the more quickly sounds above the cutoff frequency will be filtered. If you look at Figure 10.6 above it should visually illustrate what happens. All filters gradually reduce frequencies above the cutoff frequency; the higher above the cutoff frequency you go, the more the frequencies will be lowered. A 6 dB filter slope gradually removes higher frequencies and a 24 dB filter slope removes them much more quickly. There are no hard-and-fast rules as to what filter slope should be used, but with a little experimentation you will soon discover what works best for different sounds.

Each of the filter types can be used to shape the sound in different ways, depending on what kind of sound you want to achieve. If the synth you are working with has multiple filters and you can route them in either series or parallel then you have even more options. Even though your filtered sound may be tonally more what you were looking for, there's every chance it will still sound a bit boring! In order to really get some life into your sounds, they need movement. And that is where *modulators* come in.

Modulators further shape your sounds and make them more dynamic. The two types of modulators seen most frequently are LFOs and envelopes. Each is used for a different purpose. For introducing cyclic change to a sound, an LFO is the best option. If you want to introduce a more controlled and linear pattern of change, then an envelope should be what you look for. What would be typical uses of both of these modulators?

An LFO outputs a *control* signal, which is often based on the same waveforms as a normal oscillator but instead of running at frequencies we can translate as notes, an LFO works in the subsonic frequency range. Often an LFO will have a frequency range that extends from around 20 Hz at the highest down to 0.1 Hz (or lower) at the lowest. We couldn't possibly hear sounds with a frequency that low, but we *can* hear the effects of a control signal that modulates a sound parameter. The actual way in which you connect the output of a modulator to a synth parameter varies from synth to synth. There are certain modulators that are *hard-wired* to certain parameters, other times they are completely free (as in the case of *modular* synths), and occasionally you have a combination of both. It would be impossible to describe the routing possibilities for every synth here purely because they are so varied. But if you want to get the most from your sound design, the flexible modulation routing is a very good thing. Take a look at the manual for your synth(s) for instructions on how to route modulators to parameters. Once you have done that you are ready to start making some motion happen!

When a singer or a violinist adds vibrato to his or her notes, what you actually hear is *pitch modulation*. The pitch fluctuates to a small degree, and in a cyclic manner, around the center frequency (the note). This technique, when used by singers and instrumentalists, often adds a great degree of expression and emotion to what you hear. And the good news is that you can apply the same technique to the sounds you create. By applying the output of an LFO to the fine-tuning control of an oscillator, it is possible to recreate a vibrato type of effect. You would typically use a sine wave or possibly a triangle wave on the LFO to achieve this as you want the pitch change to be quite fluid. If you were to use a square wave the pitch change, although very small, would sound jumpy as it flipped from one extreme of the modulation range to another.

Another option for LFO modulation is to use the LFO to control the amplitude (volume) of the sound as a whole, or perhaps just one oscillator within a sound. This gives more of a *tremolo* effect. Or you can use the LFO to control the panning of the sound. All of these are techniques you have probably heard singers or instrumentalists use before—pitch modulation as vibrato, amplitude modulation as tremolo (often used by guitarists), and spatial/panning modulation (one of the factors that gives a rotary speaker its unique sound).

Other common things to modulate with an LFO are filter cutoff frequency and oscillator pulse width (assuming you have a pulse wave as one of your sound sources). If you modulate the filter cutoff frequency with a slow LFO then you'll have a sound that seems to rise and fall in a way similar to the sound of waves crashing on a beach. Some LFOs even offer the option to synchronize to the tempo of the track. In these cases the *frequency* of the LFO is measured in beats (or subdivisions of beats) so you can make the cyclic swelling of the sound rise and fall in time with every bar, for example. This can be a very effective way of creating subtle movement on pads and other sustained sounds without having to work too hard.

Pulse width modulation is slightly different in that it works by actually changing the harmonic content of the oscillator, rather than modifying the sound after it has *left* the oscillator. If you manually change the pulse width of the oscillator you will hear it ranging from a thin, raspy sound at either extreme to a square wave in the middle. But if you modulate this change with an LFO you get something that sounds remarkably like a traditional *chorus* effect. This warms and softens the sound and gives it a sense of depth. In fact, given that a square wave is one stage of a pulse oscillator, the sounds you can get from pulse width modulation (often referred to as PWM) are sometimes remarkably like a sawtooth oscillator.

Beyond the examples given here, there are many other options for using LFOs. If you happen to have access to a modular synth, hardware or software, then you can experiment like crazy. Mostly when messing around with modular synths and just experimenting, the results can be quite unpredictable. But every now and then you might come across a modulation routing that does something pretty special. Have fun with it. Synthesis and sound design is as much about having fun as it is about learning!

The other main type of modulator is an envelope. The most common types of envelopes (other than the ones you get your junk mail in) are four-stage envelopes commonly referred to as ADSR envelopes. Each letter corresponds to one of the stages:

- (A)ttack—The modulation signal starts from a fixed point (usually zero) and rises to the maximum at a predefined rate. It is this rate that the *attack control* will vary.
- (D)ecay—Once the signal has reached its maximum it will then reduce, at a predefined rate, until it reaches the next stage. This rate of reduction is the decay.
- (S)ustain—At the end of the decay stage the signal will stabilize at a level determined by the sustain control. Out of the four controls in an ADSR envelope this is the only one that controls a *level* rather than a *rate*. The control signal stays at this level until the keyboard key is released.
- (R)elease—You have probably guessed what happens here! The control signal dies away at a rate determined by the release control until it reaches zero again.

So, with a combination of these controls you can create envelope shapes that vary from short and snappy to longer and more evolving.

The most obvious, and common, use for an envelope is to change the amplitude or volume of a sound. If you think of a guitar string being plucked or a piano string being hit with the hammer you can almost visualize the volume of the sound. It starts up very quickly with the initial pluck or hammer strike and then gradually dies away to nothing if the note is held. In addition, on a piano, if the key is released before the note has died away naturally then, owing to the dampers on a piano, the sound will stop. If you listen to the sound of a single piano you will probably be able to visualize the amplitude envelope. Do you see the way this works now? Perhaps you want a sound that starts more gently and gradually builds up in intensity. You can achieve this by using a slower attack (usually given by a higher value on the attack control). Or maybe you want your sound to die away slowly once you have released the key. In that instance, increase the release value. If you want a sound that is more like that of a pizzicato string, then set a fast attack (low value), medium release, and low sustain value, for that snappy or percussive sound. These are all ways in which you can manipulate the volume of a sound using an envelope.

You can also control the filter cutoff frequency with an envelope. In many ways you will get a superficially similar effect to using the envelope to control the volume but, especially if you are using a low-pass filter, as well as seeming to control the volume of the sound (because the more harmonic content you have in a sound the louder it seems to be, to a point at least) you also control the tone of it. So as well as the sound seeming to swell up in volume it can also change from muted at the start to more bright and then back down to muted again. Or, if you were to apply an envelope to a high-pass filter, the sound could start as very thin and then get more body and depth as the envelope changed.

It doesn't end there. You can sometimes even modulate a modulator! You could set up an envelope to slowly build up and then apply that control signal to the speed parameter of an LFO so that whatever the LFO was controlling (let's say the fine tuning of a sound) could start off with quite a slow variation and then speed up over time. Once again, depending on the modulation routing possibilities of the synth you are working on, the possibilities can be endless. Sometimes the little things can turn a good sound into a *great* one!

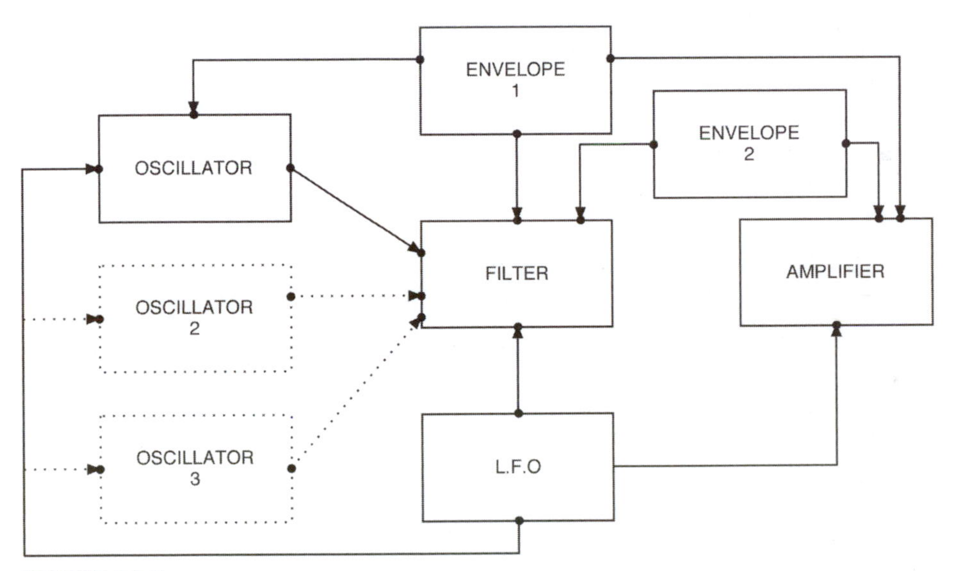

FIGURE 10.9
The signal path of a basic subtractive synthesizer. More complex synths exist, of course, but all follow this basic model.

That is pretty much all there is to the basics of subtractive synthesis. Some synths have built-in effects like reverb, delay, chorus, and flanging, but those are the fairy dust on top of the sound. The oscillators, filters, envelopes, and LFOs are where the serious sound design is done. In Chapters 11 and 12 we look at ways we can use these building blocks to create different kinds of sounds, but for now let's move on to another type of synthesis you might come across.

ADDITIVE SYNTHESIS

Not surprisingly, *additive* synthesis takes a completely opposite approach to building sounds. In the introduction to this chapter I mentioned that sounds were made up of many different frequencies. Subtractive synthesis starts with all of these frequencies already present and allows you to selectively remove them. Additive synthesis lets you to choose which of these frequencies you want to *include*. Many additive synthesizers give you control over 128 (or more) individual sine waves; you can control their relative levels and their level envelopes.

Equipped with the right knowledge, you could easily recreate a square wave or a sawtooth wave, but that really isn't the point of additive synthesis. With additive synthesis you have the power to create waveforms from harmonics that don't follow the simple 1st, 3rd, 5th, 7th... harmonics rule that a square wave has. You could, for example, choose to include only the 1st, 3rd, 4th, 7th, 11th, and 15th harmonics. You could choose to include frequencies that are *inharmonic* (that is, they're not an exact frequency multiple of the fundamental frequency), to give the sound a more metallic feel. You could choose to have the levels of the harmonics follow an inverse pattern. Usually the fundamental frequency is the loudest with all of the harmonics gradually reducing in level as they become higher. You could do the opposite and have the highest harmonic as the loudest and the lowest as the quietest.

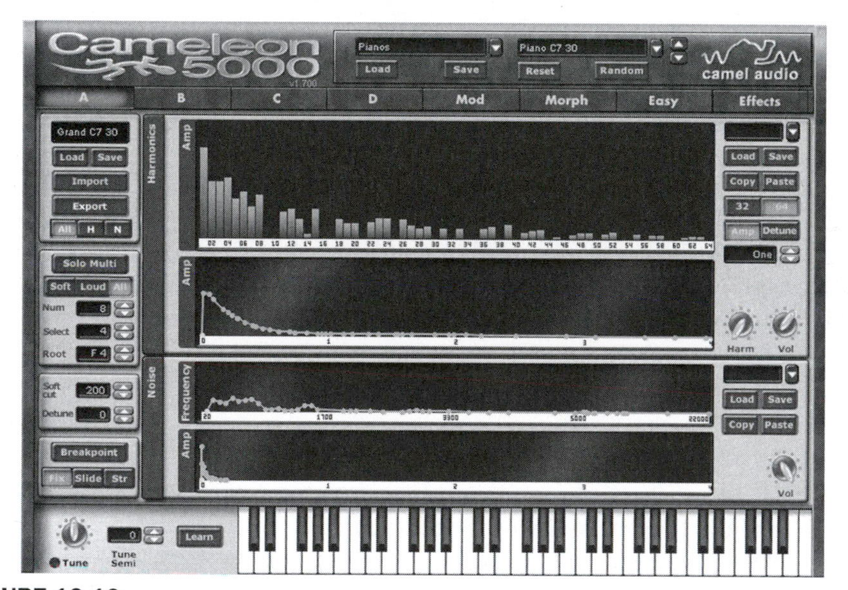

FIGURE 10.10
In additive synthesis you create your sounds by specifying the levels of a large number of individual frequencies (or *partials*) and specifying how they change over time.

Of course, all of this comes at a price. While much of subtractive synthesis is fairly intuitive after a while, additive synthesis takes real thought if you have a particular sound in mind when you set out. This really *is* something for the more experienced sound designer. It is also probably why additive synths are quite rare (in both hardware and software forms) and why some of the more advanced software additive synths include *resynthesis* capabilities. Resynthesis occurs when you load an audio file (a *sample*) into the software and it analyzes the sound and creates an approximation of it by adjusting the levels and envelopes of each frequency in the synth to match the frequency balance and change over time. This can give some very interesting results as it lets you do all manner of manipulations on the resulting additive sound that would not be possible with any form of conventional sample editing.

You could, for example, stretch the sound in the harmonic sense. When a sound is analyzed there will be separate frequencies that make up the sound. These are spaced a certain distance apart and it is the relation between these frequencies, as we have already discovered, that gives the sound its character. With an additive synth, you can make the gaps between them wider or narrower. The resulting sound is impossible to describe in words; it's best to try it out yourself if you possibly can.

You can also apply the stretching approach in the *time* domain. For each individual frequency in an additive synth there is a volume envelope. These are usually much more complex than the four-stage ADSR envelopes we spoke about earlier, and allow for the level of each frequency band to rise and fall multiple times. The reason for this is that the balance of the different frequencies in real instruments is rarely a simple affair. Some frequencies might start off low, then rise, then dip and then rise again, while others may just have a short burst of activity at the very start of the sound. A resynthesized additive synth sound tries to track these changes for each separate frequency. So, if we have a sound that lasts a total of five seconds, the total length of the longest envelope will also be five seconds. But if our synth allows us to stretch all of those envelope times equally then we have a very unusual form of time-stretching available to us! Again, remember that these sounds are only an approximation of the original sound so I wouldn't advise you to use this method as your main source of time-stretching on anything critical. It does, however, provide some interesting effects.

Because of the way additive synthesis works, it negates the need for filters, because you already have in-depth control of the sound sources to actually build change into them. Many traditional additive synths didn't have any kind of built-in filters. However, with the power of modern computers, many of the newer additive synths *do* include filters. You can use the filters in exactly the same way as described in the previous section, only in an additive synth the sounds being passed to the filters could be inherently more complex.

The same can be said of the modulators. The additive sound source itself contains a large number of modulators to control the generation of the sound, but you will often find more *global* modulators for broad changes to amplitude, filter frequency, and other common synthesis parameters. These will take the form of the usual envelopes and LFOs.

With this in mind, and assuming that you are dealing with a modern additive synth that *is* equipped with filters, LFOs, and envelopes, you can approach additive synthesis in two ways. You can look at it like a standard subtractive synth with the power to have more complex and interesting output from the oscillators, or you could look at it as a tool for really complex sound design (and even resynthesis) that happens to have filters and modulators attached at the end. Which of these works best for you really depends on how much time you are prepared to put into learning how to synthesize the sounds. As with so many things, the more time you put in, the more you will get out of it. And if you really want to get into sound design and come up with your own unique signature sounds then additive synthesis can be a very useful tool.

FM SYNTHESIS

FM stands for *frequency modulation*, a type of synthesis that has been around since the early part of the twentieth century. During the 1930s, frequency modulation was used in radio transmissions as a means of improving the range and clarity of the broadcasts. However, it wasn't until the early 1970s that John Chowning, a researcher and composer from Stanford University, started to explore the possibilities of using frequency modulation to actually *create* sounds instead of merely transmitting them. For the remainder of the decade he continued his research and then, in 1982, Yamaha released the first synth to use the FM, the GS1. The product wasn't a great commercial success, but it paved the way for one of the most legendary synths of all time, its successor, the Yamaha DX-7. Released in 1983, the DX-7 has become a legend in the world of synthesizers.

Frequency modulation, not surprisingly, has modulators at its heart. However, unlike additive synthesis where the modulators (at least the ones involved in the sound generators themselves) are usually envelopes, here the modulators are oscillators! The main difference here is that, in FM synthesis, the modulators used in the sound generation aren't *just* low-frequency oscillators, they are *full range* oscillators that can start way below the threshold of human hearing and go right through the audible range and beyond.

Remember when we talked about subtractive synthesis I was describing an LFO that was modulating the fine tuning of an oscillator to produce a *vibrato* effect? This worked because the LFO was causing a change (modulation) in the tuning (frequency) of the oscillator. That example is, in essence, the most basic form of frequency modulation. And that is the basis on which FM synthesis works. Of course, if all FM synthesis could do was apply *vibrato* to an oscillator it would be pretty pointless. In an FM synth the oscillators are called *operators* and can be connected together in a variety of ways. Different FM synths have different numbers of operators and a different number of ways they are connected together (called *algorithms*) but, at the very least, there needs to be two operators for FM to work. There are two different types of operators in an FM synth, carriers and modulators, and each oscillator is either a carrier or modulator. In some cases, and in the more complex algorithms, any given oscillator can be a carrier (that is, it is modulated by another operator) *and* can act as a modulator for another operator. It is this modulation complexity that allows FM synths to produce harmonically complex sounds and tones.

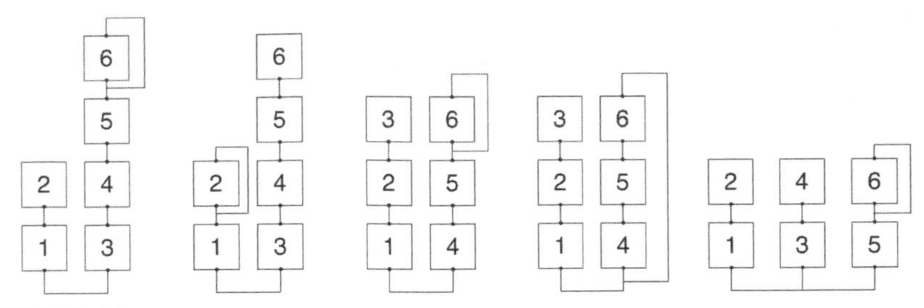

FIGURE 10.11
Some of the many different FM algorithms available on the Yamaha DX-7.

Let's go back to the most basic example mentioned earlier. If we have a simple one-carrier and one-modulator set-up, we can start to look at what happens with FM synthesis. The carrier is the waveform that is actually *heard* in the end. The modulator doesn't (normally) have an output that can be heard; rather it is just used to modulate the carrier. So if we started off with a sine wave for the carrier operator that was running at a fixed frequency and then we applied the modulator to its input, we would, at low-modulator frequencies, hear a *vibrato* effect. The depth of this effect is controlled by adjusting the level of modulation applied to the carrier, and this level of modulation can be further adjusted by an envelope applied to the modulator. So, once again, we see that complex modulation is the key to so many aspects of sound design. In this case we are modulating one modulator (the oscillator) with another modulator (an envelope) to give a vibrato effect that steadily increases in effect. If we change the frequency of the modulator operator to run at a much higher frequency, we start to hear something that can no longer be classed as *vibrato*.

As the frequency of the modulator increases, we start to get the classic FM sounds, although they might not seem especially musical at this stage. The real magic of FM starts to occur when the modulator and carrier waves are at similar frequencies. For reasons that are mathematically far too complex to explain here, FM synthesis produces a series of harmonics that are directly related to the ratio of the carrier frequency to the modulator frequency. So by changing the ratio of these two frequencies you are changing the harmonics and, consequently, the character of the sounds produced. A very simplistic explanation of the harmonics generated (called *sidebands*) is given here.

There will *always* be a harmonic generated at the carrier frequency (let's call this "C," and we'll call the modulator frequency "M"). In addition, there will be harmonics of higher frequencies generated in a sequence according to the following pattern: C+M, C+2M, C+3M, and so on. Finally there will be a series of harmonics of lower frequencies generated according to this pattern: C−M, C−2M, C−3M, and so on. To give a more specific example let's put some numbers in and see what happens.

If we had a carrier frequency of 900 Hz and a modulator frequency of 200 Hz we would get a harmonic series generated as follows:

C−4M	C−3M	C−2M	C−M	C	C+M	C+2M	C+3M	C+4M
100 Hz	300 Hz	500 Hz	700 Hz	900 Hz	1100 Hz	1300 Hz	1500 Hz	1700 Hz

Does that sequence of numbers look familiar? Do you remember when we were talking about square waves and how they occur? Square waves are composed of the fundamental frequency and only odd-numbered multiples of that frequency. So if we took a fundamental frequency of 100 Hz, there would be harmonics at 300 Hz, 500 Hz, 700 Hz, and so on, which is exactly what we have here! So FM can be used, once again, to produce similar waveforms to those used in a subtractive synthesizer.

If we were to use a carrier frequency of 400 Hz and a modulator frequency of 100 Hz this time, we would get a different harmonic series generated:

C−4M	C−3M	C−2M	C−M	C	C+M	C+2M	C+3M	C+4M
0 Hz	100 Hz	200 Hz	300 Hz	400 Hz	500 Hz	600 Hz	700 Hz	800 Hz

What about *those* numbers? Well, if we assumed a fundamental frequency of 500 Hz this would be the harmonic series for a sawtooth wave. These waveforms won't sound exactly the same as those produced by the oscillators in a subtractive synth for a number of reasons (including the relative levels of each of the harmonics) but it does illustrate that even with a very simple FM configuration there is plenty of scope for creating varied sounds from adjusting just a few simple parameters.

What if we were to change the frequency of the modulator over time? Say we applied an LFO to the pitch parameter of the modulator, which changed it smoothly from 100 Hz to 200 Hz and then back again. What would be the result? Well, we would get an output from the carrier that changed from a pseudosawtooth to a pseudosquare wave and then back again. But it wouldn't just suddenly switch from one to the other; it would *morph* from one to the other. This is a very useful technique for adding subtle (or not so subtle) movement and life into sounds created with FM.

It is also worth noting at this point that if you wanted to change the basic tuning of the sound (or transpose it) then you need to adjust the frequency of the carrier. If you have set the frequency of the modulator manually then you will need to recalculate the ratio and change the frequency manually as well. However, most FM synths work with ratios so you wouldn't have actually used a specific frequency as in the previous examples, you would simply have set the modulator to a ratio of 200:900 Hz, or 2:9 (1:9 in the second example). In this instance any changes in the carrier frequency would be automatically matched by corresponding changes in the modulator frequencies.

We have figured out that changing the frequencies of the carrier and modulator is what determines which frequencies and harmonics are created. But to control the level of the harmonics we have to vary the level of the modulator. And this is why each of the carriers and modulators has an associated envelope and can use LFOs as modulation sources. In a subtractive synth we could create a sound that had a low-pass filter controlled by an envelope so that when the sound was triggered the envelope would open up quickly and then close down slowly. This would have the effect of a sound that had a lot of harmonics at the beginning and then gradually became more muted as the envelope closed. We can achieve a similar effect with FM in the following way.

If the carrier was not being modulated at all then it would just be a sine wave and, as such, have no extended harmonics. If it was being modulated in one of the ways we described previously, it would have a series of extended harmonics. The levels of those harmonics would depend on the level of modulation being applied. If we were to apply an amplitude envelope to the output of the modulator with a fast attack and slower decay, the modulation amount would follow that envelope. As a result, the carrier would be modulated to a large extent at the beginning, producing lots of harmonics, and then as the modulation level decreased (because of the envelope on the modulator), so too would the level of the harmonics. Of course, the effect wouldn't be exactly the same as it is a completely different form of synthesis, but it shows that *similar* effects can be created.

Once you extend the principles listed here to include four or even six FM operators you start to get an idea of the potential for sound creation. You could use the six operators as three carrier/modulator pairs and create something along the lines of a three-oscillator subtractive synth, or you could stack up the six operators with each feeding into the next as a modulator.

In hardware FM synths there is a limited number of these algorithms to choose from, but in the latest software FM synths there is a modulation *matrix*, which allows you to configure your own algorithms to create a virtually unlimited range of options. In addition, more recent FM synths allow you to choose from waveforms other than sine waves as the sources for both carriers and modulators, giving you even more options for designing your sounds. With all of this in mind, FM synthesis offers perhaps the widest palette of sounds, but that comes at the expense of less intuitive programming.

PHYSICAL MODELING SYNTHESIS

From the dawn of sound synthesis, one of the goals has been to reproduce real instruments. Attempts at reproducing real instruments on the earliest synths were crude at best, but the resulting sounds would often give the feel of the instrument they were trying to recreate if heard in the context of a full piece of music. As synths became more complex and elaborate, so did the sounds they were capable of producing. Nevertheless, even the best synthesizers in the world cannot recreate the range of *expression* of a real instrument. When you hear one static note in isolation you might be hard-pushed to tell the difference between the real instrument and the synthesized one, but when a melody is played you will hear all manner of subtle differences. It is these differences and subtleties that are the holy grail of sound design. In order to do this, a totally new approach was needed.

Physical modeling works in a way totally unlike any other method of synthesis. Instead of dealing with a static waveform and then modulating it using envelopes and LFOs, a physical modeling synth instead uses a computer model to describe the actual physical *process* that happens when the instrument being modeled is played. From that model, it calculates what the real instrument would be doing under any given set of conditions. The following is an example in the form of a physical model for a guitar.

Think about what happens when a guitarist plays a guitar. What are the deciding factors in the sound produced? Well, at the very beginning there is the plectrum or pick. This is what is used to move the string. If the plectrum is harder or softer or made of different materials then that will affect the sound. A thin and flexible plastic plectrum will produce a very different sound to a thick metal one. Then we have the strings themselves. Different metals used in the strings will sound different, as well as different thickness of strings and different tension in the string. The strings then cause a vibration in the body of the guitar so the material of the body of the guitar is also a factor. If it is an electric guitar then the size, location, and other factors of the magnetic pickups will change the sound. If it is an acoustic guitar then the size and location of the sound hole will affect the sound. There are many other factors as well but these are the most common.

Physical modeling takes all of this into account and creates what is, essentially, a computer simulation of a guitar, which has parameters for all of these variables. You program in the values of these variables—plectrum type and thickness (stiffness), string material, string thickness, string tension, body material, and so on—and then the simulation will be able to calculate what the sound generated would be. This enables very accurate and natural sounding emulations of real instruments, but it doesn't allow for a huge amount of variation in the sound. That is why there are different simulations used to recreate different types of sound. The one described earlier would be called a "plucked string" simulation. There are others for "bowed string," "blown pipe," and "taught membrane" systems as well, for recreating the sounds of, for example, a violin, a flute, and a drum, respectively.

Each simulation has specific parameters that relate to the model in question, so there are no generalized parameters like you have with the other synthesis methods described. Each is unique and targeted, and produces a specific type (or group) of sounds. There are some creative possibilities with physical modeling that do go beyond the purely emulative, however. You can produce models that are simply not feasible in reality due to size or material restrictions, but they will always retain that inherently "plucked," "bowed," "blown," or "hit" tonality.

In terms of synthesizers that use physical modeling as a synthesis method, there is very little choice at present. Owing to the computational complexity of these systems, it is only recently that they have started to become more widespread. The first commercially available synth to use physical modeling was Yamaha's VL1, released in 1994 (later came the cheaper VL7). These were revolutionary instruments at the time, but failed to really connect with people owing, at least in part, to the high cost and complex programming architecture. The VL series was followed by Korg's OASYS system and its Prophecy synth. Nowadays, the same basic techniques are available in much more accessible formats. Software instruments that make use of *physical modeling* techniques include Logic's EVP88 and EVB3 recreations of a Fender Rhodes piano and Hammond B3 organ, as well as the Modartt Pianoteq plug-in, which uses these techniques to (very successfully) recreate the sound of an acoustic grand piano. There will, undoubtedly, be future developments in physical modeling technology as computing power continues to increase, but there is still some doubt as to how much it will be adopted owing to the programming complexity and the inherently limited variety of sounds that can be produced. But, as with all of the other synthesis methods listed here, it *does* have its place.

SAMPLING

Sampling isn't, strictly speaking, a synthesis method, but it is an essential part of the overall sound-design process. Sampling is, in essence, a recording of a sound, which is then manipulated in various ways. Although it wasn't recognized as such when it first came out, the first well-known "sampler" was really the Mellotron. Released in the 1960s, it was a machine that featured actual tape recordings of different instruments, one tape for each note. The sounds were triggered by the keyboard and the tapes pulled over tape heads by electric motors. There were several limitations to this system. The first was that the recordings were of a limited length (about eight seconds) so if you wanted a sound like strings to hold for longer than eight seconds, well, you were out of luck! People got around this limitation by adapting their playing style to suit, but it still wasn't ideal. The other *big* limitation was that, because of the "tape banks" used to store the sounds, a Mellotron could only be loaded with one sound at a time, and changing a tape bank wasn't an easy job! But it paved the way for the later development of much more accessible digital sampling.

Digital sampling can be traced back to the late 1970s and the legendary Fairlight system. The aim of the Fairlight was simple: to bypass the limitations of synthesis techniques in reproducing real instruments by simply recording the actual sound of them and allowing those sounds to be played back, in a melodic way, at will. The early systems were extremely limited by the technology of the time because, as the recordings were digital, they had to be stored in some kind of memory and, at the time when the Fairlight was released, the multiple gigabytes of memory that we have become accustomed to today weren't even a dream! The less memory you have available to store the sample, the shorter the length of the recording and the lower the quality of recording that you can store. But still, even with those limitations, the arrival of sampling was a real breakthrough and allowed a far more accurate representation of acoustic sounds.

The first samplers had very limited editing capabilities and were fundamentally similar to the Mellotron in the sense of being "playback" devices. The biggest difference was that these digital samplers had looping capabilities. These weren't perfect (although great improvements in looping would come later), but it meant that the notes could, if desired, be held indefinitely, bypassing one of the major limitations of the Mellotron. The other big advantage was that the storage medium for digital samplers was far more convenient. Sounds (or sound banks) were stored on floppy disks, so loading a new sound meant inserting a new disk and loading from there rather than changing a large and cumbersome tape bank. The loading times were far from instantaneous, but they were a marked improvement.

Later samplers would increase the quality of the samples and the total length and amount of samples that could be stored at any one time. Later, manufacturers began to include filters and modulation sources so that the samplers of the day were more like subtractive synths that used real recordings rather than static waveforms as the sound sources. But in its early days, digital sampling was still, primarily, sampling. Then, in 1987, Roland released the D-50, the first real synthesizer that incorporated samples. The D-50 had both static waveforms and samples available as oscillator sources, and these could then be combined before being passed through the filters and further modified by all the usual elements of a subtractive synth. Admittedly the samples within were very short and mainly limited to *transient* samples (the initial "attack" portion of a sound in which most of the identifying character—bowed, plucked, blown, and so on—is contained), but by grafting a sampled transient onto a static waveform the resulting sounds were much more lifelike and realistic.

This approach became known as the S+S method (or sampling + synthesis). The next development in this approach was the Korg M1, which used more elaborate samples as the main source for the oscillators. There were traditional square, sawtooth, and triangle waves included, but much more emphasis was put on the samples as the source of the sounds. This was the start of a trend that was to last for many years and produced some of the best known and best-selling synths of the 1990s including Roland's legendary JV series. In fact, it is only in the last five years or so that the trend has swung back toward "vintage" synthesizers (or at least their software recreated counterparts).

SUMMARY

In this introduction to sound design I have explained a little about the most common forms of synthesis you are likely to come across in everyday use.

They all work on the same acoustic principles but set about arriving at that point in completely different ways. There is much more to each of them, of course. In fact, a deeply detailed examination of any of them would probably take up more space than this entire book, but if you want to get more detailed knowledge there are plenty of great sources of information in print and on the Internet to satisfy even the most deeply technically minded among you. What I described here was the difference between them and, as a consequence of their differences, their different tonalities. Each of the synthesis methods has strengths and weaknesses, and the beauty of sound design is that we get to play with these and to mix with them.

Sometimes you might be struggling to get the sound you are imagining from a subtractive synth. It might be *almost* there, but perhaps there is a tonal element that you can't quite get right. You *could* continue trying to make it perfect on just one synth, but why struggle and possibly end up compromising a sound that is 90% right? Try layering it up with a sound from another synth, or even another *type* of synth.

More and more people are producing music "in the box" (meaning that the sounds are all produced, mixed, and mastered inside their computers) and, with the outrageously powerful computers available to us today, we have access to tens if not hundreds of virtual synths. It's much easier now to mix and match your instruments and synthesis types. You can get the sound to 90% on your virtual Moog, and then if you need a little sparkle or a harmonically interesting layer of fairy dust, you can load up an FM synth or maybe even an additive synth and layer the two together. That way you can optimize each synth to the element of the sound you are trying to create. This is something I do all the time to create new sounds. I do love getting deeply into programming synths but there is no *one* synth that can do *everything*. So learn to program the synths well and exploit them for what they are best at, but don't be afraid to layer synths together to get the overall sound you are looking for. You can always sample the combined sound and create a sampler patch from it if you prefer the convenience of being able to load it from just one place!

The next two chapters examine the different types of sounds—bass, pads, arpeggios, leads, and percussive—and we look at the general characteristics each of these sound categories exhibits. I'll show you how the different synthesis methods that we have explored here can be used to build those sounds. But before reading on, put the book down and play around with some of the synths you have available. Just explore without trying to achieve a specific sound. Change one parameter at a time to get a feel for what they all do, and after a while the process of sound design should become more intuitive.

Often people will look to some new piece of equipment (real or virtual) to solve a problem with their work when, in reality, spending time getting to know the equipment they already have can prove much more useful and effective and, in some ways, satisfying. Of course, there are always better pieces of equipment out there, but if you learn to use the ones you already have to their full potential first, it is almost certain that, when you *do* get those shiny new toys, you will be able to get far more out of those as well!

Because I have taken the time to get to know my sound generators, many times when I am listening to a track I have a fair idea of how I would program the sound without even having a synth in front of me. In fact, it can even get to the point where I *see* the synth (plug-in in my case) when I close my eyes and can imagine the positions the controls would be in. Terribly geeky I know, but great fun too!

CHAPTER 11

Sound Design: Rhythmic and Percussive Sounds

INTRODUCTION

There are three different categories of rhythmic and percussive sounds: the main *drum kit* type of sounds, traditional *percussion* sounds, and all other sounds that are used as percussive sounds without necessarily being related to anything "real world." We cover all of these in this chapter, beginning with the main drum kit sounds, as these will form the backbone of any remix you work on.

I will not go too deeply into the harmonic theory surrounding each particular kind of drum sound, as that subject in itself would be enough to fill half of this book on its own! There are many great books, articles, and websites that cover the complex physics involved in recreating drum sounds if you're looking for a more detailed investigation into the subject. For the purposes of this book, I will cover more practical applications and explain how the majority of us go about creating our drum sounds on a day-to-day basis.

Later in the book, when we look at mixing and effects (Chapters 16 through 19), we will be covering how to use compression and EQ to enhance different types of sounds, including drum and percussion sounds. Here we're dealing with actually creating the sounds, rather than how to get them to cut through in a mix or how to handle the dynamics. You can always jump forward to those sections as you are working on creating your own sounds if you like!

DRUM KIT SOUNDS: KICK DRUM

The most obvious drum sound to any dance music producer is the kick drum. In pretty much every genre of dance music this is one of the fundamental elements of a track and, sometimes, the choice of kick drum sound can make or break a groove. Kick drum sounds range from the authentic "acoustic" sounding to the totally synthetic, and a million shades in-between. Some are "pure" and are a direct sample from a real kick drum, others are 100% synthetic and have been wholly created on a synthesizer, some are hybrids, and others have been sampled from other records. These sounds are often endlessly sampled, resampled, edited, and reincorporated with others until their true origins are indistinguishable. If you're thinking about trying to recreate a favorite kick drum sound, you could be in for quite a struggle as the sound in question could, potentially, be made up of a huge number of elements and layers. Trying to recreate all of the subtlety of that could be a mountainous task!

Better to come up with your own, unique sounds! You could, if you like, use a sampled kick drum from one of the ever-increasing number of (often very good) sample libraries. It would be misleading to tell you that I have *never* done that, because I have. But more often than not I will layer different kick drum samples with synthesized sounds to create something that is totally unique to me and perfectly suited to the track I am working on.

Let's break the sound down into a few components to see how these work together and how we can come up with ways of building our own kick drum.

Kick drums, like most drum sounds, can be broken down into *attack* and *sustain* portions, with the attack part being the initial hit of the sound and the sustain part making up the body of the sound. There are four different frequency ranges: subbass, bass, midrange, and treble. Even without knowing too much about the theory behind a kick drum sound, most people would automatically recognize the sharp attack of the sound and the softer sustain. Most people also automatically know that any high frequency (and some upper midrange frequencies) only normally occurs, if at all, during that initial attack phase. After all, kick drums are also called bass drums, and that is for a reason! With that in mind, we can start to think about a kick drum in terms of four separate frequency components with different amplitude envelopes, with the higher frequency components generally dying away quicker than the lower frequency ones. But how do we go about choosing how to create each of these components and, ultimately, mixing them together into one, coherent whole? Well, in order to do that let's take a look at the differences between a sampled "real" kick drum and the almost ubiquitous Roland TR-909 synthesized kick drum.

In a real kick drum there is a constant presence of lower frequencies in addition to the higher ones (which are harmonics of the fundamental frequency generated in the vibrating drum skin and drum shell). In a TR-909 type kick drum there are only higher frequencies in the initial attack phase. The reason for this is that the body of the TR-909 sound is made up of a single oscillator outputting a single waveform. So there are higher frequencies in the attack phase when the oscillator is tuned to a higher pitch, which then drops to a lower pitch to provide the deeper sustain. There is also a filtered white noise burst to further emphasize the attack of the sound and add a variable degree of snappiness, without having to fundamentally edit the basic sound. As such, synthesized kick drums have only a very tiny delay before the low frequency body of the sound actually kicks in. This is only a matter of a few milliseconds, so it is pretty much imperceptible to the human ear. However, it is worth considering if you are building your own sounds from scratch.

There are also many things to take into account when you consider the genre you are working in, as this will have a great effect on the type of kick drum you are looking to create. Breakbeat and drum 'n' bass genres often have kick drums that have a very "real" feel to them. Funky house kick drums often have a combination of real elements and synthesized ones. Electro, progressive, tech, tribal house, and many forms of trance and techno have more TR-808/909 style synthesized kick drums. All of this is relevant.

In starting to build a kick drum of your own you *could* look at synthesizing all of the four frequency elements, and this is something I certainly recommend doing, if only for the experience and depth of knowledge and understanding of the kick drum. It can take a lot of time so it is perhaps best to experiment with different sound sources to see what works best. As a guide, and a warning, when using totally synthesized sound sources it is all too easy to end up with kick drums with an obvious pitch to them. This isn't always a problem, but it is something to bear in mind. An alternative to this is to use a combination of sampling and synthesis to create a kick drum. This can give a sound that's deep and interesting, and has a great deal of flexibility and adaptability. Intrigued? Then read on.

The best way to illustrate this is to walk you through the process I use to make a kick drum. I will start off going through kick drum samples (from sample libraries, *not* from other tracks as that would be an infringement of copyright law, even if those kick drum samples were then edited and manipulated or combined with other sounds). These may be real or synthesized kick drums. As I audition each of them I look for elements in the sound that I like. Note that I am not looking for *one* kick drum sample that does everything I want it to do. On one sample I might like the woodiness of the attack portion of the sound, on another I might like grittiness of the midrange, and on yet another I might like the weight of the sound. I will probably pick out more than three samples, having multiple options for each of the weight, midrange and attack parts. I then load them on to audio tracks and mute and unmute different tracks to give me different combinations of the sounds. I may end up using multiple treble and midrange samples. On rare occasions I may even use multiple samples for the bass component, but this happens less often as multiple bass components often don't match up, with unpredictable results!

Once I have a few options I like, I experiment with changing the balance of the different samples and then start to use EQ to further cut away frequencies from the different samples. I might also use a noise gate or simply change the length, and fade out different samples to provide different amounts of snap to the different samples. There are a few examples of this on this book's accompanying website that illustrate these points; it might be simpler to actually listen and see what I am talking about on screen rather than have me try to explain here.

The reason I initially mentioned *four* frequency ranges and in the last few paragraphs I have only mentioned three, is because there are a couple of different options for dealing with the subbass component. The first, and in some ways simplest, is to just boost the really low frequencies in the bass component. This can work, but it can also lead to a rather unpleasant boomy kick drum. It can also cause further problems down the line when trying to balance the levels of the kick drum and bass line. The second option is to find another sample to use as a subbass component. Often this involves removing all but the lowest frequencies from the sample. This has the advantage of giving you more control over the lowest frequencies in the sound; it was the method I used for quite a long time. But it still isn't the most flexible or, in my opinion, the best approach.

My secret weapon of kick drum programming is the humble sine wave. What I have been doing a lot recently is creating a kick drum as described earlier using all but the subbass component (or perhaps just reducing the level of it) and then bouncing it to a new sample for ease of use (more on this in a moment). I then have a separate subbass component, which is simply a synthesized sine wave with an appropriate amplitude envelope applied. This offers a great deal of flexibility because you can manipulate the subbass level of the kick drum independently (using level as opposed to the less predictable and accurate method of automating EQ changes to the main kick drum sound) and it allows you the option of fine tuning the pitch of the subbass component to perfectly match the track. It doesn't always work exactly as you hope, but when you get it right, it has the effect of being an almost imperceptible extra weight to the track that it also very clean.

We now come to the question of how to combine all of these into one complete new sound. We *could* simply keep them as they are, but copying and pasting upwards of three of four audio files every time you want to duplicate an individual kick drum hit is impractical. You might inadvertently miss one while copying, which would result in an unexpected change in tone of the kick drum. So what do we do? Well, I normally start by routing all of the individual samples to a bus and then make sure that the channel levels on each sample are low enough to avoid overloading the bus. Then I might apply some overall EQ to the sound. This is done mostly to help glue the sounds together. At this stage I might also go back and tweak the individual EQs of the samples to help them all blend better. I would also apply some light compression to help pull all of the components of the sound together before finally adding some very light limiting at the very end, just to make sure that nothing is clipping following the EQ and compression. From this point I would simply bounce the combined sound and then drag that combi kick into Logic's Arrange window. From here I can work with the sound while still keeping the original files intact, in case I want to go in and tweak the sound again once I progress a little further with the track.

Depending on just how many individual components were used to make up the sound I might even bounce several variations of the kick drum by muting individual components or combinations of components to give different results. Perhaps the most useful is a version where all the bass components are muted, leaving only the midrange and the treble components active. To some extent, the same effect can be achieved by applying a high-pass filter to the combi kick. This is a good alternative to have, given that it is relatively little extra work to just mute the bass components of the sound, bounce the combined sound, and drag into the Arrange window as well.

If this all sounds like too much work for what is, in effect, a very small part of the final track, remember that you are dealing with one of the most important elements of your finished track! If you want to you can always try to find everything you are looking for in just one kick drum sample. There's is a good chance that you might find what you need, but if you take the extra time and effort to design the best sound you can, you will be pleasantly surprised by the results!

There is also a final option to consider, the excellent Metrum plug-in from the Vengeance Producer Suite (*www.vengeance-sound.de/eng/VPSMetrum.html*). This does, in plug-in form, much of what I described earlier in terms of layering different samples, each with its own independent amplitude envelopes. It includes one dedicated oscillator layer along with three sample layers, plus a modulation matrix and several useful built-in effects. The Metrum has the convenience of being self-contained and, because it is a plug-in, there is no need to bounce the combined sound to work with it in a convenient manner. I have used it with great success on some recent tracks. That said, I still use my tunable subbass layer underneath the sounds produced within Metrum, simply because I like the way it can be tuned to follow the bass line.

DRUM KIT SOUNDS: SNARE DRUM

The snare drum (along with the clap) is the second in the Holy Trinity of groove-making drum sounds in club music. If the role of the kick drum is to provide a solid and regular beat for people to dance to, and the role of the hi-hats (along with things like tambourines and shakers) is to provide the pace and movement in the groove, then what does the snare drum do? Well that isn't always an easy question to answer. The snare drum can provide some extra rhythmic interest in terms of syncopation but, more often than not, it is simply used as an accent on beats two and four of the bar.

Many of the choices you make when choosing and programming snare drum sounds depend on the genre you are working in. Some genres go for quite small and tight-sounding snare drums, others for bigger and brasher sounds. Some do away with the snare drum completely in favor of a clap sound. So rather than go into detail for each genre, I will simply go over the basics of programming snare drum sounds and finish with some potential ways to modify and adapt those sounds to take it more in the direction of the genre you happen to be working in.

Once again the snare drum has separate attack and sustain components to the sound, but they might be a little harder to distinguish because of the inherent qualities of a snare drum. The attack component of a real kick drum is made when the beater of the kick drum pedal hits the drum skin. The high frequency harmonics generated in the vibrating drum skin die away comparatively quickly but the deeper frequencies continue on. In a snare drum, however, we have an additional factor to deal with. A snare drum is, in many ways, similar to other drums but differs in that it has snares drawn tightly across the bottom skin of the drum. If we were to remove these snares, the process for recreating a snare drum sound would be very similar to that of a kick drum, albeit in a higher frequency range and without the subbass considerations to deal with. There would be the snappy attack (this time caused by being hit with a drumstick rather than the beater from the pedal) and the resonating sustain caused by the drum shell and the skins.

But the addition of the snares brings in a third factor. The snares themselves are actually just a group of coiled wires under tension held against the bottom skin of the snare drum. When the top skin is hit with the drum stick, the pressure created inside the drum shell causes the bottom skin to vibrate and, as it does so, the up-and-down motion causes the wires to bounce against the surface of the skin. This alters the vibration of the snares and the bottom drum skin as well. And it is this distinctive rattle that gives the snare drum its unique sound. From a sound design perspective this can be seen as an additional mid-range to high-frequency component with a less snappy amplitude envelope and a slightly longer decay time. So to create a convincing snare drum sound from scratch we need to consider all three components: the "stick" noise, the drum shell and skins, and the snares themselves.

Given that a real snare drum is hugely difficult to model with conventional synthesis techniques, perhaps the easiest option is, once again, to use different samples layered up and edited for each of the components of the sound you want to recreate. The same techniques we would apply to choosing samples to create a composite kick drum apply here, albeit with slightly different attributes.

What if you actually *do* want to synthesize a snare drum sound yourself? The two main factors to consider are the sound of the drum skins, shell, and the sound of the snares themselves. In order to accurately synthesize the snare drum shell and skins we have to consider a complex harmonic structure. Because of the "closed system" of the two drum skins and the drum shell and the complex interactions between them, the frequencies produced aren't nicely harmonically related, so conventional oscillators wouldn't be up to the job here unless you used individual sine wave oscillators and a basic additive synthesis analysis to create the individual frequencies.

The problem lies in the fact that the frequencies produced don't follow on in a ratio system where each is a multiple of the previous one. Rather, the frequencies produced here have a (relatively speaking) fixed interval between them. By that I mean that the harmonics follow a sequence along the lines of 130 Hz, 230 Hz, 330 Hz, 430 Hz, and so on (adding 100 Hz each time) rather than 130 Hz, 260 Hz, 390 Hz, 520 Hz, and so on (which is the pattern for the harmonics of a square wave). You might be seeing some kind of similarity.

A "normal" harmonic series starting at 100 Hz instead of 130 Hz would develop as 100 Hz, 200 Hz, 300 Hz, 400 Hz, and so on. Based on that you can see that we are getting closer to what we want. All we need is a way to somehow offset the harmonics by adding 30 Hz to each. We can do this with a device called a *frequency shifter*. Frequency shifters work by doing exactly what we require: adding a fixed frequency offset to the harmonics within a waveform. So if we ran a 100 Hz square wave through a frequency shifter with an offset value of +30 Hz we would get the resulting frequencies we wanted. There are a couple of problems with this though. The first is that frequency shifters aren't exactly standard equipment in most synthesizers. They are normally reserved for the likes of expensive modular analog synths or perhaps Arturia's excellent Moog Modular V plug-in. The second problem is that you will need two of these set-ups, as the nature of the snare drum has two of these inharmonic series combined (along with other elements). It's starting to look like a great deal of work, and that's before we have even gotten to dealing with the sound of the snares themselves!

Is there a way to create something that sounds approximately like a snare drum without going through all this trouble, or should we just revert to samples? Well, fortunately for us, the clever engineers at Roland came up with an answer in the form of the (did you guess?) TR-909 snare drum. Most of us know that it doesn't sound especially like a "real" snare drum, but it gives us all of the essential characteristics of a real snare drum. And you can rest assured that they didn't use complex frequency shifters or the like to create this trademark sound.

What they actually did was strip the sound down to the essentials. Out of the many harmonics present in a snare drum sound, there are two main frequencies called the "0,1 modes." Without going into unnecessary detail about what these are or how they are calculated, the engineers worked these out to be around 185 Hz and 330 Hz. So you form the basis of your sound by having two triangle wave oscillators set at those frequencies (which work out to F#3 and E4 in case your synth doesn't allow you to set absolute frequencies) and then set the amplitude envelopes to have the attack as short as possible and a decay/release time of about 100 ms (possibly a little less on the 330 Hz waveform if you have the option of giving each its own envelope). These two combined won't, at this stage, sound very much like a snare drum. And that is why we add in the filtered white noise component next.

The purpose of the filtered white noise is to emulate the sound of the snares. If you think about it, with a real snare drum, when it is hit with anything above a medium strength, you hear more of the snare sound than the resonance of the drum skins and shell. That is what we are aiming for here with our TR-909 type snare. So we take a white noise source and then we run it through a low-pass filter to take off some of the really high frequencies (around 7 KHz cutoff frequency seems to sound about right) and then we run that through its own amplitude envelope. Whereas the tonal parts of the snare sound had a pretty short decay/release time, here we can extend that to around

300 ms or perhaps even a little more. It is this component of the sound that really contributes the most, so if you wanted to increase the tightness of the sound you could reduce the decay and release times accordingly. Once you have added this in you will start to hear something that sounds a little more like a snare drum. The TR-909 also had a "snappy" control and this added in an additional layer of white noise (this time high-pass filtered at around 600 Hz in addition to the low-pass filtering already described) but with a much shorter decay/release time of around 170 ms. The level of this could then be controlled independently to give a degree of adjustability to the sound.

You could use a compressor with appropriate settings to further emphasize the attack of the sound or to give it more body (more on this in Chapter 17) and a little EQ to make it sound fuller or brighter, but what you have now is pretty much snare drum-like in its tone. It's unlikely you'll fool anybody into thinking that it is a real snare drum, but remember, we don't always go for realism in dance music, we go for what sounds good. And the TR-909 snare drum sounds so *right* in this context!

The best part about creating a sound in this way is that it allows you to actually change the sound as the track requires it. You could, for example, change the balance between the pitched part of the sound and the noise part of the sound, or you could change the decay of the sound. In short, you have much more flexibility than you might get from using a sample. In much the same way as an original TR-909 had controls that allowed you to alter certain aspects of the sound, creating your own synthesized snare drum allows you to build a little more control and variation into the sound.

With that being said, there is nothing to stop you, once again, layering your synthesized snare drum sound with a sample of a real snare drum. You could use only the attack portion of the sample to give the sound a little more bite and still have the flexibility of making changes to the underlying sound you created. Or you might choose to use a high-pass filter to remove the body of the sampled sound.

One final point to mention before we move on to claps is that, unlike kick drums, it isn't unusual to have two (or even more) different snare drums used in the same track. Often you might find a main snare drum being used on beats two and four and one or more slightly different, perhaps softer sounding, snare drums used for offbeat hits and ghost notes. This *can* be achieved using programmed (perhaps velocity dependent) changes within a single sound, but don't be afraid to use a few different sounds if it gets you the result you are looking for!

DRUM KIT SOUNDS: CLAPS

While claps are not, strictly speaking, a drum kit sound, they are closely related (both sonically and in terms of the context of use) to snare drums so I have decided to include them here. We all know what a real hand clap sounds like but how would we go about creating that sound? If you look at the classic TR-808 handclap sound you might be surprised! Once again we have a filtered white noise source (this time with a band-pass filter set at around 900 Hz with quite high resonance) that is, this time, fed through an envelope generator that creates four, very closely spaced and very quickly decaying "spikes" at around 10 ms apart. Each of these is designed to simulate a single person clapping, so the four together gives the feel of a group of people clapping at once. In addition to this there is a very simple reverb circuit to give a bit of ambience to the sound.

Another alternative is to take the same filtered white noise source but this time use an amplitude envelope with a longer decay/release time of around 80 ms and then use a sawtooth LFO set to around 100 Hz to modulate the amplitude in addition to the main envelope. This LFO gives the effect of a repeating mini-envelope with a time period of about 10 ms, which is pretty much what we had previously but we use the main envelope decay/release time to prevent this sound from continuing *ad-infinitum*.

As you can see, in terms of recreating the classic synthesized clap sound there really isn't a lot to it. We can vary the filtering on the noise source, the spacing and envelope of the four initial spikes (or the frequency of the LFO, if we use the second method), and the decay and level of the reverb portion of the sound, but that's essentially it. However, much more can be done with effects and many of the classic clap sounds used (especially in trance music) tend to have a much higher level of reverb on them than the reverb generated within the TR-808. In addition they are often layered up with snare sounds, and there are times when the line between the two is indistinct. You can move across the full spectrum of snare drum sounds and then across into handclap sounds quite easily without a break.

I find programming my own drum sounds hugely rewarding and interesting, and I feel that it adds a unique flavor to the percussion on my tracks and remixes. But claps can be pretty hard to get as you want them. It's sometimes hard to get the initial attack right, and then the sound can end up sounding a little soft in comparison to a sample. This could be due to a number of factors including the minimum attack and decay times of the synth you are using to create the sound and even the filtering used on the raw noise source.

Not having a sharp attack may not be a problem for you. After all, not every sound needs a razor sharp transient and, in fact, there has been a trend recently toward snare and clap sounds that almost seem to reverse into themselves. There are a number of ways of doing this, including bouncing and reversing the sample and mixing it in at a lower level, and, of course, moving it forward so that the end of the reversed sample matches up with the beginning of the normal one. Another way might be to add a short reversed reverb (Creative uses of reverb are discussed in Chapter 18.) to each hit. Either way, in cases like

my production style now is more minimal than it used to be. I use fewer individual sounds and competing rhythms and melodies than I used to, and that can actually make the sound design process a little more fun. I'm working more on bigger and more unusual sounds, as there is more space in the mix to place them into.

The last thing I want to mention are so-called *signature sounds*. Certain genres of music have become inextricably associated with certain kinds of sounds. On a less macroscopic view, certain remixers and producers have become associated with certain sounds. I can see good and bad in this. If you come up with a truly amazing sound, which then becomes your signature sound, people will start identifying your work. As a marketing and branding tool it is the audio equivalent of a logo. But unlike a visual logo there is nothing to stop somebody else from creating a very similar sound. Another potential problem is that the music scene changes very quickly so what is *de rigueur* today will very quickly become *passé* so, unless you are extremely lucky and have come up with a sound that sits comfortably in an ever-evolving production style, it is unlikely that this signature sound will last you a lifetime. Nonetheless, anything that you can do to help your branding and marketing is a good thing, so go for it as long as stays current.

Sound design is one of my favorite aspects of music production and it's something I could easily spend a lot more time on were I not kept busy with meeting deadlines. I hope that what I have shared with you is helpful when you are working on your own sounds and makes the whole process a little easier.

Next we will move on to something that is a fundamental aspect of remixer's job: the manipulation of time.

CHAPTER 13

Time Design: Time-Stretching

The one thing every remixer will have to deal with on a regular basis is *time-stretching*. This is the process of changing the length of an audio file. Although the process has become known as time-*stretching* it does, of course, include time-*compression* as well. If you have a song to remix that is at 110 bpm and you want to do your remix at 140 bpm you have to somehow make the audio files you use from the original song the same tempo as you need for the remix. And while this may sound superficially simple, the mechanics of it are far from simple.

Changing the length of an audio file is easy. If you imagine playing an LP on a record deck (remember those?) and then changing the speed from 33 rpm to 45 rpm you will, in doing so, change the length of the record. A one-minute song would now last about 44 seconds. Likewise you could do the opposite and take a 7″ single and change the speed from 45 rpm to 33 rpm and one minute of this would now last about 82 seconds. Given that we said that time-stretching is basically the process of changing the length of an audio file, then surely what we have done here is time-stretching. Well, yes and no. As anybody who has ever tried this trick with a vinyl record (or adjusted the pitch slider on a DJ-type CD player) will know, doing this changes the length, and the pitch. As a remixer, in most genres at least, this isn't particularly useful, especially if you have to stretch or compress the audio by a large amount.

Using this method, if we halve the speed of the audio file the pitch will drop by one octave. So for every 8.33% we adjust the length, in either direction, the pitch will change by one semitone. If we happened to need to adjust the tempo by exactly 8.33% then the resulting file would, perhaps, be usable, but even a change of +/− one semitone on a vocal recording can sound spectacularly different in terms of the tone of the voice. And this all, of course, assumes that you want a tempo change of exactly 8.33%. If you wanted a tempo change of, say, 10% then the resulting detuning factor would be somewhere between one and two semitones and the resulting files would not be especially useful to us.

For some genres, though, this effect has developed into part of the "sound," so a technical flaw has become a creative trademark. It is far from the first time in musical history that this has happened and, in many ways, the people who originally came up with this idea have my respect because turning a weakness into a strength is not easy to do. But for most of the rest of us it is far more useful to have the vocals in the same key as the original track to avoid the risk of making your female vocal sound masculine in its tone. Equally, in the same way a male vocal can be made to sound very feminine in tone or, if pitched down, slightly demonic! It is in this respect that time-stretching and compression are hugely complex. What is actually needed is a means of digitally deconstructing the audio waveform and then, taking into account stretch/compression factor, reconstructing it with the appropriate length. To explain why this is so complex and so difficult, let's take a look in a little more detail starting with a very simple example: time-stretching a sine wave.

If we take a continuous 100 Hz sine wave we will have a waveform that completes one *cycle* of its waveform every 0.01 seconds. If we stretch out the waveform by a factor of two, then one cycle will now take 0.02 seconds, which means the resulting frequency is 50 Hz. This is what happens with the turntable speed-change technique. Now if we take a finite length of that waveform of, let's say, 0.1 seconds, we have an original waveform that completes 10 full cycles in that time. We want it to last 0.2 seconds (a time-stretch ratio of 2:1), so, again, we have the option of simply stretching out the waveform that is already there. However, this will leave us with a resulting waveform that completes 10 full cycles in 0.2 seconds, which will, again, reduce the frequency of the waveform to 50 Hz. In order to do a useful time-stretch, what needs to happen is some kind of analysis of the original waveform to determine that it is a steady 100 Hz sine wave and that information is then used in the reconstruction of the time-stretched version of the file.

In this particular instance this is a very easy calculation because we are dealing with a simple waveform of a constant frequency with a very simple (mathematically speaking) time-stretch ratio. Obviously, if there were a time-stretch factor of something like 1.34:1 then you couldn't just double the number of cycles or simply copy and paste the entire waveform. You would, instead, need to calculate what the resulting length of the audio file would be and then copy the first 34% of the file and paste it on the end. This is a major over-simplification of what goes on but it serves to illustrate the point.

Now let's look at a slightly more complex example, with a waveform of 0.1 seconds again but starts at 100 Hz and steadily increases to 200 Hz. At the very beginning the cycle length will be 0.01 seconds, but for the very last cycle that length is reduced to 0.005 seconds. Now things become a little more complex because each successive cycle is shorter in length than the one before. In fact, it is even more complex than that because even during the duration of a single cycle the frequency is actually increasing. But for the purposes of the discussion here, let's assume that each cycle has a fixed frequency. If we have a time-stretch ratio of 2:1 we could simply cut the overall waveform into single cycles and just repeat each to get a rough approximation. This wouldn't be accurate, however. To be truly accurate, there would need to be a smooth change throughout the entire time-stretched version.

If, for the sake of argument, the frequencies of each successive cycle in the original waveform were 100 Hz, 110 Hz, 120 Hz, and so on all the way up to 200 Hz, then using the method we just described, the time-stretched version would end up with cycles as follows: 100 Hz, 100 Hz, 110 Hz, 110 Hz, 120 Hz, 120 Hz, and so on. While this would be a rough approximation, it would be better to have cycles of 100 Hz, 105 Hz, 110 Hz, 115 Hz, 120 Hz, 125 Hz, and so on. While this would provide a smoother frequency change to the final time-stretched file, it comes at the price of the final waveform not being exactly double the length of the original one.

As we can see, neither of these methods provides a truly accurate representation of the original waveform that is doubled in length. In one instance the length is accurate but the resulting waveform is "stepped" rather than "smooth" in terms of frequency, and although the second is much smoother, the accuracy of the length is compromised. Are you starting to get a clearer picture of how complex this is?

If we now further increase the complexity by having two separate sine waves overlaid on top of each other with frequencies of 110 Hz and 128 Hz (musically speaking an A and a C) then we can see that the combined waveform doesn't have a clearly obvious single cycle like before that we can use to chop up and reassemble. This is where the aforementioned method falls apart. However, we *can* use this idea to help us develop a potentially better method. If you refer back to Chapter 10 and, in particular, the idea that any complex waveform can be expressed as the combination of a number of sine waves at different frequencies and at different levels, then we have the basis for an alternative method. If we were to analyze the spectral content of any given sound we could come up with an additive synthesis model of that sound. The accuracy of the sound would obviously depend on the number of separate sine wave partials we could have available simultaneously, and the number (and resolution) of the steps in the amplitude envelope for each of those partials. With enough computing power, we could come up with an accurate model of the changing harmonic content of any given sound over a period of time.

Once we have this model it then becomes very easy to time-stretch it because all you are effectively doing is lengthening (or shortening) the duration of an amplitude envelope for each of the hundreds (or even thousands) of individual sine waves that make up the sound. Say the length of the longest of those envelopes was one second with a slow build-up from nothing at the start to a level of 100% at 0.4 seconds and then slow decline to 0% at one second. If we applied a time-stretch ratio of 0.7:1 we would end up with an envelope that starts at nothing, builds up to 100% at 0.28 seconds and then slowly declines to 0% again at 0.7 seconds. All very simple, and when applied to each individual sine wave partial it gives us a very good method of time-stretching because we are, effectively, dealing with a synthesized sound. This means that any transitions can be made smoothly because the synthetic nature of the sound allows us to modulate the parameters on a very small level.

audio files that are as easily manipulated as conventional MIDI data in terms of timing, groove, and duration. The aim of all of these different implementations is to turn what have traditionally been very static audio files into something that can be molded and manipulated very easily and used in even more creative ways.

Each of the main DAW software packages will have its fans and its haters, and the same is true of the time-stretching algorithms within each of the major packages. Some will swear that Pro Tools' Elastic Audio is the best; others will argue that the Logic Flex Time is the best, and so on. In my opinion there simply isn't one *über-algorithm* that's all things to all people; at least not yet. Some are more suited to real-time manipulation while others are more suited to static time-stretching. Some are better suited to sustained pad-like sounds and vocals, while others work better on rhythmic material. And even then, just changing the sound or part you are time-stretching can vary the results. One algorithm you might not have liked might, for some reason, just work better with a particular sound.

PRACTICAL TIME-STRETCHING

I want to raise a few more general points about time-stretching now that we have discussed the technical background and implementations of it. One of the problems that can come up when you are trying to time-stretch files is actually figuring out the original tempo. In an ideal world you would receive some kind of information about the tempo of the original files, but this isn't always the case. In a vast majority of instances when I'm asked to do a remix I am sent a copy of the original track to listen to prior to accepting the job. There are several different options to figure out the tempo of the original version. The manual way is to load the track into your DAW, line up the start of the track with the first beat of a bar, and then simply adjust the tempo until you find the one that fits. This can be a bit hit-and-miss to start with, but you will get much better at estimating the tempo of a track just by ear. Until that time a good way of going about it is to have a guess at the tempo and, if it is slower than the correct tempo, choose a different one that is a good amount faster. If this second guess is faster than the correct tempo then you have two goalposts you can work in from. I find this to be generally quicker than just guessing randomly or making one guess and then increasing by one beat per minute each time.

If, for any reason, you don't receive either a copy of the original track or a written indication of the original tempo, then by far your easiest option is to just ask someone! If time or some other reason prevents you from doing that, you have to figure it out somehow. The first thing to do is take a look at what you have actually been sent for the remix. Sometimes you will receive many remix parts, which will hopefully include some kind of drum track (a kick drum, snare drum, hi-hat, or something similar, which has a regular rhythm). Even though you may not be interested in using the drums from the original track in your remix, you can use them to work out the original tempo.

You may just receive an a cappella, which, on its own, could give you a few problems working out the original tempo. The biggest problem with just receiving an a cappella, with no original song and no original tempo (and the track isn't already available on YouTube or some other similar site), is that you really have no clear idea of where the downbeat of the vocal is. Even in situations like this, however, the chorus of the song will most likely have some kind of repetitive hook, which you can use to establish a definite number of bars or a rhythmic pattern, which can then be used to work out the tempo. It's all a little bit around the houses, but you do get more adept at figuring these things out more quickly as time goes on. This last situation I described comes up very rarely, so for the most part things should be a little easier than this.

Once you have the original tempo you can then start to think about actually time-stretching the vocal. Most of the time you won't have to worry about whether to speed the vocal up or slow it down because it won't be too far from your target tempo. If you wanted to work at 134 bpm and your track was at 120 bpm, then obviously you will speed it up. If your target was 126 bpm and your track was at 70 bpm, then you would probably treat it like a half-tempo vocal and slow it down (to 63 bpm to make it half-tempo). More rarely you might have a target tempo of 70 bpm (perhaps you are doing a chillout remix) and your track is at 130 bpm, so you assume a double-tempo and speed it up to 140 bpm. Sometimes, however, you will be faced with a situation where your original track tempo is midway between your target tempo and either the half- or double-tempo feel, in which case you need to make a decision about what to do. Either way you could face potential problems and there really is no easy solution. You could be facing a potential time-stretch of 30% in either direction and that will *always*, at least with current technology, introduce many artifacts from the stretching process. You'll also have to deal with phrasing and performance problems from the tempo-changed vocal (including too fast or too slow vibrato, too many notes in a short period of time, and pitch-bending of notes, which sound way too slurred).

In fact, time-stretching vocals is such a potential minefield that I have devoted an entire chapter to it (Chapter 15), which deals with different ways in which you can get around certain problems created by large ratio time-stretching, and also highlights some of the things that, sadly, you can't do anything about. There have been huge improvements in time-stretching technology in recent years, which have made the actual audio artifacts less and less of an issue. But there will always be an issue of the basic physics of speeding something up, so I have also mentioned my ideas for future technology, which might solve some of the problems that still exist today.

The vocals always have to be there, front and center, and probably not smothered in effects, which is why you have to be so much more careful with your time-stretching. When it comes to other musical parts or instruments, there are still issues to face when dealing with high ratio time-stretching but, as a remixer at least, these tend to be less important because you can, if you need to, bury the sound back in the mix or perhaps not use it at all. The biggest actual sonic issue with all instruments and sounds is a pseudometallic "ringing" overtone to sounds that have been stretched a lot. It doesn't seem to be quite so noticeable when sounds have been sped up by a similar amount, but it is still present. You can, to some extent, cover this with effects or perhaps a combination of very clever EQ use and filtering, but depending on the actual sound, it can render the stretched sound unusable.

In the case of something like a rhythm guitar part you could always try to stretch it the other way if it doesn't sound right. For example, if you slowed the guitar part down and the ringing overtone was too much, you could always try speeding it up instead. Of course, this would mean that the part would last half as long (in terms of the number of bars) so you would probably have to cut the part up and get creative with repeating certain regions to make sure the chords changed in the right place, but it might just work and it might even give an extra dimension of pace and bounce to the remix. Synth parts, pads, arpeggios, and possibly (although much less often) even bass lines are all potential candidates for this trick. You have the luxury of trying this with musical parts because a musical part has only two dimensions: pitch and rhythm (by that I mean timing and duration together), whereas a vocal part has an extra dimension in the form of *meaning*. You can't simply cut and paste vocal parts to fill up space in the way I just described, without obscuring the meaning of the words.

Another thing that tends to suffer quite a lot with time-stretching in general is anything with a very sharp or pronounced attack transient. The first fraction of a second of most sounds is what gives them much of their character. So much so, in fact, that the development of Roland's D-50 (and other D-Series) synths were based around this idea. They worked out that, after the initial attack of the sound, a violin and a trumpet finally ended up resolving to similar waveforms (sawtooth in this case) and it was the initial scrape of the bow on the violin or the initial breath noise in the trumpet that helped us decide which instrument it was. So they developed Linear Arithmetic synthesis, which became known more commonly as the "S+S" (or "Sample + Synthesis") technique. What they did was have a sampled attack portion of the sound played back by one oscillator and then a more conventional analog-style oscillator playing a sawtooth waveform. The combination of the two gave a sound that was much more convincing than anything you could achieve with pure analog oscillators, and also didn't need to have a long sample of the whole sound (sample memory was *very* expensive and, therefore, limited in those days). As time went on, this method eventually gave way to using pure samples, but it gave a much more authentic sound than any other form of synthesis that had come before it. In addition, it proved just how important transients are to the overall sound.

What this shows is that the transients or attack part of a sound is very much an integral part of its character. The problem occurs when you try to time-stretch a sound that has a very strong transient character. When you stretch the transient, you lose the transientness of it, and it can start to sound (at best) a little squishy or (at worst) like a completely different sound. The only real way to get around this, and this is completely impractical, and quite probably not even possible to do manually, is to zoom right in on the audio waveform and manually chop every single note into two regions: the attack portion and the sustain portion. You would then have to time-stretch only the sustain portions of the sound and leave the attack portions as they are. This would, undoubtedly, give a much better result than any automated form of time-stretching, but even if it were possible to do this accurately manually, the amount of time it would take could render this approach utterly unworkable in all but the most exceptional of circumstances.

To some extent, however, there are time-stretch algorithms designed to work by detecting the transients and then leaving them intact and only time-stretching the other parts of the file. The biggest problem (and at the same time, benefit) of these algorithms is that they don't offer any kind of adjustability for the detection of the attack transients. Far more useful would be to have some kind of option whereby your audio file is analyzed and then the software showed you in some visual form where it thought the transients were and gave you a option to listen back to *only* what it had detected as transients to check how accurate the guess was. If it wasn't quite right, you could potentially have an option to adjust a threshold to make the detection process more or less sensitive to transients. This kind of transient detection and adjustable threshold exists already in software such as the excellent SPL DrumXchanger, and there are time-stretch algorithms that claim to process transients separately. What we really need is some kind of amalgamation of the two and then we would really be in business! I should probably mention at this point, that for drum and percussion recordings there is a very good alternative to time-stretching for small to medium tempo changes and we will be looking at that in the next chapter.

As I said earlier, time-stretching quality has improved greatly in recent years but it still isn't perfect. I have given you a few pointers here about what kind of limitations to expect in general from the time-stretching process, but each algorithm (and software or plug-in) has its own idiosyncrasies. I recommend playing around with the time-stretching options available to you and putting each through its paces on a variety of different material (vocals, drums, bass sounds, and complete mixes generally make the best "test matter" in my experience). Use a variety of different stretch amounts (both stretching and compression) and form your own opinions about which is the best option to use on each different kind of material, and at what kind of stretch ratio.

There really isn't a definitive answer or consensus of opinion on this, so there isn't any kind of list I can give you about what is the best time-stretch algorithm to use on a particular kind of sound at a particular stretch ratio. I sometimes wish there was, but in this case there is no substitute for trial and error, and because of the nature of it, it really pays to be prepared and not be trying to figure all of this out when you are under pressure to complete a remix on a tight deadline.

CHAPTER 14

Time Design: Beat-Mapping and Recycling

In this chapter we look at a couple of different techniques that should be a part of the remixer's arsenal of techniques. The first, beat-mapping, isn't related to time-stretching in any way and doesn't attempt to do the same thing; instead it offers something that might be used alongside time-stretching to help knock problem audio files into shape. The second, recycling, does offer a potential alternative to time-stretching in certain circumstances but can also be used in other ways that are much less problem solving and much more creative.

BEAT-MAPPING

The principle behind beat-mapping is simple. The idea is to take an audio clip that doesn't conform nicely to a click track and do one of two things: either warp the actual timing grid of the software to match the clip (by means of a series of tempo changes) or, certainly more useful in a majority of situations, warp the timing of the clip to match a precise grid. While they differ in outcome substantially, both techniques require the audio to be analyzed first and markers to be positioned in the file representing the points in time that will be used as a reference. The more of these there are (assuming of course that they have been accurately positioned) the better the accuracy of the beat-mapping and the better the final result will be.

One of the best known, and most developed in this regard, pieces of software for this type of application is Ableton's Live, which has exactly these techniques at its very core. As such, Live has found a growing number of fans who praise it for everything from beat-mapping (effectively, requantizing) old records that were not recorded to a click track to putting together full DJ sets. In fact, in its more recent versions, it is maturing into a very capable platform for the creation of music and DJ sets and for regrooving old audio files. It now comes with an increasing number of "included" virtual synths and effects and is also capable of loading VST and Audio Units plug-ins so that, when combined with its unique compositional options, is making it one of the faster growing audio production platforms.

The Remix Manual.

not recorded to a click track and then make all of those stems sound like they *were* actually recorded to a click track. Nothing is perfect of course and without a great deal of time being spent on getting the warp markers in *exactly* the right place in Live, there would be minor timing errors. The resulting "warped" file would probably be more than suitable, and any further timing errors would be quite easily correctable using more conventional means inside your DAW.

We can see that the biggest problem with the manual method lies in making sure there are no gaps or overlapping regions once we have requantized everything. Obviously if you have sound that has a long sustain portion and where notes run into one another, this is very important. But what if each individual region contains a short sound, one that doesn't actually reach the end of the region anyway? In this instance the exact start and end points aren't as important because there doesn't *need* to be that continuity of one note into the next. In this instance it might actually be more beneficial to *not* time-stretch the regions (assuming that the tempo change isn't huge) because any gaps will still be acceptable and any overlaps will probably be unnoticeable. This is something very similar to what happens with recycling.

The actual term *recycling* is derived from the software originally created to do exactly this. Propellerheads' ReCycle was first released in 1994 and was targeted mainly at users of loop libraries. At its core is the ability to take an audio file, detect transients, slice that original audio file into separate pieces and then assign those individual pieces to a sampler instrument so that each "slice" plays on a separate key on the keyboard. Then, as the final step, it creates a MIDI file to accompany the sampler instrument that contains the MIDI note and timing information necessary to completely replicate the original file. If you were to "recycle" a drum loop, for example, and then do nothing to the resulting MIDI file, there would be no change to the sound of the loop. If this is the case, you might be asking what the point is. At the most basic level, now that this audio file had been recycled in this way it would follow any change in the tempo of the sound, whereas the original audio file would be static in this respect. To put it another way, in the original audio file the first snare drum hit might occur at exactly 500 ms into the file. Whereas in the MIDI file the slice representing the first snare drum hit might occur on the second beat of the bar. If we now slow the tempo down by 10 bpm, what happens? Well, in the audio file the first snare drum hit *still* happens at 500 ms, which is now too early, but in the MIDI file version the snare will *still* happen on the second beat of the bar.

That is just the beginning for recycled files. In addition, because each slice is a MIDI note it means that requantizing the drum loop is now as easy as requantizing any other MIDI data. Want a little more swing on the loop? Just change to a 16B quantize. Want less swing? Change to a straight 16 quantize. Of course, for all of this to work as hoped the original recycle file needs to have been created properly and this means those good old transient markers again. But you soon get used to making sure they are where they should be and it becomes second nature.

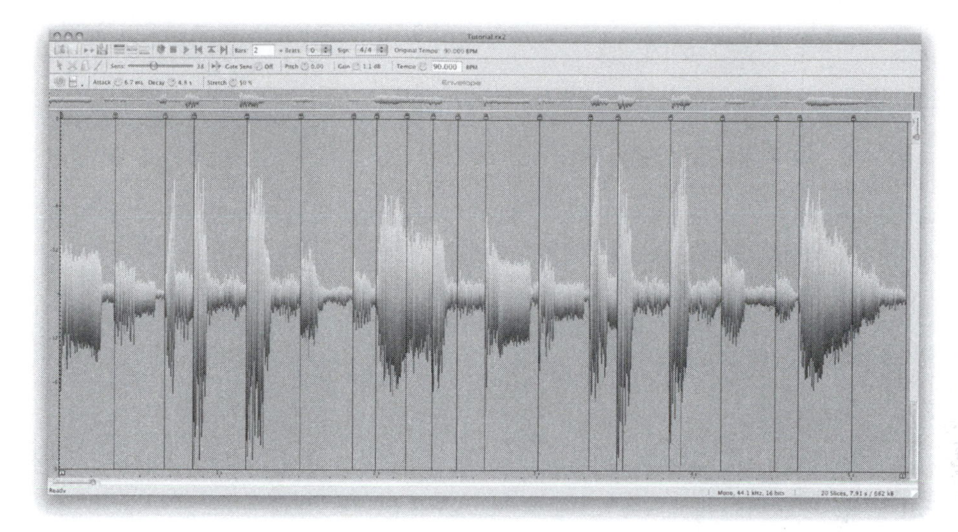

FIGURE 14.2
Propellerheads' ReCycle was one of the first tools that enabled manipulation of the timing and groove of samples.

There is also an additional benefit to recycled files, which is that it is now easy to actually get editing and completely change the original file. If we stick with the example of a drum loop for now (although we will look at more creative uses for recycling shortly), we have the option to remove certain slices completely by muting the MIDI notes that represent them. Or we can increase or decrease the volume of individual slices by changing the velocity of the MIDI note (assuming that the sample instrument is set up to have the amplitude modulated by velocity of the note). But the real fun, and creativity, comes from the fact that we can now change the order that the notes, and therefore slices, play back in. This could be anything from creating a double snare drum hit at the end of every eighth bar to act as a turnaround, to completely restructuring the loop. And because it is all set up as a sampler instrument, it means you can do it in a more interactive way by playing the new groove on a MIDI keyboard rather than being forced to move things around on screen with a mouse. Actually *playing in* a new groove will give you something that sounds more natural and will be much more fun to do.

In Logic 9 there are new features that offer much the same facilities as ReCycle, only built in to Logic itself. There is an option to "Convert audio file to sample instrument" that does much the same as ReCycle. If you select this option, you are given the choice to use the transient markers detected by Logic (which may well be sufficient for your needs) or to use separate regions to create the individual notes in the sampler instrument. If you choose this second option you will have to precut your audio file at the points where you wish to create the separate slices in the sampler instrument. It is obviously a little more time-consuming to do it this way, but it does allow for a more accurate (possibly) and more controllable (certainly) result. Once you have chosen which of the two options you would like to use for creating the separate slices in the sampler instrument, the process is pretty quick and a new instrument track is created, a new EXS24 sampler loaded onto that track, and the sampler instrument created and loaded for you. In addition, the original audio region(s) used is automatically muted.

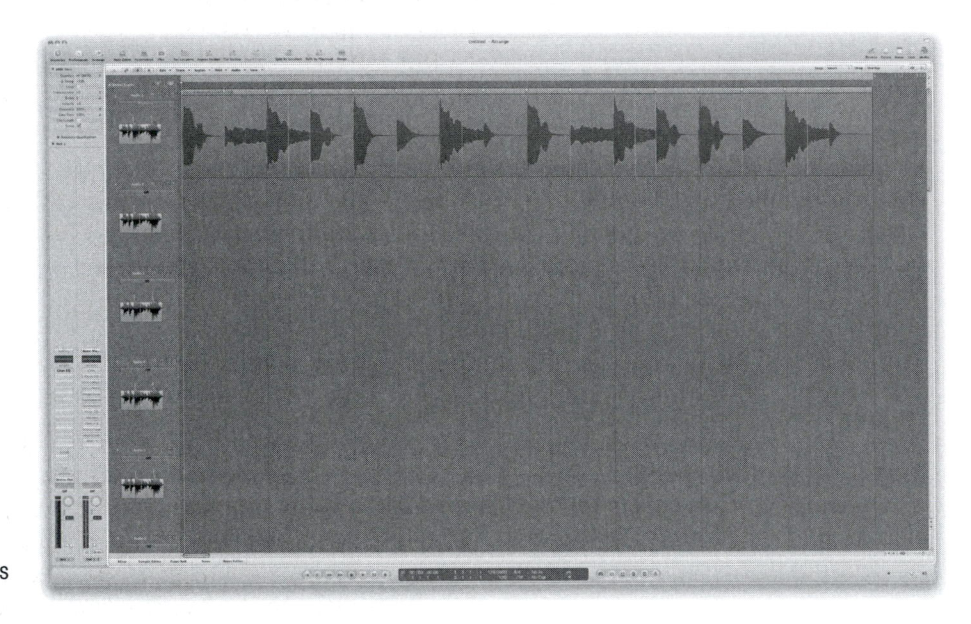

FIGURE 14.3
Logic 9 showing the ReCycle-like features it now features to manipulate audio files. The transient markers used by Flex Time offer the same functionality as those in ReCycle.

I have found this to be one of the standout new features in Logic 9, not because it is revolutionary but because of the massive amount of time it can save me. It doesn't offer anything truly new because these features (and in some ways more) have been available in ReCycle for a long time, but it does mean that it is included as a part of the Logic package, which means no need to buy a separate piece of software to achieve this. In addition, it speeds up the workflow because you don't need to leave Logic, open up ReCycle, load the relevant audio file into ReCycle, set up all the markers, export the ReCycle file and the MIDI file, close ReCycle, go back into Logic, create a new instrument track, load in an EXS24, import the ReCycle file into the sampler, locate the associated MIDI file, and then import that into the relevant track in Logic. Even just reading through that, you can see how much more complicated the process is, so it is most certainly a welcome addition to the Logic feature set in the new version.

While we are on the subject of new features in Logic, I will again make reference to the new Flex Time feature, because it does offer some of the functionality that recycling does, with even less effort. If you remember, two of the advantages of recycled files were that they would follow any tempo changes easily and could be easily requantized. In this regard Flex Time matches that functionality because any audio files that have been prepared for Flex Time use and have been analyzed will automatically follow any tempo changes and can be quantized as easily as any MIDI data. And, if anything, the workflow to do this is even easier. All you need to do is highlight an audio region (or regions), make sure that the "Flex" button at the top left of the screen is visible, then choose the appropriate algorithm from the list that appears in the track header when you zoom in far enough, and the analysis will automatically happen. The speed of the analysis will depend on the specification of your computer, but it is generally pretty quick. Once the analysis is complete, the audio is ready to be requantized or follow tempo changes. It is really *that* easy. And once again, as with Ableton Live and ReCycle, the quality of the results depends on the accuracy of the markers, but Logic, like those other two programs, gives us the option to go in and add or delete markers as we see fit.

The third benefit of ReCycle, that of being easily able to reorder the slices by moving MIDI note data around, isn't present in Flex Time, so if that is something you feel you might want to do then ReCycle (or the "Convert audio file to sampler instrument" feature within Logic) is probably your best bet. That's not to say you can't reorder the individual slices in a Flex Time file, you can, but in a roundabout way. To do this you need to choose the option to "Split file at transient markers," which will then take the Flex Time file and split it into individual slices, which you can then move around on screen into the order you'd like, copying, pasting, and deleting as you wish. The two disadvantages here are that, first, doing the reordering on screen with a mouse is far less intuitive and fun than playing the new groove on a MIDI keyboard, and second, once you have split the Flex Time region into separate slices you lose the automatic stretching of each slice to fill any gaps.

In the end, though, any of these techniques will help you bring new life into any sampled drum (or percussion or even synth) loops you have in your library by allowing you to fit them into any groove and even change the actual pattern of them. This will make them useful in far more contexts than they perhaps originally were.

It doesn't end there, however, because ReCycle has another trick up its sleeve. Most people use it primarily for drum and percussion loops, and this is where it is perhaps the most useful. But it can be used to cut up any audio file and allocate the resulting slices to keys on a MIDI keyboard through the sampler instrument it creates. As such, you can load all manner of audio files into it and chop them up for retriggering purposes. Let me give you just a couple of examples. The first one that comes to mind is to use it for a vocal track. Ordinarily, a vocal track would be allocated to an audio channel and just played back. But if you wanted to get a little more creative with your vocals you could load, say, a verse vocal into ReCycle and then use the software to cut it into slices for each word (or even small parts of words), and then, on the sampler instrument channel, retrigger those rhythmically to create a new musical part from the vocals. I have done this a number of times myself, but possibly the best-known example of a vocal treatment like this is on Nightcrawlers track "Push the Feeling On." If you don't already know this track, check it out on YouTube or somewhere similar and listen to the chopped up vocal to hear how effective it can be. I don't know if they actually used a technique like this to create that vocal, or whether it was a purely manual cutting up that made it happen, but the ReCycle method could certainly achieve that effect quite easily if you wanted it to.

As a variation on this theme I have, on more than one occasion, used the "Convert audio file to sampler instrument" feature in Logic to chop up a vocal, let's say a chorus, into equal 16th-note regions, and then mapped these to the keys in an EXS24 sampler instrument. I then started the track playing and just hit random notes on the keyboard in a rhythmic pattern and, after a bit of trial and error and sometimes a bit of tweaking, the result is an almost percussive sounding part but with very clearly defined notes (being made from a vocal performance) and a clear vocal texture to it. It certainly wouldn't work in every context, and it isn't something I would want to overuse because I am sure it could very easily become tiring to listen to, but on the occasions I have used it, the effect has been great. On one occasion I actually saw a post on a web forum where somebody was asking where the new vocal in the remix that I had done had come from. High praise indeed!

Another example might be if you had a piano part in your track that was playing simple chords. You could bounce that down to an audio file and then load that audio file into ReCycle and cut each individual chord to a new slice and then have those available on a sampler instrument to trigger as required. Just by messing around and hitting random notes, you might come up with something really nice as a rhythmic chord pattern that you might never have come across if you were playing the full chords in on a MIDI keyboard.

Another thing to consider with this particular technique is to have a very short release time on the envelope on the sampler. This way, even if the original piano (or pad or whatever) sound that you have recycled had a long decay/release, you can cut it prematurely short, which can give a really choppy feel that is almost impossible to replicate with virtual instruments. If you bounced the piano chords with reverb already on them, for example, not only will you be cutting the actual piano sound short using this technique, but the reverb part of the sound would be cut short as well. Once again, it's not necessarily going to be an effect you use on every track, but it's certainly something worth trying out and experimenting with when you have the time as it might offer you a spark of creativity in moments when you need it the most.

In summary, how often you would use either of these techniques as a remixer depends a lot on your own production style and the genre you work in. If you work in more of a minimal genre there is an increased likelihood that many of your beats will be programmed from individual drum and percussion sounds, so you will be less likely to be using sampled loops. If this is the case then the main purpose of ReCycle is somewhat defeated, but, as I mentioned, there are alternative ways in which you can use it. Beat-mapping, on the other hand, is more useful to a DJ as a way of making tracks more mixable, but it does still have its uses to the remixer and, in certain cases, it provides a way of doing things that would be extremely difficult (although not impossible) to do any other way. Time is often limited when you are commissioned to do a remix, so anything that can save you time (and not compromise on quality—this distinction is important) is definitely worth looking at.

Beat-mapping comes at a price though, and in this instance I am talking about a financial one. Most DAWs have limited functionality in this respect and even the new Flex Time feature in Logic 9 cannot readily deal with audio files that have substantial changes in tempo throughout the track. In this instance Ableton Live might be (currently at least) your only option. And, while very good value for what you get in the package, it isn't exactly pocket money price either. If you are a DJ as well as a producer/remixer then Live may already be on your shopping list (or even already installed); if not and you use another DAW as your main workhorse, I would find it hard to say that it is a must-have piece of software in the way that top quality time-stretching software would be. However, it does have its uses as an audio manipulation tool and it does do some things that other DAWs can't really do in anywhere near as concise a method. It also offers some unique tools for improvisational composition and live performance. But if you are only going to use it for beat-mapping and the very occasional remix parts that weren't recorded to a click track, then I would have to classify it under the "great if you can afford it" heading. Personally, I do have it and I do use it, more and more in fact, but it was something I purchased once I had bought everything I considered essentials first.

Both of these techniques, and pretty much every other one I mention or describe in this book are open to interpretation and interesting methods of (ab)use. New genres of music may come up with currently unimaginable ways of using these techniques and turning them into trademark sounds in the same way that Auto-Tune became "that Cher vocal" sound or, more recently, "that T-Pain vocal sound."

When you have the time to experiment with these techniques, try to stop thinking about how they can help you achieve what you want to achieve and start thinking about what they are truly capable of when pushed to extremes. You never know what you might discover.

CHAPTER 15
Time Design: When All Else Fails

In Chapter 13 I mentioned that there are times when "conventional" time-stretching simply doesn't get us close enough to a good sounding vocal. Even though everything is in time and at the correct tempo and even though, with the latest and best time-stretching algorithms, there are few enough actual audio artifacts, there are still issues with the vocal because of two main factors: the vibrato is way too fast (or slow) and there are too many (or too few) notes over a given period of time for it to sound natural. One of these we can do something about and one we can do very little about. We'll start by taking a look at the one we can't really improve upon.

Part of the problem we have when time-stretching vocals by a large amount is the issue of trying to make the performance sound natural or, at the very least, plausible. In most cases this part of the process pretty much takes care of itself. Given the natural flexibility of the human voice, a huge range of vocal performances can sound natural. Some are slow, languid, and fluid while others are much faster and choppier. Because of this we can take a vocal that was, originally, quite slow and speed it up quite a lot before it starts to sound physically impossible. The same also works in reverse. However problems can occur when an otherwise slow and fluid vocal features a section that has, compared to the rest of the performance, quite a few notes or words in a short space of time. This kind of vocal gymnastics makes things extremely difficult. We could quite plausibly make the vast majority of the vocal sound natural when sped up to this degree, yet parts of it will sound utterly impossible for a vocalist to sing.

The problem we have when dealing with this situation is that rapid notes occupy a (comparatively) short space of time and they usually don't occur in isolation. More often than not they are at the end of a word or the beginning of another word. My first reaction would be to just slow these parts down again at least a little, but if they occur at the end of a word and lead straight into another one, as they often do, stretching them out at all would make them overlap the next word, so we can't do that. Another possibility is to take them out completely, but often that will result in an unnatural end to (or start of) a word, which can be very hard to cover up. It can be improved if you have another version of the same word that can be copied and pasted onto the offending part of the word from the unaffected version, but that's not often the case. A third possibility is to try to edit out some of the notes from the phrase and leave yourself with only enough to fill up the space, but I can tell you from experience, this is almost impossible to do because vocalists tend to "bend" from one note to the next. If you simply cut a part of it out the transition from one note to the next in your final version will sound horribly unnatural and will probably draw more attention to itself than the plain time-stretched version would.

The final solution is to simply do nothing and leave it as it is. Given that it is only a relatively minor part of the whole vocal and that (hopefully at least) it doesn't happen that often during the song, you might just get away with it. However, if you do decide to do this, be prepared to have to defend yourself against the "it sounds time-stretched" argument from your client. Whatever your choice in this particular matter, the result will always sound less than perfect. Don't feel bad though; there really isn't anything that you could do better. No matter how much the technology of time-stretching and audio manipulation improves, there will never really be an answer to this particular issue, as the problem isn't the technology, it's a performance issue.

The problem of too many notes in too short a space of time shouldn't dog you if you're slowing a vocal down. It might not sound especially natural because the transitions from one note to the next will sound rather lazy, but at least there won't be the thought in the listener's mind that there is no way on earth that the singer was physically able to sing that part. Also, as well as not having as much of an issue to deal with in the first place, there are also more options available to fix things if we choose to. The biggest problem with a sped-up vocal is not being able to slow the offending section down again to make it sound more natural because it will overlap what comes next. When we're slowing down a vocal, however, we can do exactly that (well, the opposite). We *can* speed the section in question back up again because there will be a bigger gap than we need.

We could run into problems if the vocal gymnastics run from the end of one note into the beginning of the next note because there will be a gap when we speed the difficult section up. In that case I would probably choose to leave it as the stretched version and then try to fix any issues with it sounding unnatural through messing with the pitch transition rate a little in Melodyne. This could help to at least minimize the unnatural sound, if not get rid of it completely.

You'll probably never get something perfect. No substantially time-stretched vocal is ever going to sound 100% natural (a fact that worryingly few A&R execs, artists, and management seem to understand) and while there is a lot we *can* do, there are still these few little areas that hold us back. Perhaps, if I had much more time to spend, I might be able to find a better way to deal with this issue but, after 10 years of trying, the best solution I have come across is to pretty much leave it as it is, do the best I can in Melodyne to make it sound as natural as possible, spend the time on doing a better job of cleaning up the

95% of the vocal that I can work on effectively, and then ready myself for *that* discussion with the client if it comes up.

So far our discussion hasn't been especially confidence building, but things are about to look up. Let's now consider the vocal time-stretching problem that we *can* do something about.

The task of trying to make a heavily time-stretched vocal sound *not* heavily time-stretched is the technical holy grail of the modern remixer. I have tried pretty much every method and piece of software there is to get the best results I can. And after all of that research and experimentation, the best method I have found is, basically, a fairly time-intensive and laborious semi-manual method. I suggest that you give it a try yourself to see how much more effective it can be. I will warn you, however: don't expect quick results. It really does take a while to make this work, especially when you consider that you will have to run through the process with each individual track of vocals. In other words, if you have a lead vocal, a lead vocal double track, and three or four tracks of backing vocals or harmonies you will have to run through the procedure I am about to describe for each and every line of each and every one of those vocal tracks. It isn't uncommon for me to spend a full day just sorting out the vocals and "prepping" them if the original vocal is a particularly difficult one. Having said that, do remember that this full process is only really necessary if the singer has a pronounced vibrato. If the singer has a subtler vibrato, it might not be necessary at all. If that is the case then count yourself lucky and move swiftly along!

Let me start by explaining the thinking behind this method. The time-stretched vocal will, for the most part, be acceptable. The parts we are looking at here are the long, sustained vocal notes that have a noticeable vibrato on them. How many of these, and how noticeable they will be, depends on the individual singer. In the worst cases, and especially with more ballad-type songs, you could be looking at having one or more of these notes to fix in every single line of the song. But even then, it's not the whole line that you need to fix, just the part of that line that has the sustained vibrato.

At this point I should probably say that when you are time-stretching vocals (or any audio part for that matter), always—and I really cannot stress this enough—save a copy of all the original parts you were given so that you know you have a copy of the original files to go back to should you need them at any point.

The first thing I do is solo the time-stretched vocal track I am working on in my DAW and then just listen through the vocal. I have the "scissors" tool selected and each time I hear a note that has an excessive vibrato, I cut a new region out that is just this note. It doesn't matter at this stage if it is cut in exactly the right place as this is purely designed to give a rough position of the parts of the vocal I need to work on. When I have reached the end I will normally do another run-through to check that I have all the sections I need to cut out as separate regions, and I will often assign the "blocks" a different color so they're

If you want to avoid going into that much detail, the better option is to try cutting out a couple (or more) of shorter sections spaced evenly along the length of the sustained note. By doing this you stand a better chance of not having to deal with large changes in level or even of tone. It might be necessary to make several small cuts and nudge the vocal along a little each time until you reach the desired length. The more cuts you make (and the shorter the section removed), the better and more natural the edit will sound. However, there is an obvious tradeoff between quality of sound and the amount of time it is going to take you to fix. In theory, if you needed to cut out half the length of a sustained note, the best way to do it would be to cut out every other cycle of the waveform, but this would mean doing 128 separate edits for every second of a sustained vocal note of middle C. This would be insanity! Striving for the best quality is one thing, but rarely would that level of edit be justified. Of course, the more often you do this kind of thing, the quicker you will get at it, but even so, it can rapidly start to eat up your day. Vocals are very important, so try to find a balance between quality and effort where your extra time investment is justified.

With each of these edits, and any others you make, it's good to try to make them work so you don't need to crossfade the files at all. The best way to do this is to make sure that all of the edits occur at *zero-crossing* points. These are the points in the audio recording where the waveform "line" crosses from below the zero-line to above it. Any edit made at one of these points, assuming of course that the direction of the waveform continues (meaning that if the line was moving from below to above on the preceding file then it should start from zero and move upwards in the following file), shouldn't need a crossfade to avoid any little clicks or pops. To make sure this is the case, you will need to zoom right in, probably to maximum zoom, and then adjust the start and end points of the regions you are editing to make sure they all happen at a zero-crossing point. In general, I make sure that all of the edits I do are moving upwards from zero at the beginning of the region and then moving upwards toward zero (from below the zero line) at the end. All that is left then is to move each region around until there are no gaps and overlaps between them so that one region starts exactly where the previous one finishes. When this happens you will be able to play back the edited vocal and you shouldn't hear any clicks or pops at the edit points.

From this point on it is just repeating the process for each of the problematic lines, or words, in the song:

- Find the problem
- Locate the same line in the original vocal
- Paste the original vocal in place of the time-stretched one (making sure they start at the same point)
- Edit down the length of the original vocal either by doing a simple fade-out or by cutting out sections throughout the duration of the note

The only additional complication I can think of here is if the difficult note is one continuous sound that changes to another note halfway through. This can be overcome by treating them both as separate notes and dealing with the "sustained" part of each separately.

There shouldn't be any need for crossfades if you have followed the steps I outlined previously, but if you feel the need, you could give every single region a very short crossfade on the beginning and end. Assuming the crossfade time is very short, there shouldn't be any squishiness going on at the transition points.

As you can see, there isn't any alchemy or any other kind of magic going on in fixing excessively wobbly vibratos, just a basic and simple (but slightly in-depth) editing job. You *are* degrading the faithfulness to the original recording, but the results are nearly always good and far better than I have achieved from any other method. Just make sure you allow the time for it, that's all.

One final thought, before I finish up: I have often wondered if there *is* any way in which software could do as good a job or even a better job of this task. And while I haven't found anything yet that can do it, I do have some ideas about a way in which it could be implemented in the future, if any software developer decided to take up the challenge.

As a starting point, I would take the technology behind the already excellent Celemony Melodyne and then look at ways of developing that to make it more useful in this context. What Melodyne does is analyze the audio it is presented with and then provide, for each individual note, a center frequency and an amount of deviation from that center frequency at any given point in time. Visually in Melodyne this is represented by a blob whose vertical center on a piano-roll–type display represents the center frequency, and superimposed on top of that is another thin line that represents the instantaneous pitch of the note as it deviates from this theoretical ideal note. In the system I am proposing, that is most of the hard work done already. What remains is a simple mathematical analysis of that instantaneous pitch line and then some post-processing to bend things into shape for us.

How I propose it would work is this. Once a vocal had been analyzed in the normal way, there would be an additional stage of analysis that would track the instantaneous pitch. When a regular deviation from that pitch is detected (a sine-wave–type oscillation in the pitch value over a period of time) the software would mark that section as having vibrato. There could, of course, be adjustable controls to determine the minimum level of depth required before it was marked (a kind of *threshold* control), along with controls for the minimum actual frequency of oscillation of the pitch and the minimum length of time over which it must occur. With these three controls we could hopefully successfully mark out all of the sections of the vocal that had vibrato on them.

The next stage would be to analyze the parameters of that vibrato and determine the average frequency of the vibrato, the average depth (the actual amount of detuning and amplitude variation applied during the vibrato), and whether the singer gradually applied the vibrato and, if so, over what period of time. The analysis of all these parameters could either take place on a per-note basis for each marked section or as a global average calculated for the entire song. Once all this analysis had taken place, the software could then remove the vibrato completely from the marked sections (as you can in Melodyne with the Pitch Modulation controls). The vocal could then be time-stretched inside the software while still retaining the position of all the marked sections. Once the time-stretching had been completed, those marked sections could then have pitch modulation reapplied (perhaps with a variable depth of application just in case we wanted to play with it a little), according to the parameters (rate, depth, and fade-in) that had been determined by the analysis.

In this way, if there was a note sustained for, say, 2 seconds in the original file, and the singer's vibrato was at about 5 Hz, that would mean around 10 complete vibrato cycles over the length of the note. Using conventional time-stretching, if the stretched version was only 1 second in length, there would still be 10 full vibrato cycles, which would raise the rate of the vibrato to 10 Hz. Using the method described here, the vibrato would be reapplied to that 1-second version so that the rate would remain the same at 5 Hz, but there would only be five full vibrato cycles instead of 10. There would, of course, be a bit of a tradeoff. If you have a note with strong vibrato and you remove the pitch modulation completely in Melodyne, you can still hear a cyclic modulation going on in the note; however, it is a subtle tonal modulation rather than the obvious pitch modulation that was there before. Unless the technology improved a little more, our idea here would still suffer from that. So even though the vibrato itself were at the right speed and depth, there would probably be a slightly odd overtone to the vocal in that section because of the tonal modulation consequence of flattening out the original vibrato. It's certainly an interesting idea, and I believe that something like this would be of great interest to remixers, as it would save an incredible amount of time. Celemony… are you listening?

CHAPTER 16
Mixing: Introduction

WORKFLOW

The art of remixing, while it does have many similarities with production, differs in a few ways, one of which being the extremely tight deadlines. As such, anything that can be done to speed up the process can be a huge help. Most modern DAWs allow you to save and use multiple templates. I have already touched on this briefly, but I will expand on it here and give a few examples of different possibilities, depending on your preferred way of working.

In Chapter 3, I mentioned the idea of using "templates" to make it quick and easy to load up your DAW and be creating in a matter of minutes. This is obviously a very useful timesaver, but there are other, perhaps less obvious, things you can do to improve workflow as well. It might be easiest to illustrate this by showing you how my standard remix template is organized.

I like working in a traditional way. I tend to group sounds that are related into submix buses, for two main reasons. The first is to allow for processing (EQ, dynamics, reverb, etc.) to be applied to the whole group of sounds. With modern computers having ridiculous amounts of processing power it's not always necessary to do this as there is usually more than enough CPU power to put the processing on each sound individually, but sometimes, by applying the processing to the group of sounds it can help to pull the sounds together and seem more coherent and focused. This might not be the effect you want all of the time, but by routing the sounds in this way it doesn't prevent you from applying processing to the individual sounds but *does* give you the option to apply it to the whole group.

The second reason is that you can use the submix buses to make broad mix changes quickly and simply. For example, if you're layering up a few different bass sounds and applying EQ and balancing up the bass sounds with each other until you're happy with the combined sound. Later you could change the kick drum sound and then might feel that the bass sound was now too quiet. If you had three different sounds layered and not routed into a submix bus you would need to adjust the levels of all three sounds. It's not necessarily a major headache to do, but as the decibel scale (and hence volume of sounds) isn't linear then simply increasing the volume of each sound by 3 dB won't actually do what you expect it to do if the different layers have different volumes. The increase from −3 dB to 0 dB is very different from the increase from −18 dB to −15 dB and so simply reducing the volume of each sound that way will actually alter the relative balance of the sounds, whereas routing through a submix bus will keep the balance of the sounds constant while adjusting the overall level.

It might seem like a lot of fuss, especially if you are working in more minimal styles with fewer separate parts, but once it has been done, it is just as easy to load up a template based around this set-up as any other, and there are many other reasons why this tried-and-tested mixing approach is still used by many professional mix engineers. So with all of that in mind, Figure 16.1 shows how my template is set up in terms of signal flow and routing.

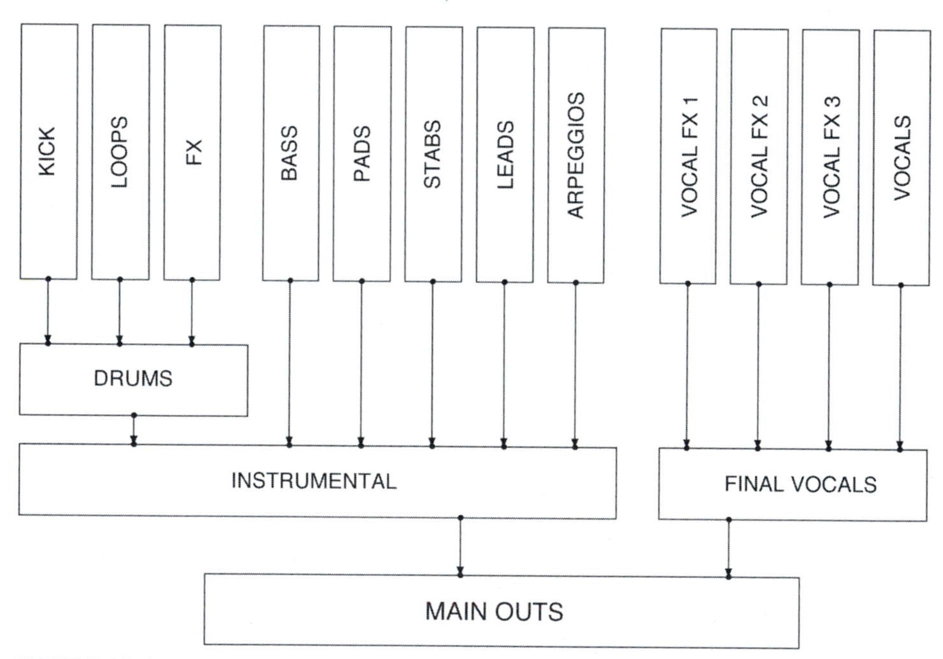

FIGURE 16.1
By using submix buses you can make broad changes to the mix more easily and apply effects to groups of sounds simultaneously.

Whether this would work for you is uncertain; you might prefer a more simplified way of working. But if you have Logic, you can load up the sample files from the website that accompany the remix walkthrough in Chapters 21–24 and you will be able to experiment with a remix that is mixed and routed this way to see how it feels to you.

Having covered those few potential shortcuts, we now take a look at some other important mixing basics. Possibly the most important is the subject of *gain staging*. To have clean and powerful mixes you should avoid distortion as much as possible, unless, of course, you are talking about deliberately applying a distortion effect as a part of the sound. This has *always* been good practice, but if you are working in a fully digital environment it is even more important because, while analog distortion can sound warm and actually add some nice harmonics to a sound if used in moderation, digital clipping distortion sounds very harsh and nasty. To avoid this, it is best to work with lower levels to allow plenty of headroom.

In a traditional analog studio there is always background noise, and every additional channel of mixing and every additional outboard device that is added to the mix compounds the noise levels. To overcome this, signals would need to be as *hot* (loud) as possible to provide the greatest signal-to-noise ratio while still trying to avoid the levels being so loud that they run into distortion. With today's digital systems, things are a little different. Background noise, while not completely a thing of the past, is much less of an issue because the digital processing does not add any additional noise of its own (unless 100% by design in an attempt to mimic vintage hardware!). Of course, any noise present in any original recordings will continue to be a factor, but much less so. This greatly reduced signal-to-noise ratio (and this is *especially* true if using a 24-bit format) means that signal levels can be generally much lower, allowing for greater headroom, which minimizes the risk of digital clipping while still maintaining a more than adequate signal-to-noise ratio.

The first step in the signal path is the sound source, and this can be either an audio recording or a plug-in generated sound source. When dealing with

FIGURE 16.2
Comparison between a signal recorded too *hot* (top) and one recorded at a good level (bottom).

remixes you normally have at least some audio parts from the original track, and there is little you can do if these have been badly recorded (you're unlikely to run into clipped or distorted recordings, though). However, if you are going to be adding recorded parts of your own to the remix, it is important not to record them *too* loud or compressed or processed because once this is done, very little can be done to undo it, short of actually rerecording the parts—something to be avoided if at all possible. Once the audio parts have been recorded, it is impossible for an audio recording to actually overload the individual channel strip in your DAW (assuming no plug-ins at this stage) but the same cannot, strangely, be said of synth plug-ins. I have found quite a few that, when set to maximum volume (or sometimes not even as high as maximum), can actually overload the channel strip.

There are, as always, a few different ways of dealing with this. You can lower the output volume of the synth itself (assuming it has a volume control… not all do!), which will prevent overload of the channel strip and prevent distortion occurring within the plug-in itself. Some plug-in synths, when the output level is too high, will actually distort their outputs so that, even if the channel gain is lowered and the peak signal level is way down, they still have audible distortion artifacts in the sound. Another obvious thing is to turn the main channel fader down, but, as we will see shortly, this doesn't prevent overloads and clipping occurring before the signal actually reaches the fader, as this fader is the very last component in the chain. A better option is to use a plug-in such as Sonalksis' excellent and, as the name suggests, *free* plug-in FreeG (*www.sonalksis.com*). FreeG provides a great way of adjusting the gain and level of a signal at any point in the chain. The advantage of using a plug-in like this is that it can be inserted *between* other plug-ins to provide an indication of the signal level going into each plug-in rather than just the overall output. This is important because it is quite easy to overload the input to one plug-in in a chain while still keeping an output level that isn't clipping the channel. It might seem like a lot of messing around, but it is worth inserting a plug-in like FreeG after each plug-in on a channel to make sure the levels are under control and have an easy means of adjusting them if they aren't. There are more creative uses for this kind of set-up as well, if you are feeling more adventurous. These will be covered later.

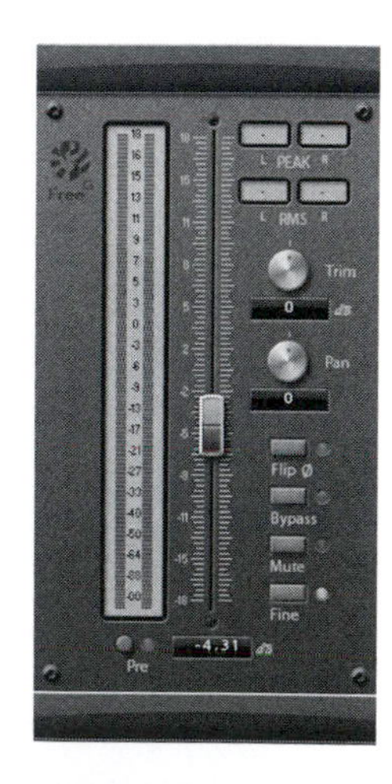

FIGURE 16.3
Sonalksis' FreeG plug-in is invaluable for controlling levels at various stages of the mixing process.

If you are using a multiple submix bus type system as I described earlier, then you also need to consider that, while each channel may be below clipping levels, when two are more of these are fed into a submix bus the cumulative level might overload the bus. In this situation, the FreeG plug-in will be useful as a global level trim on the bus, rather than having to reduce the levels of the multiple channels feeding into the bus. And, finally, the same thing applies when all of the buses are fed into the master output channels.

If this is all set-up correctly, you should end up with a final signal level that has plenty of headroom and allows for decent dynamics in the track as a whole. The final mastering process may well squash many of those dynamics out of the mix in an effort to keep up in "The Loudness Wars" (the constant battle to make tracks louder than those that have gone before so they sound bigger and more impressive) but it is always better to leave such things until the mastering stage rather than building such heavily compressed dynamics into your original unmastered version.

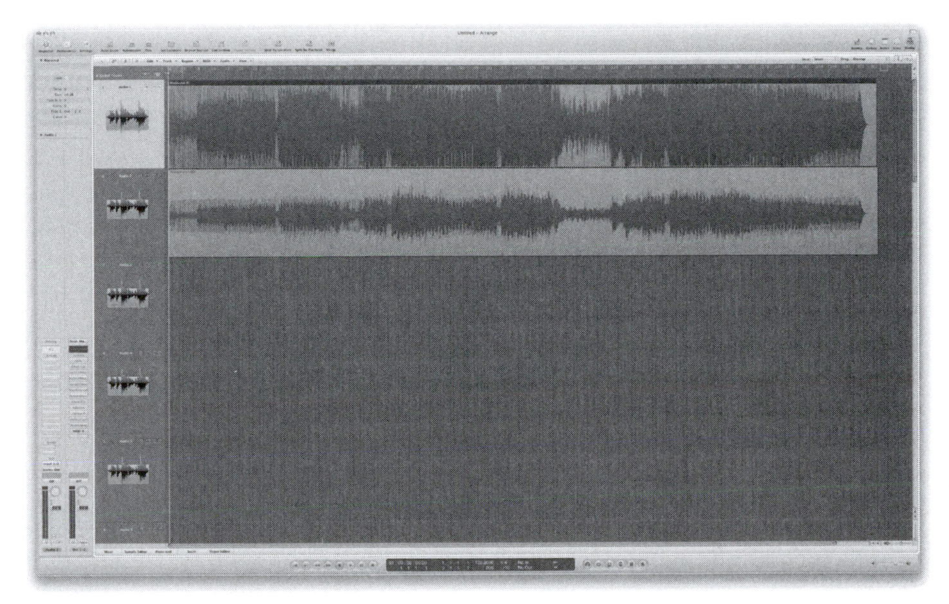

FIGURE 16.4
A comparison of a mastered track (top) against the same track unmastered (bottom) shows the very obvious increase in overall level and the corresponding decrease in dynamics.

BUILDING A MIX

It is more than likely that you will be working on the mix (levels, EQ, effects, etc.) of your remix as you go along, because there is not usually the time available with remixes to get to a point where all of the musical content is in place and then pull all the faders down and start mixing the track again. As you approach the end of the remix you will probably find that the mix itself is pretty close to being finalized. While the traditional approach of pulling everything down to zero and starting again might not apply, the principles applied during that mixdown process still do. There may be differences in the way you apply them or the order in which you apply the process, but having a good knowledge of what to do and, even more importantly, why to do it will only do you good in the long run. There are as many different approaches to working on the mixdown of a track or remix as there are producers, remixers, and mix engineers, and you will, eventually, find a method that feels comfortable to you and becomes your own.

In spite of this diversity, some common threads run through many of the methods. In almost all forms of dance music, one of the most fundamental elements of the mix is the kick drum, followed closely by the bass line. These two sounds, both in isolation and in their interaction, are crucial to your production. In fact, some genres of dance music are minimal in content and, other than the kick drum and bass line, may contain very few additional elements. In which case the importance and focus on them is even greater. With this in mind, it is quite frustrating that these two elements are often the hardest to judge and sculpt effectively on most smaller monitoring systems. Most remixers, and especially those starting out, will be working on relatively small monitors, and the bass extension of those monitors won't even come close to that offered by a decent-sized club PA system.

There are a number of different ways to compensate for this, the most obvious being to get a decent subwoofer to work with your monitors. This will go quite a way toward rectifying the lack of bass, but unless set up very carefully, a subwoofer can give a misleading impression of the real amount of low bass in your mixes. Another possibility is to make sure you have at least a couple of different monitoring options available. Some good-quality headphones can have very decent bass output, which will give you a pretty good idea of what to expect, but because they are headphones, you never really get that physical sensation of loud bass that you get in a nightclub. Possibly the best thing you can do is simply spend a good amount of time getting used to *your* monitors and how they sound in *your* studio. The human ear/brain combination is very adept at compensating for audio discrepancies, and once you have trained your ears to the sound of your monitors you will find yourself compensating for any lack of low bass response without really thinking about it. A very good way to do this is to play tracks that you know have a great bass end to them through your monitoring system. Play them loud, play them quiet, play them a lot, and you will start to learn what your own mixes should sound like through your monitors.

If you have gotten used to a particular set of monitors and then, for whatever reason, you change to different ones, it can *really* confuse you until you have grown accustomed to the new ones. This has happened to me a number of times in the past and has been such a source of confusion that now, if I am going to be working at another studio for anything that involves mixing a track down, I put my monitors in the back of my car and take them with me! Even if the main monitors are much bigger or better (probably better to call them "more accurate") than mine, I still like to have mine there as a familiar point of reference to give me a sound I am used to and one on which I can make a sound judgment (excuse the pun!) of the subtleties of the mix balance and EQ.

As a more general point about monitoring, it always pay to have a variety of different monitoring options available to you while you are working. Monitors can be expensive, so I am not suggesting that you need multiple pairs of main monitors available to you. What I am referring to are the cheaper options and, ultimately, what a good percentage of your end-users will listen to the remixes on. Apart from a big club PA system, most people will listen to your remixes on one of three playback systems: a car stereo, a home hi-fi system, or an MP3/CD player. If you can test your mixes on those three systems and your main monitors it will give you invaluable insight into how your mixes will translate and sound to the majority of your listeners.

Connecting a set of hi-fi speakers (probably through a hi-fi amplifier of some kind) isn't a major job and might even be as simple as connecting them up to a secondary output pair from your sound card (if it has multiple outputs). If your sound card has only two outputs, you can connect these to either a small mixer or a simple speaker-switching box. If your budget allows, there are a number of very reasonably priced monitor controllers that allow you to connect multiple input sources and multiple sets of speakers and switch between them instantly for A/B comparisons.

Similarly, listening back to your mixes on earbuds is as easy as connecting them to the headphone out on your computer/sound card/mixer and is a very quick and easy way to check for any glaring obvious faults in your mix. If your budget allows for a good quality set of headphones and a decent headphone amplifier then this can be a great investment, as a way of double-checking the mix and a way to work late at night in environments where working loud might not be possible.

One thing to be wary of: as well as a very different sensation of listening on headphones compared to speakers, there is a potentially drastic difference between the stereo image on headphones and speakers. As such, it can be difficult to make decisions regarding the panning of sounds on headphones. There are a few devices on the market, some software (for example Redline Monitor by 112 dB, *www.112db.com/redline/monitor*) and some hardware (such as the SPL Phonitor headphone amp, *www.spl.info/en/hardware/headphone-amps/phonitor/in-short.html*), that aim to give headphone listeners a more natural sound in their headphones by replicating some of the physical effects of using speakers. Because speakers are placed at a distance from our ears there is always an element of the sound from the right speaker reaching our left ear, and vice versa. This is called crosstalk, and the different systems available work in similar ways by introducing elements of this crosstalk into the headphones to give a more familiar-sounding stereo image. In effect this brings the sound out of your head and positions it at a point in front of you where it feels more natural and realistic, and where it is possible to get a greater sense of depth from the mix as well.

The ideal scenario would be to have, as mentioned before, a number of different monitoring options. There will always be one that is your main monitoring set-up and this is the one you really need to learn and get used to properly. No room is acoustically perfect and no speakers have a perfectly flat frequency response, so learning how your monitors sound and how they convey the music is vitally important. There are ways of getting a more accurate picture of the music no matter what monitors you use. One of these is to work on the there are certainly a few models of EQ that seem to be generally accepted as being up there and having that certain *je ne sais quoi*. It's probably true to say that most of the really sought-after models are vintage hardware EQs. Most of these types of EQ are used for subtle or broad tonal changes to a sound because that is what they excel at.

For more surgical and corrective requirements, there is a lot to be said for more modern digital (software based) EQs as they can offer precision that an analog (hardware) can find hard to match. Of course, even within the digital EQ world, some are regarded more highly than others because of the smoothness of the results. There is also a recently emerging third category: software recreations of classic hardware EQs. I would classify those alongside the genuine hardware EQs, because, to a greater or lesser degree of success, they aim to recreate all of the characteristics of the units they model.

With any kind of EQ, hardware or software, analog or digital, it is good to always try to achieve what you want to do by cutting frequencies instead of boosting them. This general rule came from the time before digital EQs, and was born, like so many things, out of necessity. You see, each time you boost a frequency with an EQ, you will also be boosting the inherent background noise in that part of the sound. While each individual boost might be very small and would only result in a small increase in background noise, the cumulative effect of all of these little EQ boosts in a whole mix could be very noticeable. With digital technology, this isn't as much of a problem because the amount of background noise in digital systems is generally much less than in analog ones. However, there will still be noise on any recordings made through microphones, preamps, and any kind of analog input signal, so there *is* still noise to be boosted by EQing. Even if there isn't so much of a technical necessity in a digital system, cutting rather than boosting is still a good habit to get into even on a digital system.

WHAT IS COMPRESSION?

Compression is a way to make loud signals quieter. It can be a little confusing because much of the time, compression is used to make things *louder* rather than quieter. Compression is actually a two-stage process. The loudest parts of the input signal are squashed down (compressed) and then the overall level of the signal is raised to compensate. If we assume that the loudest parts were reduced by 4 dB in the compression stage and then the overall signal was raised by 4 dB, the effect would be that the loudest parts still remained at the same level but the quieter parts, which weren't affected by the compression process, were then boosted by 4 dB and were now 4 dB louder than they were before.

This is a somewhat simplified description of the total process, but it gives an idea of one of the aims of compression, at least. With appropriate settings, compression can also be used to emphasize or subdue either the attack or sustain portions of a sound, and if the compressor in question has a side-chain capability you can also use it to create a pumping effect, like that used (some would say over-used) on a number of dance tracks in recent years. A compressor (of both the single-band and multiband variety) is also an essential part of the mastering process. Another related piece of equipment is the limiter, which is a compressor with infinite compression ratios.

To comprehend the different uses of compressors, we need to understand the main parameters of a typical compressor and then figure out how to set these to achieve the different scenarios we described earlier.

FIGURE 17.5
Logic's Compressor plug-in is useful and versatile.

COMPRESSION PARAMETERS

Threshold

Put simply, the threshold control on a compressor is the level of incoming audio signal above which the compression starts to take place. Any signals that are below the threshold level will pass through the compressor unaffected and any that are above the threshold will be reduced by an amount determined by the ratio control. The threshold level control normally starts at 0 dB and can be adjusted downward from there. The lowest value you can set the threshold level to varies from compressor to compressor, but most will go down to at least −40 dB, to provide a wide range of adjustability.

Ratio

The ratio value is the amount by which any signals above the threshold will be affected. This is expressed as a ratio that defines how strong the compression is. For example, if a ratio of 4:1 is selected, that would mean that if the input level is 4 dB above the threshold level then the compressor would reduce that to just 1 dB above the threshold. If the input level were 8 dB above the threshold then the output would be 2 dB above the threshold. Likewise if the ratio were 10:1 then an input level 5 dB above the threshold would result in an output level that was only 0.5 dB above the threshold. Because of this, the higher the ratio selected the stronger the compression effect will be. It is normal for compressors to have a ratio that is variable between 1:1 (no change) and 20:1 (strong compression). There do exist ratios higher than 20:1, but at that point the process becomes more like *limiting* than compression, and this will be explained in more detail later in this chapter. There is also the option to have ratios lower than 1:1, but, again, this isn't compression as it serves to *emphasize* levels that are above the threshold and is known as *expansion*. Once again we will look at this later.

One further variation, which is far less common, is that of *inverse ratios*, such as −1:1. This would mean that if the input signal level were 5 dB above the threshold level, the output level would actually end up being 5 dB *below* the threshold level. This puts us very much into "special effects" territory, and it wouldn't be something you would use that often. For the sake of completeness, we will look at possible applications for this shortly.

Attack/Release

These two controls, in combination, determine how a compressor's effects vary over time. The *attack* control determines how quickly the compressor will react to changes in the input signal. The *release* control determines how quickly the effects of the compressor will die away once the input signal has dropped back below the threshold. If a short attack time is chosen, the compressor will very quickly respond to any changes in the input level that take the signal above the threshold. Whereas with a longer attack time is it is entirely possible that very short dynamic spikes will sneak through without the compressor squashing them down, as would normally be the case. Similarly if a short release time is selected, as soon as the level drops back down below the threshold the compressor will stop compressing. But if a longer release time is chosen, it is possible that the compressor will continue reducing levels even though the input signal is below the threshold level. This has the effect of reducing the level of a more significant part of the signal and not just the peaks above the threshold, and can make the compression sound gentler and less "pumping." In combination, the adjustment of the two parameters can result in a wide range of different effects, which can be used for anything from gentle overall dynamics control to more specific control of transients.

Knee

Knee is the rate at which the compression takes effect. The choices are normally *hard knee* and *soft knee*, although sometimes there is a variable control rather than a switch between the two. The practical difference between hard knee and soft knee is that with hard knee compression the ratio kicks in as soon as the signal level crosses the threshold. If the input signal is even 0.1 dB above the threshold, it will be compressed by the amount set by the ratio control. Whereas with soft knee compression there is a more gentle transition into compression.

To give you a comparison between the two, let's look at an example. If the input signal was 0.5 dB over the threshold and the ratio was 10:1 then with hard knee compression the 0.5 dB would be compressed to just 0.05 dB; with soft knee compression it might be compressed less, so that the output level was 0.25 dB, which gives an *effective* compression ratio of 2:1. When the input signal increases to being 2 dB above the threshold, hard knee compression sticks to its rigid ratio and gives an output of 0.2 dB above the threshold. This time the soft knee compression would take more of an effect, so it might perhaps give an output of 0.4 dB, which gives an *effective* compression ratio of 5:1. If the input signal was 5 dB above the threshold, then both the hard knee and soft knee would give the same compression ratio of 10:1, as we would expect. As such you can see that the knee control only really has an effect when the input signal level is fairly close to the threshold level.

Peak/RMS Detection

We have seen so far that a compressor deals with reducing the level of an input signal when it exceeds a predefined threshold. There are actually two different ways in which a compressor can determine whether the signal has crossed the threshold level: *peak detection* and *RMS detection*. I won't go into too much detail about how the two different systems work on a technical level (there is plenty of information about that on the Internet if you are technically minded in that way), but will just say that one works by detecting the absolute peak value of a signal at any time (peak detection) and the other works by detecting the average signal level over a period of time (RMS detection).

The simplest advice I can give you in this respect is that peak detection is better for applications where you want fast or instantaneous control of the signal level, such as catching loud transients from drum sounds to avoid them clipping. In use, this style of signal level detection is more likely to sound jumpy because the input signal might be constantly crossing and recrossing the threshold level, and therefore moving in and out of compression. Depending on how you set up the other parameters, this could either sound quite subtle or very obvious. Neither of these is a bad thing depending on the context in which you want to use it, and it certainly pays to experiment with different settings.

On the other hand if you want to achieve something that is more like an overall leveling effect, such as keeping a more consistent vocal level throughout a track even though the singer varies in volume quite a lot, then RMS detection is more likely to be the most useful. Because it works on an average level, there is less likelihood of the average signal level jumping around wildly and moving the compressor in and out of compression. As a result, RMS detection generally sounds much smoother and less obvious, and is useful for situations where you almost don't want to hear the compressor working. Of course, like most things music-production related, you can take it to extremes, and even RMS detection can be pushed to the point where it sounds heavily squashed. As with peak detection compression, although the method of detecting the input level is different, there is a wide range of possible sounds to be had depending on how the other parameters are set. And, of course, using peak detection for transient limiting duties and RMS detection for more subtle duties is not a rule as such. Those are the roles they seem to work most naturally in, but that doesn't mean it is impossible to use them in other ways.

Sidechain

The sidechain part of a compressor isn't really a control as such, but it can have a very dramatic effect on how the compression works, so it is worth describing. Under normal circumstances, compression occurs when the level of the input signal is above the threshold. In this way, any level peaks in the input signal can be controlled. However, if you use the sidechain input, the compression process can be triggered by the level of the signal coming into the sidechain input instead. If the uses of this aren't immediately obvious to you then don't worry, there are a few *very* useful ways in which sidechain compression can help us when we are mixing tracks and one of them has actually become very much a trademark of certain genres of dance music. Even if you wouldn't use it in a really obvious way, it can be a great problem-solver to help us deal with excessive bass levels in a track, or, with a little more effort, with sibilance in a vocal recording. The sidechain input may well have very simple EQ controls in the form of high- and low-cut filters. These are there to help us fine-tune the response of the compressor to the sidechain input by giving us more control over the particular part of the frequency spectrum the compressor will be listening to, in order to determine whether the threshold level has been exceeded.

Using Compression for Level Control

One of the most common uses for a compressor is to make things louder. This seems to go against what I have already said about compression being used to reduce the level of sounds above the threshold level, but it isn't really the compressor that is making the sound louder. What the compressor is doing is reducing the size of the peaks in the audio signal, which means that we can now increase the overall level of that sound without it clipping. Perhaps an example will help clarify this.

Assume we had an audio recording with quite a few short peaks in it, which had a maximum level of −3 dB. If we wanted to increase the level of this recording by 5 dB those peaks would clip because they would, theoretically, end up at 2 dB above the maximum possible level. This would end up causing distortion to the inverse ratio but the elysia mpressor, which is available as both a hardware compressor and as a plug-in, is one (very good) example. Used on a full drum submix, the effect can be devastating and it can squash the peaks of the drums down in such a dramatic way that it feels like they have become a living and breathing creature straining to burst out of your speakers! It really can be quite an intense effect; not something you would use all the time, but certainly something worth having in your box of tricks.

Before moving on, I would like to give you a little advice. As a general rule I think that it's always better to compress too little rather than too much, unless you are aiming for a very specific effect, in which case they sky is the limit. I always try to make sure that a compressor is giving no more than 5–10 dB of gain reduction. If I find myself needing more than this, I would more than likely place another compressor in series and use this second one to reduce the peaks some more. For some reason two compressors placed in series, each giving 5 dB of gain reduction, generally sounds cleaner and clearer than a single compressor giving 10 dB of gain reduction. Many top producers and mix engineers use multiple compressors in series. In addition to the fact that each compressor has to work less, there is the possibility that you can set the different compressors up differently to do different things to the sound. The first could be set up with a very fast attack and release, peak detection, quite a high threshold, and a high ratio, to essentially act as a peak limiter. From there the sound could move to the next compressor, which had a slightly slower attack and longer release, a lower ratio, and lower threshold, but still with peak detection, providing an additional stage of gain reduction. Finally you could add in a low-ratio, low-threshold compressor with RMS detection to add some density to the track overall and give a little more gain reduction. There really aren't any hard-and-fast rules here, and using combinations of different compressors for different purposes and then using them in sequence can give some truly great results. You can, if you wish, still stick to using just one compressor if that is giving you the effect you want, but certainly don't be afraid to try this multiple compressor technique.

Using Compression as an "Effect" (Sidechain)

Sometimes you want your compression to happen without it really being discernible. If all you want to do is control the peaks in a sound, then you don't really want it to sound like you are doing anything, you just want it to seem like the sound in question was played quite uniformly. On other occasions, such as with parallel compression, you might want to have the compressor take a more upfront role and be more dominant in the sound. Even then, it wouldn't necessarily be obvious that a heavy compression was giving the effect. There is one usage of heavy compression, however, where the effects are *very* audible and very identifiable as being compression. And that is with *sidechain compression*. This has become a deeply ingrained part of the whole French house sound, and was used to immense effect on Eric Prydz's huge hit "Call on Me" where it came to define the whole sound of the track.

sound and wouldn't be good. But if we apply a compressor to the recording with a threshold level of −13 dB and a ratio of 5:1, then the compressor would take those peaks and, seeing as they were 10 dB above the threshold, would reduce them to 2 dB above the threshold (5:1 ratio), which would mean that they would now be at −11 dB. We could then happily increase the level of the track overall by 5 dB because, even then, the loudest peaks would still be at −6 dB, which is actually 3 dB *lower* than they were before even though we have *raised* the level of the track by 5 dB.

This kind of compression is used commonly with sound sources that can have transient peaks, which are much higher than the overall average level. Instruments such as drums and guitars often exhibit peaks like these (which can easily be 10 dB or more higher than an average level), so a well set-up compressor can make the resulting sound much more manageable and flexible to mix with. But there are other ways in which we can use compression to control the dynamics of our tracks. We can equally use compression on a whole mix rather than just an individual sound. We could use our compressor in a similar way to just squash down the peaks of our whole mix but, being a whole mix rather than just one sound, the peaks are likely to be much more regular as they will be a combination of the individual peaks of all the different sounds in the mix. Moreover, the overall average level will probably be much higher given all the sounds mixed together, which means that the peaks will probably not seem as "peaky" compared to the average signal level. Because of this, if you try to use a compressor on a full mix in the same way you would on an individual sound you will probably need to use a very high ratio to really hear any reduction of the peaks. And high ratio compression is, essentially, limiting. This does have its place when used on our final mix, but isn't really the best way we can use a compressor in this context. You see, what we are doing with limiting is basically a "top down" process in that we are trimming from the top and then raising the overall level to compensate. We can, however, tackle things from a different angle with a more "bottom up" way of compression. Let me explain what I mean.

With high ratio, high threshold compression, or limiting you notice the effect on the peaks, but most of the signal remains unaffected. However if you use a much lower ratio and a much lower threshold, the compression will have an effect on most of the signal but it will be a much smaller effect. Once again we will use an example to clarify.

If you had a track with peak levels of −5 dB in certain places, but most of the song had peaks of about −20 dB (that's quite an extreme example but serves to clearly illustrate the point) and we set a threshold of −50 dB and a ratio of 1.5:1, what would happen? Well the peaks in the track that were at −20 dB would be 30 dB above the threshold, which means compression would leave them being 20 dB above the threshold at −30 dB, so these peaks would have been reduced by 10 dB. The peaks that were at −5 dB would be 45 dB above the threshold, so would end up, after compression, being 30 dB above the threshold at −20 dB, representing a reduction if 15 dB. We could then apply 15 dB of *makeup gain* to bring the level of the loudest peaks back to where they were before we started. In this case we have applied quite a lot of gain reduction, but, as a lower ratio was used, it sounds to the ear less like you are cutting off certain loud peaks in the track and more like a general increase in loudness and density. This type of low ratio, low threshold compression is often used in the mastering process.

Using Compression to Shape the Sound

I mentioned earlier how adjusting the attack and release times of a compressor can be useful in emphasizing certain characteristics of a sound. Here I will go into a little more detail. Let's use a kick drum as a starting point. If you think about a kick drum sound and what parts it has to the sound, you can pretty easily split it into two: the initial hit and then the resonance of the drum shell. If you look at the waveform of a typical kick drum recording you will see that there is a very short initial peak followed by a more steady decay to nothing. This initial peak represents the sound of the beater hitting the drum skin, and the more steady decay is the resonance of the drum shell gradually dying away to nothing. Compressors give us a few options here.

If we set a very short attack time and choose an appropriate threshold and quite a high ratio then the compressor will react almost instantly to the peak at the beginning of the sound and squash this down. But when we get to the drum shell part of the sound, this will be below the threshold so the compressor won't change this at all. This has the effect of making the decay part of the sound louder in relation to the attack part, so we have, in a small way at least, rebalanced those two parts of the sound. If we set a longer attack time, the first part of the sound will be able to pass through unaffected while the second part of the sound *will* be affected. We might need to adjust a few parameters to get the required effect, but we can, if we set this up right, use the compressor to allow the peak of the beater hitting the skin through, unaffected, and actually reduce the level of the drum shell part of the sound. Once again we are rebalancing the parts of the sound, only this time we are doing it in the opposite direction.

If you have a compressor with inverse compression ratios then you can take that kind of effect even further. Let's stick with the kick drum and see what would happen if we set up a ratio of −2:1. If we use a fast attack time, the compressor is triggered by the initial beater hit. Let's assume that it peaked at −5 dB, with a threshold level of −10 dB set. With a conventional ratio of, say, 5:1 that would mean that the sound peaked at −9 dB after compression, but with an inverse ratio of −2:1 it would mean that the peaks were now reduced to −12.5 dB. For as long as the input signal remained above the threshold level, the output signal would remain below it. It is quite hard to explain the full effect of this in words, so I recommend that you listen to it yourself if you possibly can. There aren't many compressors that have an

Sidechain compression is actually a very general term, but it has com
associated with a very specific type of usage. If a regular 4/4 kick drur
into the sidechain input of a compressor and a reasonably high ratio a
threshold are used along with a fast attack and a medium-to-long rele
effect on the sound will be to "duck" the sound when the kick drum is p
because the kick drum will be triggering the compressor to reduce th
every time it sounds. The combined effect is to make whatever is bein
pressed sound like it is pumping or sucking. If this compression is
to the whole track (minus the kick drum of course), the effect can be
intense.

It can also be used to take advantage of a natural phenomenon to g
impression that our tracks are louder than they actually are. The hum
like any kind of listening device or recorder, has limits to how loud
can be before it will distort and overload. What happens when those lin
reached is that our ears just compress (or limit) any sound louder than i
inbuilt threshold. In the case of kick-drum-heavy club music, the kick
will most likely trigger our ears built-in limiter with its huge bass ener
els. When the sound of the kick drum dies away, everything else will s
get louder for a second until the kick drum comes again. And this is
the effect we have been describing with sidechain compression. So if
this sidechain compression up in the right way and on the whole mix
have the same effect on the sound as listening to it extremely loud woul
ears hear the effect and naturally make a connection between the "overly
effect and the "sidechain compression" effect and it goes some way towa
ing the impression that we are listening to the track really loud, but at th
time quietly. Of course, nothing will ever truly give the same impression
tening on a huge club PA system would, but it is a useful side-effect of th
chain compression process that can work in our favor to help make our
sound somehow bigger.

Parallel Compression

Parallel compression, like sidechain compression, is a generic term for
cess that has been adopted by a very specific production technique.
called New York compression, parallel compression is, pretty much
explanatory. Rather than the compressed sound replacing the origina
with parallel compression the two versions run side by side and can th
blended in different proportions to give different effects. Sometimes
possible within a single compressor if there is a wet/dry control, but in
cases, you will need to make a copy of the sound you wish to process i
way, apply the compressor to the copy of the sound, and then blend th
by adjusting the levels of both versions. Another option might be to
the sound to a bus and apply the compressor to the bus. This is the pre
method for using parallel compression on drums because several sounds
of the drums in the drum kit) can be routed simultaneously to the sam
and the compressor can act on all the sounds at once.

with the kick drum, but that would also have
of the rest of the track as well, which gives ris
effect.

This result can certainly be useful as an effect, b
ally want it. Without multiband compression
to apply any significant amount of compressi
when you start using a multiband compressor
over the compression process as a whole. In it
ply set up two separate bands—one that only
kick drum and low bass sounds (maybe up t
covers everything else—and set them up identi
ilar effect to having a single band of compres
you will clearly see that the kick-drum band v
it than the other band. If you did this, you w
thing was wrong because now the balance be
cies in your mix will be all messed up! Fortu
that I am aware of include *makeup gain* (or s
independently, so you can now use the makeu
the level and bring things back into line. Th
independent band can be very useful at this p
least a ballpark figure of how much makeup
to how they were.

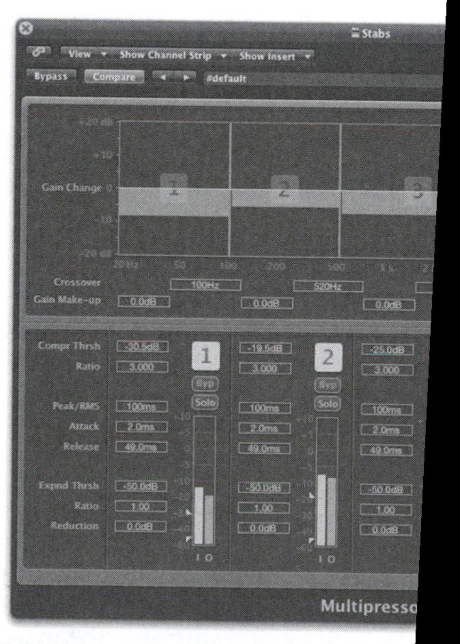

FIGURE 17.6
The Multipressor plug-in in Logic is one example of a mu
different amounts of compression being applied to each

But herein lies the first problem. Once we have applied the makeup gain, the approximate frequency balance of the input signal is returned. But because we have changed levels with the compression and then changed them again, we risk changing the frequency balance and spectral content of individual sounds within the mix (assuming you are processing a full mix through the multi-band compressor). These changes might only be subtle, but can be enough to start to make things sound a little strange. You may be thinking that much the same thing happens when you EQ sounds, so why worry about it here. To some extent that is true, but in this situation you are changing the frequency response of the sounds and the dynamic response. If you reduced the level of a particular frequency range with an EQ, that change is constant (unless you are using dynamic EQ, which is in many ways similar to multiband compression), but with multiband compression there might be points when the level of that frequency band is reduced and other times when it remains the same. It is this kind of effect that makes multiband compression hard to set up correctly. Fortunately, if used subtly, these effects are generally minimal and often considered acceptable.

The possible applications of multiband compression go well beyond the simple two-band possibility we described earlier. Most multiband compressors include at least three bands, and as many as six in some cases, which allows for some fairly specific dynamic control if you need it. Another possible use is in isolating the very high frequencies in the same way we did for the bass, to avoid the vocals and other midrange sounds being "clamped down" by an especially loud cymbal hit. Once we have tackled these two potential problem areas, we can split up the remaining frequency range still further (assuming there are enough bands left) to single out other sounds. You have to bear in mind that the midrange frequencies are where many different sounds occur. Pianos (when playing chords at least), strings, brass, guitars, synth pads, lead, and arpeggiated sounds all sit in this midrange area. Last, but not least, so do vocals. Consequently, the midrange of a final mix can become quite crowded. Trying to isolate particular sounds or instruments this way with a multiband compressor could prove quite difficult.

The mention of vocals brings us to one quite specific application for a multi-band compressor. You have probably already heard of *de-essing*, and chances are that you will have a need for it on a regular basis. The process of getting rid of harsh sibilance ("sss" sounds in a vocal) is not easy, at least achieving it in a way that doesn't make the rest of the vocal performance sound strange. You could simply EQ out the harsh frequencies, but if that EQ change was a static one it would probably make the rest of the vocal sound quite odd because we *do* need to be able to hear those "sss" sounds, but just in a controlled way. If you apply enough of an EQ cut to get rid of the harsh sounding ones, then the rest of the "sss" sounds in the track will be reduced in level as well, resulting in the vocalist sounding suspiciously like he has a lisp! One possible option here is to automate the EQ cut amount, but, depending on the vocal of course, this could be a laborious task. Another possibility would be to cut all of the sibilant parts of the vocal out as separate regions, move them to another track, and then apply a static EQ to that track only. In practice this would sound very clumsy and unnatural because there would be a sudden change from one vocal tone to the other, which would not be good. What we are looking for is a way of changing the amount of reduction of a particular (narrow) frequency range in an automatic way, and that is where the multiband compressor comes into its own.

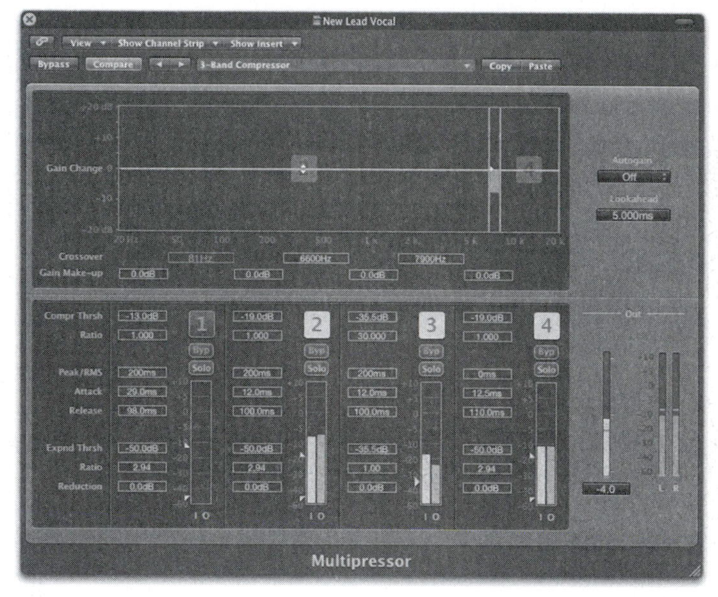

FIGURE 17.7
A multiband compressor set up with strong compression on a very narrow frequency band can, if set up right, work very well to reduce sibilance.

If we set up a 3-band compressor with a very narrow frequency range on the middle band and then set the low and high bands to have no compression whatsoever, then we can use that middle band to reduce the level of *only* those frequencies (as we would with an EQ) and *only* when the level in that band passes a certain threshold. If the frequency band is well-chosen (it varies by singer but is generally between 5 KHz and 8 KHz) and the threshold and ratio are set carefully, it is possible to have the vocal sound totally unaffected in every part of the song except for the very specific parts where the sibilance is too loud. It is certainly possible to try to do this on a final stereo mix of a track, and you might be able to go a little way toward fixing it if you really have no choice, but something like this should be applied to the vocal track(s) *before* mixdown. It's probably something you would notice early on anyway.

There are dedicated de-esser units (both hardware and software), and some of these work very well. If you have one, certainly try it, but I would also recommend the multiband compressor as a possible alternative, because you might even be able to perform some leveling on the vocal at the same time. Whether you prefer the convenience of an all-in-one solution in a multiband compressor or prefer to use a separate de-esser and compressor, is something you will decide as you go along. If nothing else, give this method a try, because it might help to give you a greater understanding of how the process works. An appreciation of *why* certain things happen is equally as important as understanding *how* to make them happen.

WHAT IS EXPANSION?

Unlike compression, which decreases the dynamic range of an input signal (that is, reduces the difference between the loudest and quietest parts), expansion serves to actually increase those differences. And whereas compression has as many uses as there are compressors, expansion is perhaps more limited in its scope. That doesn't mean it is more limited in its usefulness.

Expansion Parameters

Many of the parameters for an expander are the same as those for a compressor, because they work in a similar, if opposite, way. So the *threshold* is there, as are *attack* and *release* time settings. You might even find an option to select peak or RMS detection, but this is less common in expanders, which normally just use peak detection. The biggest difference lies in the fact that the "compression ratio" is now called "expansion ratio," and offers ratios that start at 1:1 and then *reduce* from there; instead of going from 1:1 up to 20:1 they might go from 1:1 down to 0.1:1. When the input signal crosses the threshold (i.e., when it drops *below* the threshold), every decrease in signal level below that threshold of, say, 0.5 dB will result in an final output decrease of 1 dB. A 5 dB decrease in input level would equal a 10 dB decrease in output level.

When to Use an Expander

The uses for an expander might not necessarily be as immediately obvious or, for that matter, as diverse as those for a compressor, but they can still be good problem-solvers for us. Like pretty much any audio processor, an expander does have a number of well-established uses, but there are probably other ways of using them that haven't yet been fully explored. Let's start with the obvious and then move on to a couple of the more specialized. The two main uses that spring to mind are to reduce the level of headphone spill in vocal recordings and to effectively quicken the decay of something like a crash cymbal or a similar FX sound.

Headphone spill on vocal recordings is an inevitable fact of life. The degree to which it is a problem depends on many different factors including the type of headphones being used, the monitoring level the singer wants, the type and

FIGURE 17.8
An expander does the opposite of a compressor in that it aims to increase the dynamic range instead of reducing it.

sensitivity of the microphone, whether the singer had only one headphone properly on his or her head, the music that is being played back through the headphones and, finally, whether any compression was used at the recording stage. In most cases, the headphone spill is negligible, but there can be times when it's quite obtrusive. When the vocalist is singing it generally isn't so much of a problem, because, to some extent at least, it is masked by the (comparative) loudness of the vocal itself, but when the vocalist isn't singing it can quickly become a problem, especially if that vocal recording is subsequently being quite heavily compressed. A noise gate (see below) is often used to try to clean up the background noise, but it isn't always the best option with a vocal recording, at least not on its own, because of its inherent on/off behavior. With vocals, there are often parts at the end of words or phrases that gently fade away, and a noise gate isn't particularly well-equipped to deal with this. You have to make a choice between setting it up so that it removes all the headphone spill completely, at the risk of prematurely cutting off the end of some of the notes, or setting it up so that it only "closes" right at the very end of the vocal notes, which might risk leaving some headphone spill audible at the end of those notes.

An expander can be a useful alternative, or, at the very least, an additional step. What we try to do in this situation is set the threshold high enough so that most of the vocal passes through unaffected, with only the very ends of the notes—the points where the headphone spill is starting to become audible again—being affected. Once we have figured out a good average for this level, we then use a high expansion ratio to make sure that anything that drops below this level is reduced quite substantially. When the vocal level drops below the threshold, the remaining decay of the note does continue, but at a quicker rate. As a result, we get something that is halfway between the two extremes we just described for the noise gate, and sounds much more natural than simply cutting off the end of a note. If you wanted to, you could then add in a noise gate with a very low threshold to make sure any residual background noise was taken care of by the noise gate.

Using similar principles, we can make crash cymbals or other FX sounds seem to die away a bit quicker. Of course, you can do this with either volume automation or a fade-out on the audio file itself, but using an expander can be a better option because it doesn't require you to remember to replicate the automation if you copy a region to another location. In addition, it isn't a destructive edit on the audio file, which helps because the original file can stay as it was in case you have used it in another project and don't want the new changes to reflect in the other project. One other benefit is that if the original sound has a decay that isn't a simple linear fall in volume, the unique shape of the decay is still maintained (in general, at least) after being processed by the expander.

Another possible use of an expander is to reduce the level of reverb on a sound. If you were given a recording that had reverb applied already and wanted to reduce the comparative level of that reverb, you can possibly use an expander to help with this. Be warned, however, this isn't always going to work. It relies on the original sound having a relatively clear and quick decay in order to be effective. There has to be a noticeable change in level between the end of the original sound and the reverb part for the expander to be able to distinguish between the two. If the sound has a long natural decay, like a sustained piano chord for example, this method won't really work, because the expander will simply lower the volume of the decay part of the original sound as well as the reverb. This can be a useful effect in itself as we saw with the crash cymbal example earlier, but it doesn't really have the effect of removing the reverb from the sound.

Something like a vocal or, especially, a drum kit that has reverb present will be a better candidate as the threshold for the expander. It can be set at a level where the majority of the sound has decayed naturally. The expander would then take effect and reduce the level of anything that follows, which is, in this case, reverb. When the next drum or vocal happens, the expander threshold will be crossed and the reduction effect of the expander will stop, allowing the sound to pass unprocessed. You would probably need quite a high expansion ratio for this to be noticeable, and even then you might need to use two expanders in series to get the desired effect. We are in similar territory here as dealing with headphone spill, so you might choose to use a noise gate after the expander(s) to truly get rid of reverb. As with headphone spill, a noise gate alone would probably sound too harsh to use.

What Is a Noise Gate?

A noise gate is a type of dynamics processor that works in a very simple way. If the input level is above the threshold level, the signal passes through unaffected. If the input is below the threshold level, the gate either stops the signal completely or, less commonly, reduces the level by a predetermined amount. In fact, if you use a noise gate that has a reduction "amount" control, it is more like an expander, which we covered in the previous section. Here we will focus on the pure gating effect. In terms of uses, we have already mentioned many of the main ones. You'll recall that any time you have a recording—particularly one made with microphones rather than through DI boxes or direct from the instrument—there will be some degree of background noise. While this might be minimal and not obviously audible, it is still good practice to use a noise gate to minimize the noise, because multiple tracks of background noise playing at the same time soon build up and the noise could then become noticeable.

Noise Gate Parameters

The good news is that noise gates are generally very easy to set up. There are many of the same parameters we see in other dynamics processors, with the addition of a few new options. As usual, the threshold control is probably the most important as this determines how early the gate will close. The attack and release parameters also play an important role because they determine how quickly the gate will react to changes in the sound. Normally the attack time would be set fairly quickly as you wouldn't want to miss any important transient detail in the sound being gated. Occasionally you might set it a little slower, because you might want to soften the sound of the gate opening a little. The most obvious example I can think of for this is on a vocal track where there might be some breath noise, or a word that begins with a softer sound such as "sss" or "mmm"; in this case, a slightly slower attack would preserve at least some of that smooth transition into the following sound. The release control is perhaps more important in shaping the overall response of the gate. If you were using the gate with quite a low threshold for quite hard gating effects, you would most probably use a quicker release time to make sure nothing

FIGURE 17.9
In many ways a noise gate is like a very extreme version of an expander that reduces the level to zero when the threshold is crossed, rather than just reducing it by a ratio.

slipped through. If the threshold is set higher, you might again want to soften the sound of the gate closing and use a longer release time. There is one specific example I can think of where the release time plays a crucial role in shaping the effect, and I will come to that shortly, but before I do, I want to run over some additional controls you will most likely see on a noise gate.

There is an addition to the attack and release controls in the form of a *hold* control. The combination of these three and the threshold control is fairly self-explanatory, but can be summed up like this. Once the input signal becomes louder than the level set by the threshold, the noise gate will open at the rate set by the attack control. It will continue as it is until the input signal drops back below the threshold level. At this point the gate will stay fully open for an amount of time set by the hold control, and once that time has passed, will then close at the rate specified by the release control.

Slightly more complex is the *hysteresis* control. The term *hysteresis* means, in a general sense, the rate at which a cause relates to an effect. In a noise gate, this is a control that determines the behavior of the noise gate when presented with a signal that fluctuates quickly around the threshold level. Without any kind of hysteresis, a noise gate can react in an unpleasant way to signals that hover around the threshold level. The gate can open and close rapidly in response to an input signal, and the result is often referred to as *chattering*. This rapid fluctuation between open and closed states can be dealt with in a couple of different ways. One way is to adjust the attack and decay times to combat this rapid opening and closing. This can definitely work but if you lengthen the times so that the chattering has disappeared the noise gate may now not react to the louder sounds passing through it in the way that you wanted. The other, more useful, way is to use the hysteresis control. This usually takes the form of a control that is measured in decibels and, in simple terms, it sets the difference in level between two different thresholds: one threshold for opening the gate and another, lower, one for closing the gate. Sometimes there won't be an actual hysteresis control but will, instead, be two separate threshold controls. There isn't really a technical difference in the process whichever method is used, but in some ways, the hysteresis control is more useful because it allows you to set a difference in level between the two, and then this difference tracks any changes you make to the main threshold level without you having to remember to readjust the other threshold to compensate.

This method of having two different thresholds works extremely well to compensate for the chattering because it means that the signal will have to rise again by a reasonable amount for the gate to open again once it has closed. In most situations, you will want to set the threshold of a noise gate as low as possible to allow as much of the sound to pass through as unaffected as possible. When you reach the lowest point there may well be background noise or microphone spill that is much more consistent than the main sound you are gating. If your threshold level is right on the limit of the level at which this unwanted noise occurs, then even a change of 0.5 dB up and down in the background noise can cause the gate to keep opening and closing. But if you set a hysteresis value of, say, 2 dB, any fluctuations in the background noise of even as much as 1.5 dB won't cause the gate to reopen.

Using a Noise Gate versus Manual Editing

A noise gate is a very useful tool for cleaning up messy recordings, and, in the case of most multimicrophone recordings of live drums, is a necessity if you want to get anything workable as a result, but it doesn't solve every problem. Sometimes you may set up a noise gate that works perfectly on 90% of your track, but in a few places there are little glitches or sounds that the noise gate won't deal with properly. In situations such as these you have to rely on manual editing to solve the problem. In an ideal world I would rather deal with *all* background noise and microphone spill/headphone bleed problems by manual editing, but this is simply impractical. Imagine how long it would take to manually cut a vocal track into separate regions every time there was a space between words, and to then delete those regions and perform fade-ins and fade-outs on every region that was preceded or followed by one of these gaps. I have done it, and trust me it takes a substantial amount of time to do properly, so it isn't always feasible.

While manual editing could be seen as the better, if more complex and time-consuming, method for situations like this, there are other situations when a noise gate is the far better, and possibly only, choice. One such example is the classic *gated reverb* effect. This is probably best-known for being applied to drums, and if I had to pick out one example to illustrate this more clearly than any other it would be the sound of the tom-toms on the Phil Collins song "In the Air Tonight." The idea behind gated reverb is that quite a long reverb (a large hall or plate effect) is applied to a sound, and this has the effect of almost sustaining that sound. Reverb can take a relatively short and percussive sound and make it seem to last much longer. For reverb to really be effective on a percussive sound, you would need to set a reverb decay time of anything over 5 or 6 seconds, and mix it quite high in volume in relation to the original sound. Reverb on its own would render the sound almost unusable, so we use the noise gate to close off that reverb after a set period of time by using a combination of the hold control and the release control. Care has to be taken with this kind of effect, both in setting up the reverb itself to sound obvious enough without overwhelming the original sound and in getting the hold and release times right. The times and rates needed for those parameters will depend on the tempo and the complexity of the song. To achieve that "classic" gated reverb drum effect I would use a hold time of somewhere around 150–250 ms and a release time of 50–100 ms as a starting point, and then adjust the times while listening to the effect in the context of the whole track, rather than in isolation.

Of course, this effect isn't limited to just drum sounds. I have achieved some truly remarkable effects by applying a very large sounding reverb to some synth stabs and then setting up the noise gate so that the "hold" time was calculated to be exactly a 16th note or an 8th note. This combination gives a very organic and shimmering rhythmic quality to a very average and unimpressive synth sound, and the effect can become a crucial and integral part of the sound itself. I am sure that this is just one example of many ways in which noise gates can be used creatively rather than correctively.

SUMMARY

As stated at the beginning of this chapter, the importance of EQ and dynamic effects on the final shape and tone of a mix should never be underestimated. As a producer, in general you will find yourself using them all the time and, hopefully, in ever more creative ways. As a remixer you will probably not have to worry so much about cleaning up vocal recordings, as quite often the vocal parts you receive from the client have already had this work done to them at the production stage. But every now and then you might get thrown a very raw vocal, which does require cleaning up to make it workable for you. This is, perhaps, even more relevant to remixers because headphone spill is usually masked by the music in the track. In the original track the headphone spill will match up with, and be in time with, the original music. In a remix, however, the music will be changing so the sound of the headphone spill can sometimes be heard distinctly if it isn't "cleaned up" effectively. On a remix you may be changing the key of the track, changing a chord over which the vocal sits, or even just changing the groove slightly, and any one of these changes can make headphone spill sound messy in the background.

There are so many options for EQs and compressors available now, that it is sometimes very easy to get bogged down in the decision of which EQ to use or which compressor to use for a given sound. You will undoubtedly begin to develop favorites, perhaps because of their ease of use, their tonal characteristics, or perhaps both, and you will automatically reach for those favorites. However, it is certainly worth trying different options, but try to avoid too much going back and forth comparing extremely subtle changes, as it's not the best use of your time. Remember, the technical side of production is not the only thing you have to worry about when you are a remixer. Being able to deliver a quality product *on time* is also very important. Your end product does need to be of a high quality, but you might have to make some compromises along the way. Making those choices can be hugely frustrating for you as a remixer because it may well go against any perfectionist instincts you might have, but it is something you have to learn to deal with as you go along.

CHAPTER 18
Mixing: Time Domain and Modulation Effects

INTRODUCTION

If EQ and compression are the foundations upon which a good mix is based then time domain and modulation effects are definitely the fairy dust sprinkled over a mix. Time domain effects can take an otherwise flat and uninteresting collection of sounds and put them into what seems like a three-dimensional place, and that has the effect of bringing everything to life. It's like the difference between looking at a photograph in a holiday brochure and actually being there. The same thing applies to sounds in a mix that have some kind of time domain effects applied to them. The combination of reverb effects (even if they are very subtle) and delay effects provides us with depth and even height in some instances, to complement the width that we (hopefully!) already have in our mix. These extra dimensions make everything sound more natural.

Every place we find ourselves in, big or small, indoors or outdoors, gives us some kind of sense of an acoustic "space" we exist in. Subtle reflections of sounds against surfaces define this space, giving us an impression of how big or small it is, what shape it is, and where sounds are coming from. If you have ever had the chance to stand in an anechoic chamber you will know what a weird sensation it is, and how it almost makes it feel like your voice is sounding only inside your head. It's a truly strange experience. Similarly, songs that have no trace of *ambience* about them (and this could easily be the case with electronic music, if the sounds are all generated artificially) sound totally unnatural. If we are listening on speakers, at least the position of the speakers and the acoustic characteristics of the room we are listening in will give us some sense of space, even if it is very narrow and isolated. Listening on headphones to a track with absolutely no reverb or delay anywhere, however, is very odd. It's like every sound seems to be coming from inside your own head rather than existing in the world outside you. To some extent headphones can do this anyway, but the normal narrowing effect of headphones is massively compounded if there are no time domain effects to help push the sound out of our heads and into space.

Modulation effects, on the other hand, are used more to give a sense of movement to a sound, and by that I mean movement in a tonal/spectral way rather than in a spatial way. Almost all modulation effects (such as chorus, flanging, phasing, and ensemble effects) work by layering the original sound with a copy of that sound and then adjusting something about the relationship between those two versions of the sound in a cyclic manner. This doubling up of sounds gives a thickening effect to the sound, and if the effect is a true stereo effect, possibly a widening effect as well. When the layered copy is modulated in some way, there is also a tonal variation over time, which can give a sense of spectral movement to the sound.

TIME DOMAIN: REVERB

Reverb is almost certainly one of the most impressive and useful effects we can have at our disposal, but is also one of the most misunderstood, misused, and generally overused effects. Reverb in one form or another is essential to any track we are working on. Even songs that sound like they don't have any reverb on them probably do, even if it is extremely subtle and only used on a few of the parts in the mix. The biggest problem with reverb is that it is extremely easy to get carried away and end up using far too much of it. Another problem, in my opinion, is using the *wrong* reverb on certain sounds. That might seem strange considering that reverb is just a collection of echoes made by a particular sound against the walls and surfaces of a particular space. How can one reverb be *right* and another *wrong*?

Before the days of artificial reverb effects (hardware and software) one kind of reverb was made by a sound occurring in a space. Of course, there were an infinite number of different sounds occurring in an infinite number of different spaces, but the mechanism by which the reverb was created was always the same: the reflection of sound waves in a real physical space. When we started to make artificial reverb effects, we began using different means to create them, and as a result, each had different sonic characteristics over and above the minor differences between each individual design or unit. Two early hardware methods used for creating artificial reverb were *spring reverbs* and *plate reverbs*. Their two different methods for creating reverb (or, more accurately, a reverb-type effect) gave noticeably different end results, and, as such, led them to be used primarily for different applications. Spring reverbs have become more or less synonymous with guitar amplifiers, while plate reverbs are used mainly for vocals and percussive sounds. The way each of the two methods responds to different frequencies and different dynamic changes makes them more suitable for one application than another. The usual caveat applies: just because a spring reverb is *normally* used for guitars that doesn't mean it can't be used on drums or vocals or anything else you choose.

When digital (algorithmic) reverbs came along, they further complicated things by offering up their simulations of different kinds of physical rooms ranging from large halls to medium-sized stages and rooms, all the way down to small rooms and chambers. Some also offered plate reverb simulation algorithms. Each manufacturer developed its own algorithms and, as a result, the sound of a Lexicon "hall" algorithm might be very different from the sound of a Yamaha "hall" algorithm even if both units were set up with ostensibly the same parameters. Another type of reverb is a *convolution* reverb. This type allows us to process sounds using recordings made in real rooms and spaces from the *captured* acoustics of that space. This really was a remarkable development and allowed for reverb effects with a degree of realism never previously heard, assuming that the recordings (or *impulse responses* as they are usually known) were well done in the first place.

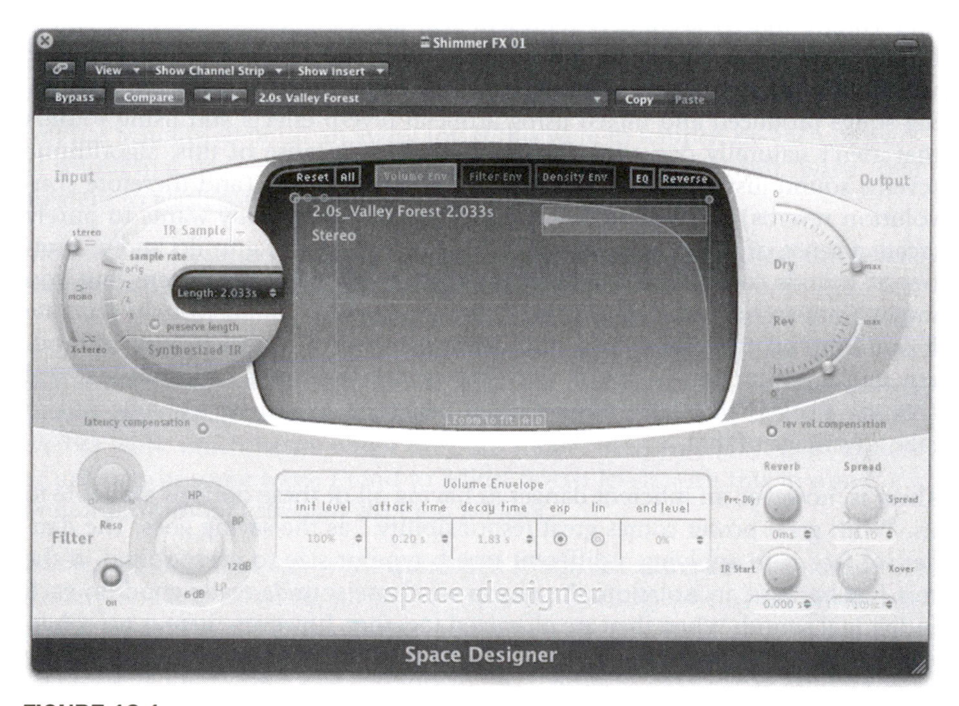

FIGURE 18.1
Space Designer is a very capable and user-friendly reverb plug-in based on the idea of convolution reverb using reverb "impulses" recorded in real acoustic spaces.

With all of these different options available to us, how can we possibly know which is the right reverb, or perhaps let's just say the *best* reverb, to use in any given situation? In order to find out, there are two factors you have to consider: the first is the length or size of the reverb and the second is the tonal character. If you want to apply reverb to a sound to make it feel like it exists in a real space while still not sounding obviously like reverb, you are probably looking for a realistic sounding reverb with a short decay time, which will end up being quite low in the mix. You might want to apply the reverb on a bus so that several different sounds or instruments can be passed through it at once giving them all the impression of existing in the same space. A convolution reverb

listen to music you may have heard many times before in a very critical and microscopic way, perhaps just focusing on one individual sound within the track or listening to just the reverb or just the overall balance of instruments, you will often hear the music in a new light and hear things you never really noticed before. I am not saying that a professionally recorded and mixed track will never have large amounts of reverb, but if they do, they will generally be used sparingly only on one or two instruments (possibly including vocals, of course).

Reverb Controls

It is quite hard to give an overview of controls on a "typical" reverb unit in the same way we did for EQs and compressors, because the different methods of generating reverb may well have quite varied parameter sets. I can think of only a single control that every different type of reverb unit has in common.

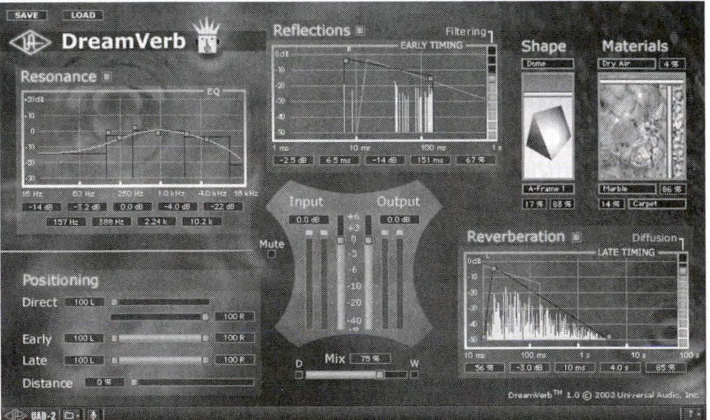

FIGURES 18.2 AND 18.3
The Sonnox Oxford Reverb (left) and Universal Audio/UAD DreamVerb (right) are both "algorithmic" reverb plug-ins. They offer different visual approaches to the same goal.

Wet/Dry Balance

Another way in which reverb can be helpful is to give our recordings *depth* in a near/far way. Panning sounds moves them left or right relative to our listening position and applying reverb will put those sounds into an acoustic space outside of our heads. Using reverb creatively and well can add a front/back dimension to this space, and allows more separation in our mix. If you imagine each sound in your mix as a person and your task is to fit them all into a small room, obviously if they are all positioned side-by-side in a straight line across the room they will be more crowded and close together than if you split them up into rows and spread those rows across the length of the room as well as spreading the members of each row across the width of the room. Using reverb allows us to push certain sounds more into the background and bring others more to the front in a simple and effective way. What's more, if we are using reverb in our track anyway then it doesn't mean using an additional effect but merely using the ones we already have in a particular way.

If you think about a real-world situation for a moment, it might help to explain how the wet/dry balance (or separate level controls for wet and dry levels) can help us achieve this depth. Imagine yourself at one end of a long tunnel or a tall stairwell. We have all been in places like that and maybe clapped our hands or shouted something to hear the long reverb that follows. Now imagine yourself standing at the end of this tunnel with a friend. Your friend shouts something and his shout causes a long reverb tail, but first and foremost, you would have heard his initial shout and then the reverb would have followed. Now imagine your friend at the other end of the long tunnel and shouting the same thing. This time you would still hear his initial shout and then the reverb, but the shout itself would seem much lower in volume compared to the reverb. This phenomenon is something we automatically associate with a sound being distant. Because of this we can use the balance between the wet and dry signals in a reverb unit to go some way toward simulating a sense of the sound being some way in the distance.

In fact, if you use a reverb as an insert effect on a "feature" sound in a track, you can, by using automation to change the wet/dry balance along with a few other parameters, create the effect that a sound is actually moving away into the distance as it is playing. This can be extremely effective in the right context. However, like changes in left/right position (and surround-sound positioning, if that applies to what you do), the generally accepted wisdom is to not get too fancy with moving sounds around because it can be quite disorientating for the listeners, especially if they are listening on headphones. It's fine to have a little movement, especially if it is a fairly slow movement, but having different sounds flying around left to right and front to back will only be distracting. Just because you *can* do something doesn't mean you *should* do it!

Reverb Decay Time

The decay time of a reverb controls the impression of how big the acoustic space is and what materials it is made of. But it isn't a control that we would find on every reverb unit. In the early days of artificial reverb there was no *reverb decay time* control, as such. If you were using spring reverb, you could change to a different sized spring to alter the decay time, but that hardly counts as a control. It was much the same with plate reverbs, because the reverb decay time was an inherent feature of a reverb plate of a particular physical size and the only way to change the decay time was to change to a different plate. And now, again, with convolution reverbs, there isn't always a reverb decay time control. The sample of the room itself will have a fixed decay time, so the only ways to physically alter the decay time of the sample is to time-stretch or compress the sample (or alter the playback *pitch*); that would result in a loss of quality and fidelity to the original room. Many convolution reverb plug-ins get around this by giving you an amplitude envelope similar to what you might find on a synth, which just controls the level of the reverb output in a dynamic way as opposed to just the static wet/dry balance. Of course, the overall level of the reverb is still controlled by that wet/dry control, but the amplitude envelope allows additional control of the amplitude *up to* the maximum level set by the wet/dry control.

It is only really in *algorithmic* reverbs that the reverb decay time becomes a truly freely adjustable parameter of the overall reverb sound. Different algorithmic reverbs offer different degrees of complexity in terms of which parameters are available for editing, but decay time is almost always one of them. Sometimes there won't be a control actually labeled "Decay Time" but there might be one simply labeled "Size," which, if you adjust it, has the effect of lengthening or shortening the decay time, perhaps along with a few other changes to the sound as well.

The combination of reverb decay time and wet/dry balance, as well as the actual reverb method or algorithm choice is what really makes up the fundamental character of the reverb sound. In addition, there are often quite a few more parameters that can further customize the sound of the reverb. I will go into a little more detail here about some of them, but please take into consideration that not all of them are included on every reverb unit.

High/Low-Frequency Damping/Rolloff

High- and low-frequency damping rates or amounts are controls that determine the change in tone of the reverb over time. Some reverbs include traditional EQs, which are (automation aside) static changes to the tone of the reverb, but damping controls allow us to change the tone of the EQ as the "tail" of the reverb decays. If the high frequency damping controls are set to a high *amount* value (or low frequency), the higher frequencies in the reverb will die away more quickly than the others. Likewise if the low-frequency damping controls are set to a high *amount* value (or high frequency in this case), the lower frequencies will die away more quickly than others. Using these controls, it is possible to customize the late reflections of the reverb in such a way that the reverb starts off sounding full-bodied and natural, and then, as it decays, becomes thin, light, and ethereal. It is also possible to set the controls up so that the late reflections become more muffled, dark, and ominous. In many ways these controls can mimic the way certain materials in a room can affect the tonal qualities of the reverb sound. We look at this more later in this chapter.

Pre-Delay Time

As reverb is basically a very large number of individual echoes arriving at our ears at different times and from different directions, if the sound source were directly in front of us and the room were sufficiently large, we would expect there to be some kind of delay between the original sound and the reverb starting. Sound travels at roughly 1080 feet (330 meters) per second, so the time between the start of the sound itself and the start of the reverb should be based on the distance the sound has to travel from its source to the nearest reflecting surface and back again. If that nearest surface were, say, 15 meters away from the sound source then the total roundtrip distance would be 30 meters, which would mean roughly a 0.1-second delay between the sound and the reverb. The greater the distance, the greater the delay. In reverb unit terms this is known as *pre-delay*. From a technical perspective, pre-delay can be very important in giving a real sense of size to larger reverb settings as it recreates a very real physical circumstance in larger rooms.

With algorithmic reverb units there doesn't necessarily have to be a direct correlation between room size and pre-delay. It is possible to use a fairly short pre-delay with a shorter reverb decay time to simulate the sound of a real room, or it is possible to use a fairly short reverb decay time with a longer pre-delay to give an effect that wouldn't occur naturally. Likewise you could use a very long reverb decay time with a very short pre-delay.

Pre-delay has another use, though. If you find that the reverb you are using is smothering the original sound a little, as well as adjusting EQ and wet/dry balance levels you can also try adjusting the pre-delay time. If this is set high enough, the pre-delay can be sufficient to put a sense of physical distance between the dry original sound and the wet reverb sound. Once again, this isn't strictly speaking a very natural sound, but it is certainly a useful way of trying to achieve a vocal sound that has clarity and presence and sounds "upfront" in a mix, while still having a noticeable sense of space attached to it. In many ways this is one of the most difficult tasks in mixing vocals well, especially in certain modern styles where the vocal sounds very dry and yet still feels like a part of the track as a whole, rather than sitting on top of it. The right choice of reverb in the first place obviously plays a large part, and getting the wet/dry balance right is also critical, but getting a good setting for pre-delay time can also help in this respect.

If your reverb plug-in doesn't have adjustable pre-delay and you are using it as a *send* effect rather than an *insert*, you can easily fake it by inserting a simple delay plug-in before the reverb on the auxiliary send bus and setting the delay output to 100% wet and 0% dry. At this point the delay time on the delay plug-in becomes the pre-delay setting on the reverb plug-in, but you might have to look around a little to find a delay plug-in that is adjustable at a fine enough resolution and one that can actually be adjusted all the way down to 0 ms delay for it to function properly as a pre-delay to the reverb.

Early/Late Reflection Balance

We have just explained that reverb as we hear it is the cumulative effect of many individual echoes. If you were to analyze the pattern of these echoes, you would see that they start out sounding like a number of closely spaced but still separate echoes that then build up into a more generalized reverb sound as the many thousands of individual echoes end up so close together that we can't distinguish them as being at separate times. In general, if two things happen less than 50 ms apart, the human ear cannot discern them as being separate events. As such, the echoes at the beginning of a reverb aren't heard in the same ways as the ones that occur later. These two different parts of the overall reverb are called *early* and *late reflections*. A good number of reverb units offer control over the balance between these two aspects of the sound. By adjusting the balance, you can vary the sound of the reverb from giving you a sense of being in some kind of room but without hearing a reverb as such (purely early reflections) to, at the opposite extreme, getting more of a sense of an expansive sound without specific localization (purely late reflections).

Reverb EQ

This is fairly self-explanatory, and will usually offer just basic EQ facilities to shape the tone of the reverb. Often this is limited to simple two-band *shelving* type EQ, but in spite of this simplicity, some very useful effects can be had. If, for example, you reduce the level of the bass frequencies and increase the level of the treble frequencies you can give the EQ an airy, almost shimmering quality, which makes the reverb sound more "open" and "near." In some ways this would be more like the sound you would get in a room with a lot of very hard and reflective surfaces. In contrast, if you do the opposite and boost the low frequencies while cutting the high frequencies, you will get a much more muted reverb sound, which can give the impression of the sound being further away. This also goes some way toward replicating the sound of a room with softer, less dense surfaces, such as a primarily wooden room or a room that has a lot of soft furnishings in it.

These EQ settings can also be used to thin out the sound of the reverb itself, to help make it less noticeable compared to the dry sound I mentioned earlier. Reducing the wet level of the reverb in comparison to the dry level will also make the reverb seem less prominent, but using EQ to achieve this has a very different sounding effect, so it is certainly worth trying if your reverb is sounding a little too dominant.

Room Shape

The shape of a real room will have an effect on the final sound and tone of the reverb, but adjusting this control, if it is present, won't have an immediately discernible effect. You may well notice a small difference in the tone of the reverb, but you most likely won't be able to immediately tell which setting has been chosen. You won't often see a room shape control on a reverb unit, and perhaps this is why. For die hard "tweakers" it is certainly a nice option to have, but few real rooms would conform to a perfect geometric shape anyway, so I wouldn't consider it an especially useful control to have.

Room Materials

Once again this isn't a particularly common control to have on a reverb unit, but in contrast to room shape, changing the materials of a room can have quite a noticeable effect, and one that is more easily identified. Harder materials such as glass, stone, marble, and tile will make the overall reverb sound brighter, more resonant, and are also likely to make the reverb sound like it is lasting longer. At the other end of the scale, softer materials like wood and fabrics tend to soak up the high frequencies more quickly, leading to a softer and warmer reverb sound. Once again, real-world experience can qualify this. The sound of the natural reverb in a large indoor swimming pool is very bright, very resonant, and sometimes almost "ringing" in its tone, whereas the reverb in a large wooden hall, perhaps a school gymnasium or a small church hall, tends to be softer and darker sounding.

Although we have already said that EQing the reverb sound can achieve a tonal diversity, changing the materials of the room does so in a more realistic way, as the rate at which low and high frequencies die away relative to each other varies with different materials. Simple EQing of the reverb output won't recreate this. An alternative to having a "materials" section on your reverb is to use the high- and low-frequency damping/rolloff controls we just spoke about. Having separate high- and low-frequency damping allows you to achieve many of the same results as changing the room materials, but you need to know a little about the acoustic properties of different materials if you want to use the damping controls to mimic a particular kind of room.

Stereo Width

The stereo width control usually works by changing the stereo width of the actual reverb output. Some of the more complex reverb plug-ins allow you to define the width of the reverb *source* within the virtual room. This can have a surprisingly big effect on the overall reverb sound. A single mono *point source* type sound in any given room will give a noticeably different reverb stereo image than two widely separated sources in the same room. Even if a "stereo width" control is available, in 99% of cases the reverb algorithm will assume that the sound source is in the middle of the room (in a left/right sense), or, if there is a "source width" control, will assume that both channels of the stereo source are equidistant from the left/right center of the room.

There is at least one plug-in in development that will take things a little further. The plug-in is called Spat and is part of the IRCAM Tools suite of plug-ins by Flux. This plug-in works by allowing you to set up traditional reverb parameters, which define the characteristics of the room itself (decay time, diffusion, EQ, etc.), and then allows multiple point sources for sounds; you can then define a two-dimensional position for each of them within the room relative to the "listener." The plug-in offers the possibility of setting a direction the sound source is pointing in, along with a *dispersion angle* to represent the directionality of the sound source. All of this adds up to a final sound that is leaps ahead of anything that has come before it in terms of creating a realistic acoustic space. Admittedly its applications lie more in audio-visual postproduction for movies than in music production, but it would certainly be a nice option to have. It is very new technology but the results are still spectacular. My only fear is that, should it become widely used, it would end up being yet another production tool that was overused simply for the sake of it.

Diffusion

Diffusion can have a significant effect on a reverb sound because it deals with the spacing between the individual echoes. Given the two-part nature of reverb and what those two parts are, diffusion tends to have more of an effect on the early reflections part of the total sound rather than the late reflections, mainly because the early reflections can be discerned as separate echoes, which means we are more likely to notice changes to the spacing between them. In some reverb effects, the diffusion control will only affect the early reflections, while in others it may affect both early and late reflections. In some others, although relatively few, there will be a separate control for late reflections diffusion.

In more general terms, the diffusion control is like a *randomize* function for the echoes that make up the reverb. Lower diffusion values mean more regular sounding echoes, while higher diffusion values seem to randomize the timing of the echoes and give a much smoother sound.

Density

The density control is fairly straightforward. Its main purpose is to adjust the spacing between the echoes that make up the reverb. If a lower density is set, there will be a greater spacing between the echoes and therefore a greater chance that the reverb will sound coarse or even resonant. Higher density values yield more closely packed echoes, which, as we have already discovered, means less likelihood that the reverb will sound like echoes and more chance that it will sound like what we typically think of as reverb.

If you use a reverb with a low density setting then you will almost certainly need to increase the diffusion amount to compensate, otherwise you could easily end up with an effect that sounds more like a delay with a high feedback amount and very short time. Conversely, as you increase the density you will probably be able to reduce the diffusion setting more should you so wish.

DELAY

Delay is a little simpler to grasp than reverb. There are, once again, many different variations on the basic theme, but these variations are more about the tonal quality of the final effect rather than fundamentally different ways of achieving it. Among the variations are mono delays, stereo delays, cross delays, ping-pong delays, multitap delays, and tape delays. Some of them are extremely simple in use and setup, and others, especially some of the more recently released ones, have a large number of adjustable parameters. They are all, basically, echo effects. As such the main parameter they are based around is the actual delay time.

FIGURE 18.4
A stereo delay such as this one (Logic's own aptly named Stereo Delay) can be extremely useful for many things, including widening the stereo image, creating interesting rhythmic effects, and even creating melodic effects.

In some of the earlier echo/delay effects, this time was set manually and often by ear, rather than having a clearly defined and accurate delay time. This was especially true of tape-based delay effects, which relied on the relative positioning of two tape heads to give the delay effect. To set up a specific delay time accurate to, say, 5 ms would have meant positioning the playback head of the delay unit to an accuracy of 1 mm or so. And that would have also assumed that the tape speed was 100% accurate and consistent as well. In fact, tape speeds often weren't that consistent, but instead of being a weakness, that came to be a strength for reasons explained in a moment.

Things got a little more controllable with the development of digital delay units, because they came with a delay time setting that was usually accurately adjustable down to the nearest millisecond, and they were also very consistent from one echo repeat to the next. Later developments, and specifically the advent of plug-ins for DAWs, saw the delay time setting become adjustable to correspond with musical note lengths (1/4 note, 1/8 note, 1/16 note, etc.), which, although it doesn't offer anything that manual delay time selection doesn't already do, can certainly make it quicker to set up rhythmic delay effects without having to resort to some kind of calculation every time.

Another control seen on a vast majority of delay effects is the feedback control. This takes the delayed output and mixes it back in with the input signal, with a varying amount. Where the output is fed back into the input, there is an echo of the echo, and this is how multiple repeats are set up. This feedback control is often expressed as either a simple *amount* type of control or as a percentage. There is no defined calibration of this control, so it really is a matter of trial and error. What is consistent is that the higher this control is set, the greater number of repeats you will get. In tape delay effects, pushing this control high enough will make the echo repeat indefinitely and push the fed back signal into overdrive. This is because, in a real hardware tape delay, once the fed back signal went beyond a certain level it would overdrive the recording heads on the tape to give a very characteristic sound; one that has become a trademark of tape delay effects. Digital delay effects, as a rule (other than tape delay emulations, of course), tend to be calibrated differently, so that a feedback control set to maximum will simply repeat the sound indefinitely but won't add any *coloration* like a tape delay effect would at extreme settings. The feedback time is also generally the same as the delay time itself, for obvious reasons, as you are feeding a sound back into the same processing loop so the feedback *time* will be the same. However, in some complex multitap delays (such as the Waves SuperTap) there is an *overall* feedback system, so that the feedback time is the same for each of the delay "taps," even if they each have different delay time settings. It seems to me that it would be far more useful to have each delay tap as its own self-contained system, so that multiple rhythmic effects could be combined. Many multitap delays *do* actually work this way, but some, as I mentioned, don't.

There may also be basic tone controls for the delayed signal in the form of basic low- and high-pass filters. These serve to thin out the sound of the delayed version of the sound, because, in some cases, if you overlap the full frequency range original sound with a full range delayed version things can start to sound a little cloudy. By thinning out either the top end or the bottom end (or both) of the delayed version, it gives some separation between the two versions and often allows you to have the delayed version a little higher in the mix should you want it. In addition, if you have any substantial amount of feedback on the delay repeats, each subsequent pass through the delay *circuit*

FIGURE 18.5
The Universal Audio RE-201 is a software recreation of arguably the most famous tape delay unit of all time, Roland's Space Echo.

will thin the sound out even more, giving the effect of a sound that echoes away into the distance in a pleasant sounding way.

If the delay is a tape delay type effect, there may well be a few additional controls to mimic the wow and flutter of a hardware tape delay. *Wow* and *flutter* are terms used to describe the inconsistencies in the speed of tape and the audible effects they have. Wow describes lower frequency "wobbles" of up to 4 Hz, while flutter describes those above 4 Hz. The combination of the two and their quasirandom nature is one of the most commonly used ways of recreating the true tape delay effect. There is much more to a real tape delay than simply that, of course, and some of the more recent tape delay effects actually go quite in-depth, attempting to model the saturation effects of recording to tape itself (a hugely complex process) before they even get to the delay part. But in a pinch, a good wow and flutter effect can go a long way toward giving the vibe of a hardware tape delay. Plug-in tape delay effects might use a single LFO to modulate the delay time or might use a combination of two separate LFOs with different frequencies, to give a less noticeable and less repetitive effect. The depth of the wow and flutter is controllable and I would recommend subtlety when using this, because too much can quickly make you feel a little seasick when you are listening, especially on headphones.

Finally there will always be some kind of wet/dry control or separate dry and wet level controls, as we would expect, and those really make up the basic controls for any delay. If we look more specifically at the stereo variants, we naturally have a few more parameters to consider. Ignoring multitap delays for a moment, we turn our attention to channel delays for now. Some have a mono input fed in parallel to each processing channel to give a stereo output; others have a true stereo input with each side of the stereo input processed by the relevant channel of the stereo delay effect. Each side of the stereo delay normally has similar controls to a single mono delay, with the possible addition of a *crossfeed* control and a *feedback* control. Crossfeed works in a very similar way to feedback except that, rather than feeding the output of the delay back into itself, it feeds it back into the input of the opposite channel. This allows the possibility of some very interesting rhythmic effects if quite different delay times are set for left and right channels. Caution should be exercised because it is easy to get into a sonic mess very quickly if high feedback and crossfeed levels are set for both channels.

An additional variation on the stereo delay theme is the so-called *ping-pong* delay. This has similar controls to a mono delay (in that it only has one set of controls) but makes it a stereo delay by panning alternate repeats hard left and then right so that the delayed sound appears to bounce from left to right and back again.

Ping-pong delays, and stereo delays in general, are especially useful for adding width to a sound. The perceived width of a single sound can be manipulated by differences between the left and right channels. A purely mono signal, when panned centrally and played back on a stereo system, will sound identical in each speaker, whereas a stereo pad sound might be fundamentally the same in both channels but with small differences and movements in each channel. If we use a stereo delay subtly, we can introduce further variation between the two channels by setting up different delay times for each channel, which give the ear the sense of there being interesting things going on on each side. Of course, there is a fine line between the effect being subliminal and it being an obvious echo, but even if it is loud enough to be perceived separately, it does still help to widen out the sound in question.

Moving on from stereo and ping-pong delays, let's look at multitap delays. These are delay units that have an increased number of taps (anywhere from three upwards), with each tap having its own configurable controls for delay time and stereo panning position. Each tap may also have its own feedback control and tone controls, but it is possible that either or both of those exist as a "global" setting, which applies to the overall effect. With multitap delays like these, especially those with six or eight or even more separate taps, it is possible that you can set up delay effects with huge rhythmic interest and have the ability to turn a single short, percussive sound into a rhythm pattern of its own. If you add to this the fact that each separate tap can have its own position in the stereo field, you can see how some really intricate effects are possible. On the downside, once again we are in that increasingly familiar territory of thinking that this could be a little too much. Like rapid and regular panning position changes, regular and quick changes of front to back depth in the

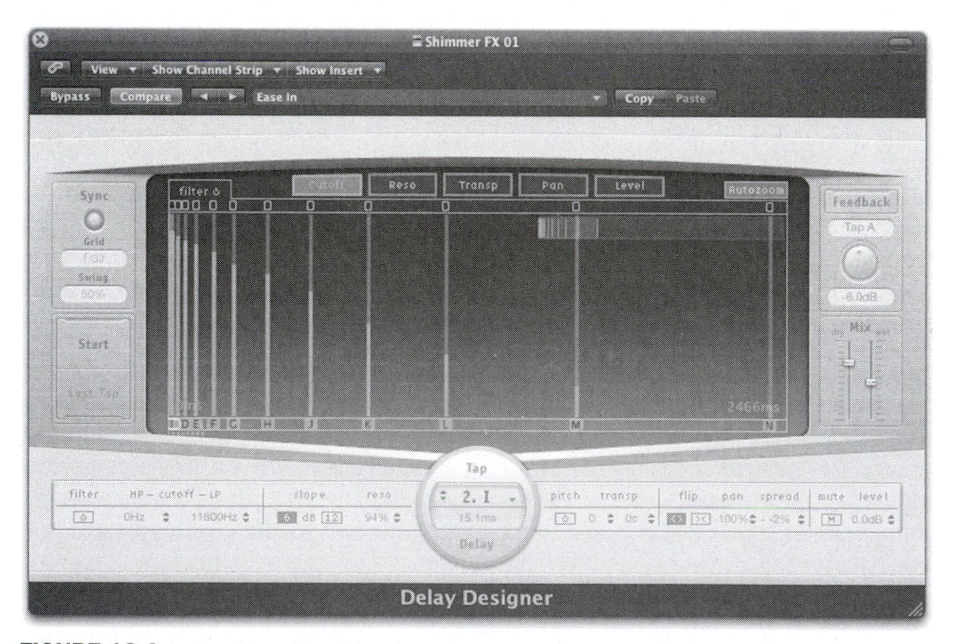

FIGURE 18.6
Logic's Delay Designer represents a new breed of delay units that goes way beyond simple echoes and pushes the boundaries of what is possible with delay effects.

mix, and like anything else that would affect the perceived position of a sound, the ability of multitap delays to add a lot of positional movement to a sound within a track should be used sparingly and with caution. It *is* a great effect and can seem to add a lot of space and width, but the overall effect can be a bit too much if, for example, you had a similar effect going on with a few different sounds at once and if that also ran through most of the song. After a while it would get tiresome and, worse, possibly even disorientating. It is certainly an effect that can be very good, but it should be used subtly if you really want to get the best possible results out of it.

Finally we come to the latest generation of delay effects. In some ways I am reluctant to call them delay effects because they often include so much more than just delay and are capable of things that go way beyond what any other delay effect can achieve. But as they are, fundamentally, based around delay effects, with additional capabilities added on, I guess that's how they should be classified. Perhaps the best example of the kind of delay plug-in is the excellent, if a little overwhelming at times, Delay Designer included with Logic. On the surface, and at the most *basic* (I use that word loosely) level, Delay Designer is a multitap delay unit with a maximum of 21 separately configurable taps. If you were *only* considering this functionality, that in itself is quite impressive, because I can't think of many, if any, other delay effects in either hardware or software form that offer this many separate delay taps.

For each delay tap there are controls for the delay time, filter cutoff frequency (selectable, per tap as either high-pass or low-pass filtering), filter resonance, level (there is also an overall effect level as well, but these level controls allow for delay fade in or fade out effects), stereo panning position, and, uniquely, transpose. Yes, in Delay Designer each tap also has its own attached pitch shifter! This really does take Delay Designer way beyond the realms of just being a delay effect. Using the transpose options, it is easily possible to turn a single note into a complex arpeggio with elaborate panning and filtering. The possibilities with just this one plug-in could take you a very long time to explore. In truth, for most typical delay effects, I would normally choose something much simpler in functionality, but on those occasions when I want something more intricate and interesting, Delay Designer is a great choice.

In the future, we'll probably see some different combinations of existing technologies in delay units. We could see a multitap tape delay that combined the sonics of a tape delay emulation complete with saturation simulation and wow and flutter effects in a highly programmable and complex framework similar to Delay Designer, although I wonder how many people would really find that useful. After all, a tape delay effect can be quite distinct within a mix, so a multitap version might be a little overwhelming. It might be nice to have the option, however, even if it was one that wasn't used that often. At the very least I would like to see a true stereo version of a tape delay plug-in, rather than having to go the long way around and "make" one by some elaborate routing within Logic and the use of two separate submix busses. All this to achieve an

the delay time it provides (*depth*), as well as a control for the base delay time (*delay*), and the obligatory wet/dry mix or level controls. The chorus effect works because of a fundamental property of delay effects. Once the original signal is in the delay effect processing, the playback speed can be changed by altering the delay time. If the delay time is sped up, the pitch rises; if it is slowed down, the pitch drops. Of course we are talking about very small variations in pitch (just a few cents really) but the effect can be quite dramatic. By modulating the delay time we get a copy of the original sound sweeping up and down in pitch relative to the original. This is the classic chorus effect.

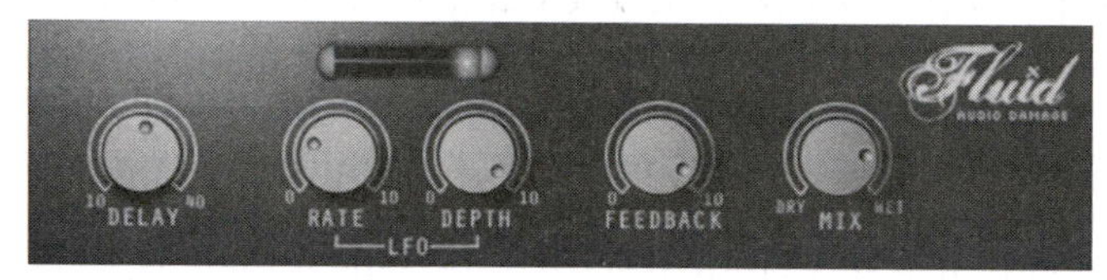

FIGURE 18.8
Audio Damage Fluid is an especially smooth sounding chorus effect.

There might also be a control for the number of *voices* of the chorus. If this is present, it will usually have the same basic parameters, with the different voices being alternate delay channels modulated by the same LFO, and taking their modulation value from different points on the LFO waveform, therefore giving each voice a different amount of variation from the basic pitch. Stereo chorus effects usually split these voices between the left and right channels to give a wider and more shimmering effect.

FLANGING

Flanging is also based on delay effects, but the application of them and the resulting effects are very different. The effect originated from reel-to-reel tape machines back in the 1960s, and was a result of tiny variations in the playback speeds of multiple tape-machines. If the same signal was recorded onto two different tape machines at the same time and then those machines played back at the same time, the differences in playback speed would create a sweeping effect on the sound in question. This happens because when two identical copies of a sound are played back simultaneously and one is delayed by a *very* short amount of time (just a few milliseconds), there is a complex interaction between the two sounds, which causes some frequencies to be cancelled out completely and others to be reinforced. This effect is known as comb filtering because of the deep "notches" it creates in the frequency spectrum. On its own this doesn't create what you and I would call flanging, but when that delay time is modulated, the varying difference in phase between the two signals causes those notches to move around. And if the variation in phase occurs in a cyclic manner then so too does the movement in the notches.

FIGURE 18.9
u-he's Uhbik-F is one of the more elaborate flanger plug-ins available. The added complexity gives a much broader scope of possible applications.

There are many similarities between chorus and flanger effects in terms of their controls and parameters. So much so, in fact, that you might wonder how they sound so different. The fundamental difference between chorus and flanging effects lies in the actual delay times used. While a typical chorus effect will have delay times of between 10 and 30 ms, a typical flanger will have delay times of between 1 and 10 ms. It might not seem like a lot, but that difference in delay time leads to the difference in effect between chorus and flanging. As a result, all of the main controls are similar. You will usually find controls for the base delay time ("delay"), controls for the depth speed and depth of the LFO modulation applied ("rate" and "depth"), and the wet/dry mix, and you might also see a control labeled "feedback." This takes the output from the flanging process and feeds it back into the input of the effect. Often this control has a +/−100% range allowing you to either *add* the flanged signal back in or *subtract* it. The differences between the two can be quite noticeable, but in either case, adding feedback in this way emphasizes the flanging effect, and can, at more extreme settings, result in a resonating and metallic "drone" effect. It is worth noting that flanging effects generally work better with lower "rate" or "speed" settings than chorus effects. Often the range of values for the rate will be similar in a flanger to a chorus, but to get that classic jet flanging effect you would probably end up using a slower rate than you typically would with a chorus effect.

Finally, as with chorus effects, stereo flanging effects usually rely on simply modulating two different delay channels with the same LFO, but at different points on the LFO phase. This gives a "wider" effect as a result of the flanging effect being at different points on the cycle in each channel.

PHASING

Phasing shares a lot of sonic ground with flanging, but it is achieved in a very different way. Where a flanger uses a modulated delay line to create the effect, a phaser uses something called an *all-pass filter*. Based on what we know of high- and low-pass filters from Chapters 10, we would probably deduce that an all-pass filter lets *all* frequencies pass through unaffected. And that is actually correct, at least in the regard that the frequency response of the sound doesn't change. What *does* change, however, is the phase of certain frequencies.

Phasers are often described as having a different number of *stages*, and these are usually a multiple of two because each two stages in the phasing process create one frequency band in the all-pass filtered signal, which has an inverted phase. When this all-pass filtered signal is mixed with the original, the effect is to create a deep notch in the frequency response where the inverted phase signal is mixed with the noninverted phase version. This isn't especially useful to us, but once we apply the LFO modulation we start to hear the expected effect. The more stages we add to the phasing process, the more notches we get, and, importantly, the spacing of the notches is different between flangers and phasers. This accounts for much of the different tonality of the two, essentially very similar (modulated comb-filtering) processes. In a flanger the notches are all equally spaced (100 Hz, 200 Hz, 300 Hz, 400 Hz, etc.) whereas in a phaser the notches are harmonically related (100 Hz, 200 Hz, 400 Hz, 800 Hz, etc.).

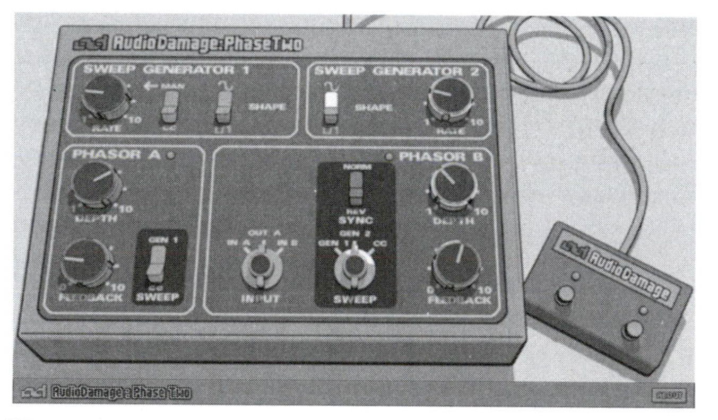

FIGURE 18.10
A phaser such as Audio Damage PhaseTwo (pictured here) has a similar sound to a flanger but there are definite differences.

In terms of controls, we find a lot of familiar ground here, but also some new things. The modulation controls will be familiar, but many flangers offer a slight variation on the theme by offering the option to control the *sweep* of the LFO modulation within a particular frequency range. This means the effect can be confined to a subtle swish in high frequencies, for example, rather than a big, dramatic sweep across the full frequency range of the sound. Instead of there being a delay time largely in control (in the case of the flanger) of the tone, here we find a control for the number of *stages* as the fundamental tonal control over the phaser. There may, if it is a stereo phaser, also be some kind of "stereo width" control. In addition there is a "feedback" control, as we might find in a flanger, but it works in a different way in a phaser.

ENSEMBLE

So-called ensemble effects can vary a little in how they achieve the effect, but the aim is to go a little further than a chorus effect. We have seen that a chorus aims to replicate the sound of another "player" who has very subtle differences in pitching and timing playing an identical part. If you think of this as a duet, it is easy to imagine how such ensemble effects might be created. If you were to have a number of parallel chorus effects, it would give the effect of a number of different players doubling up the part. In its simplest form, that is what an ensemble effect is. As such, you often have controls for the number of *voices*, which represents the number of simultaneous chorus effects happening. However, if all of the chorus effects used exactly the same modulation source, the sound wouldn't really be any different from a single chorus effect, as the differences between the original and the processed sounds would be the same in every voice. To get around this without having to provide separate LFOs and associated controls for each voice (which would, in my opinion, be a nice option to have), what usually happens is that the same LFO is used but the actual modulation amount is taken from a different point in the LFO cycle. For example, if the LFO were a sine wave shape, the first voice might be taking its modulation value from the very beginning of the cycle, the second from 90 degrees out of phase, the third from 180 degrees out of phase, and the fourth from 270 degrees out of phase. As a result, each separate voice would be modulating at the same rate but would all be at different points in their modulation cycle at any given time.

FIGURE 18.11
Ensemble effects such as this one (Logic's built in Ensemble) have a way of really thickening up sounds, but as with all "modulation" effects, the best results are often achieved when used subtly.

In addition to this, the actual modulation source is sometimes made slightly more interesting by using the combined output of multiple LFOs. If there is more than one LFO (assuming they are at different speeds), the resulting waveform will be much more complex and random sounding than a repeating sine wave. In the real world, this would translate to a sound that was more realistic in the sense that multiple "players" or "voices" in a vocal ensemble would not have timing and pitching differences that were clearly cyclic in nature. By having a waveform that is cyclic, but in a much more complex way than a simple sine wave, we can preserve much of the character of chorus effects and give it even more of a shimmering effect.

VIBRATO/TREMOLO/AUTOPAN

Often the terms *vibrato* and *tremolo* are used incorrectly as people mistake one for the other. Technically, vibrato is purely pitch modulation and tremolo is purely volume (or amplitude) modulation. What is generally called vibrato in a singer's voice is actually a combination of vibrato and tremolo as both pitch and amplitude variations take place simultaneously. In the studio it is easy to separate the two, as they can be processed with individual effects. That said it is pretty rare to get a vibrato plug-in on its own. Vibrato effects within a synth, however, are *very* common because all you have to do is route an LFO to control the pitch of the oscillators and there you have it: vibrato!

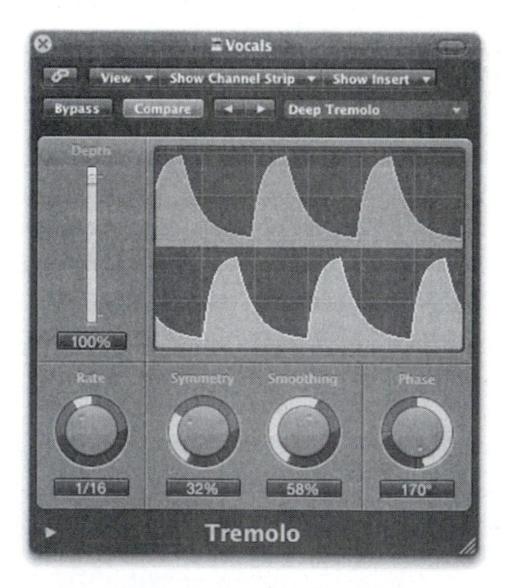

FIGURE 18.12
A tremolo effect like the one included with Logic (pictured here), is extremely useful in many ways, despite being deceptively simple. For that reason, it is often overlooked in favor of something more "exotic."

Tremolo effects, as audio plug-ins, are much more common and are fairly simple to figure out. As with all modulation effects, there is an LFO at the heart of it, which in this case provides amplitude modulation, as we have already discovered. Beyond the usual rate/speed and depth controls (which can often sync to your DAW tempo), there may be a shape control, or something similar, which will soften the edges of the tremolo effect from being much like a square wave (in that the input is either on or off after processing) to something more gentle and closer to a triangle, sine, or sawtooth wave. This could be called "shape," or something similar, but remember that it won't always be present and the only choice you might have is the square wave shape. If the effect is a stereo one, there might also be a phase control that allows a phase offset between the left and right channels.

If the two channels on a tremolo unit are 180 degrees out of phase, the sound will move from one channel to the other, which leads us to autopan effects. I often use the built-in Logic Tremolo for exactly this purpose. I set the rate to give me the required cycle time, set up the phase to be 180 degrees different between the two channels, change the shape control to be as smooth as I can make it, and then adjust the depth control. Using a tremolo in this way means that, at 100% depth, it is a very *hard* panning effect. In other words, at the extremes of the LFO cycle, the sound will, momentarily, only be coming out of one channel or the other, whereas if you reduce the depth slightly, then at no point will the sound *only* be coming out of one channel. This results in a slightly softer effect, especially when listening on headphones.

ROTARY SPEAKER/LESLIE SIMULATIONS

A rotary speaker effect is an emulation of a real-world electro-mechanical effect that has some complex properties. It is much more than simply applying a short delay to an input signal. A convincing emulation of a rotary speaker (also known as a "Leslie," after its inventor, Donald Leslie) has been sought after for quite some time, partly because a real Leslie speaker isn't the most portable thing in the world. The Leslie is large, heavy, and cumbersome, but for many,

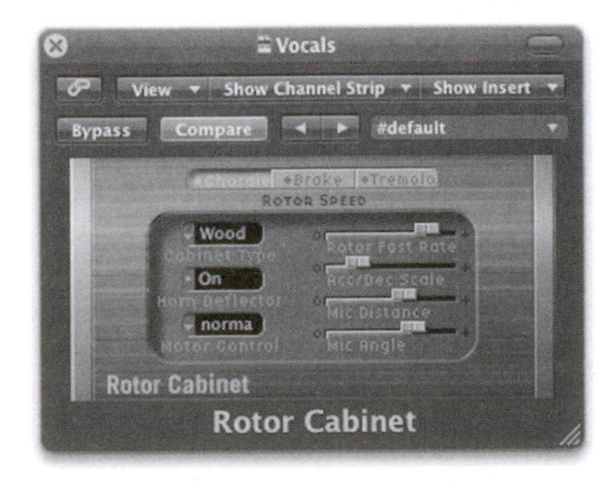

FIGURE 18.13
Logic's Rotor Cabinet is one of relatively few dedicated rotary speaker plug-ins on the market today. The sound they have is unique and surprisingly flexible if you are prepared to experiment.

it has a sound that is absolutely irreplaceable. Obviously an accurate emulation would make life much easier. The main use for Leslie speakers has been on electric organs. The legendary Hammond Organ and Leslie speaker combination has graced thousands of hit records over the last 50 or more years. Even though that is the most common use for them, Leslie speakers have also been used on any number of other instruments including electric pianos, guitars, and even vocals. The distinctive tones they produce combine elements of tremolo, vibrato, auto-panning, and chorus to give an effect that is quite unlike anything else. It might help to give you a better idea of what exactly is going on in the sound of a Leslie if I explain a little about the mechanics of how it works.

Inside the speaker cabinet are two separate speakers, one bass/midrange speaker and one horn tweeter. Some models only use a single full-range speaker, but these aren't considered proper Leslie speakers, and if anything, are easier to emulate as they only require half of the process of a full Leslie emulation. Based on that alone, we have to split the incoming signal into two frequency bands and process each separately. Each of the two speakers has a rotation mechanism. For the bass speaker, this consists of a fixed speaker that directs its sound into a drum that has a vent on one side. This drum rotates, so that the sound leaking out of the vent seems to rotate. The most likely reason why a rotating drum was used rather than actually rotating the speaker is that a rotating speaker of the size used in a Leslie cabinet would have required a lot of effort to get it rotating and then, because of the large amount of momentum it would obtain, would take quite a long time to stop again. Not so in the case of the treble horn speaker. Tweeters are inherently much smaller than bass or midrange speakers and, even with the addition of the horn, it still has a small enough mass to be fairly easy to accelerate into and decelerate out of rotation. The horn in a Leslie cabinet actually has two horns fixed in an opposing manner, but the sound only comes from one of them. The other is used purely to balance the whole assembly when it is rotating. So what actually happens to the sound waves coming out of a rotating speaker?

There are a number of things happening at once. The most obvious is that the sound appears to move from left to right and then right to left again as the speaker rotates in front of you. So the first thing we need to emulate is auto-panning. For mathematical reasons that I won't go into here, if we want to simulate a circular rotation of a speaker in a purely one-dimensional aspect (left to right only), we have to use sine wave modulation. If we add an auto-pan effect (or tremolo) to a sound and use a sine wave to modulate the panning, we have taken care of one aspect. Of course, we wouldn't want to use a panning depth that was too strong because the panning of the sound in front of us wouldn't be "hard" left/right. So there is an element of subtlety to this depth setting. The closer we put our head to the speaker, the stronger the panning would be. But unless you actually had your head *inside* the speaker, you would never get that "hard" panning effect.

Let's consider that same rotational movement we just converted into left/right panning, but shift our viewpoint by 90 degrees. As well as left to right, the sound also appears to be moving front to back. The same principle applies in that the circular rotation of the speaker would need a sine wave modulation to convert that to a one-dimensional movement (in this instance front/back rather than left/right), but that modulation would have to adjust the amplitude of the sound instead of the panning position. It doesn't require in-depth knowledge of physics to work out that sounds get quieter the further away you get. Therefore, we need to apply some modulation to the amplitude of the sound, making it quieter when the "virtual" speaker is at the back of the circle farthest from us and louder when it is at the front of the circle. The actual fluctuations in amplitude caused by this effect alone would be minimal, so we need to factor in the direction of the sound at the different points. Again calling on experience, let's consider a person who is a short distance from us, facing away from us and is shouting into a megaphone. If, while he continued shouting into the megaphone, then turned to face us, the sound would get louder even though the distance hadn't changed. This is also a factor that only serves to increase the changes in amplitude throughout the rotation of the speaker. That's the amplitude modulation (or tremolo) taken care of.

Next we consider the speed of the sound itself. Because of something called the *Doppler effect*, when an object emitting sound is moving toward us, the apparent pitch of the sound increases; when it travels away from us, the pitch of the sound seems to decrease. Again we don't really need to look deeply into the science behind it, as we have all heard this effect when a car or motorbike traveling quite fast passes us and the pitch of the sound of the engine seems to drop as it passes. This also happens with a Leslie speaker. As the speaker rotates it is alternately accelerating toward us (raising the pitch), traveling across our stereo field (pitch static), traveling away from us (pitch descending), and then traveling across again (pitch static). Of course, there is a smooth transition between these four states as the motion is circular rather than square, but the end result is a cyclic rise and fall in pitch, which is, you guessed it, vibrato. And once again we would use a sine wave to replicate the two-dimensional motion into a one-dimensional modulation value.

The one good thing here is that, because all three of the attributes stem from the same source (the rotation), they are all being modulated at exactly the same rate; only the phase relationship is different. When the left to right movement is momentarily stationary on the left side of the circle, the front to back movement will be at its greatest rate and the pitch will be dropping at its greatest rate. In contrast, when the front to back movement has momentarily stopped at the back of the circle, the pitch will also be static but the left to right movement will be at its highest rate. As a result, the front/back modulation and pitch modulation will be in phase with each other (highest amplitude and highest pitch occurring at the same time), but the left/right panning is exactly 90 degrees out of phase with that. It complicates matters a little, but not too much.

These three factors together contribute much of the character of the Leslie sound, but there are two separate speakers to consider. Each speaker rotates in the opposite direction (one clockwise and one anticlockwise), and at different speeds. This will require the use of two distinct systems to emulate correctly, but both of those systems would be identical functionally just with different LFO settings and different input signals. Both of those signals then need to be combined back to give the finished Leslie sound. Once this is done, you have, on a very rudimentary level, completed your Leslie emulation. The reality is, of course, very different because the actual speakers themselves (with their differing frequency responses) will tonally change the input signal. The same is true of the amplifier driving the speakers, and it is especially true of the actual physical cabinet the whole assembly is in. The size, shape, and materials used all have their own influence on the tonal resonances of the cabinet, and this has a bearing on the final sound. You could also consider the fact that, within the partially closed system of the speaker cabinet there are lots of sound waves bouncing around, and therefore lots of complicated phasing issues to consider as well. This is probably why it has taken so long for digital technology to even come close to an authentic Leslie simulation, and why even the best Leslie emulations still don't seem to capture the full essence of what a Leslie cabinet is about. Don't get me wrong, they are good, they are *damn good* in fact, but there is just something missing. I have actually achieved what I believe are my most realistic Leslie simulations by taking the output of my rotary speaker plug-in and feeding it through a convolution reverb set on quite a small "room" setting, just to put the Leslie effect into a *space*, rather than it existing in a full-width stereo image.

Because it is emulating a very simple (mechanically speaking) physical device, the controls on a typical Leslie or rotary speaker effect generally mimic those on a real Leslie speaker, and as such are limited to speed controls for each of the two speakers (selectable between off, slow, and fast) and then possibly a control that affects the acceleration and determines how quickly the change happens between each of these speed changes. In a real Leslie the changes don't happen instantly because it takes time for each of the speakers to get up to speed or to slow from one speed to a slower one. There may be optional controls for *drive level* to simulate the sound of really pushing the amplifier and speaker hard and to the brink of distortion, and you might even find controls for virtual microphones, which allow you to alter the width of the virtual stereo microphones to give a stronger or a more subtle effect.

I am sure there will be small improvements on the Leslie simulations in the future, and I look forward to those because I really do think it is an underrated effect. The uses on electric organs are extremely well-documented and are a part of musical history. Even its use on electric pianos, other live instruments, and vocals is quite well-established. But it can be used on any sound. Adding a Leslie effect to a synth pad and choosing the slow speed setting can give an incredibly lush sound, similar to chorus, similar to flanging or phasing, but different somehow. Adding a Leslie to a synth lead sound with the high speed setting can really help to soften the edges up, while still retaining a lot of clarity and cut in the sound. Sometimes lead sounds can be hard to mix, because, on the one hand, they need to be heard, but on the other, you often don't want them to be harsh or abrasive. By using a Leslie effect you can make the sound different enough from anything else in the track that it will stand out anyway without you having to turn it into a chainsaw of a sound!

RING MODULATION

In Chapter 10 we looked at FM synthesis, which uses a combination of two waveforms to create various *sideband* frequencies related to the two original ones used. Ring modulation uses a similar technique, but this time modulating the amplitude of one waveform with the other, rather than modulating the frequency. The result of this modulation is two frequencies, one the sum of the two input frequencies and one the difference between the two. If, for example, we had a waveform of 500 Hz being modulated by one of 80 Hz, the output from the ring modulator would be one waveform of 420 Hz and one of 580 Hz. This might seem familiar to those of you with a keen eye as being the same as the "C + M" and "C-M" sideband frequencies produced by FM synthesis. The key difference is that, while FM synthesis produces a series of these sideband frequencies, ring modulation only produces the two we mentioned, and the original two frequencies are (normally) not present.

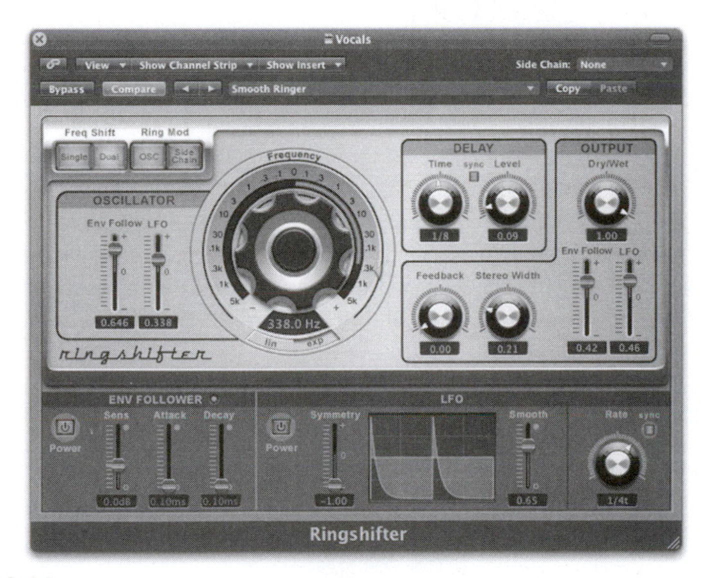

FIGURE 18.14
Ring modulation plug-ins such as Logic's RingShifter (pictured here), aren't for the faint-hearted, but they do offer fantastic possibilities for experimental sound creation.

As you might expect, the results have much in common with FM synthesis, in that the output often has metallic, ringing overtones, and a generally inharmonic feel to it. But that's not to say it *has to* sound that way. If the two frequencies are carefully chosen, the resulting *sum and difference* output from a ring modulator can be more "musical" and less harsh, but it certainly isn't something that is easily controllable. Speaking of the controls, there really is very little to control. If the ring modulator has an internal oscillator to use as the modulation source, you will have a control to set the frequency of that oscillator (and perhaps the phase of it) and that's really all there is to it. The other option is to have a modulation source input, which would use a different signal as the modulation source. More often than not these days, that would take the form of a sidechain input in the plug-in. There are other options that involve having one channel of a stereo input as the *carrier* wave and the other as the *modulator* wave. I think that the sidechain input is far better; it allows stereo signals to be ring modulated and is far easier to set up.

Ring modulation is a great effect for the more experimental programmer, but it certainly isn't something I often think of using. There is, perhaps, a better implementation of ring modulation to be found. One of the problems I have with it is that you can find a really great sound by adjusting the modulator frequency while holding down a certain note, but when that note changes, the relationship between the carrier and the modulator frequency changes, so whatever harmonics were being generated have now changed. If the pitch of the modulator inside the plug-in could track MIDI notes so that, once you had found a good position for it relative to an input note, the final frequency of the modulator could then be "played" by a MIDI sequence, that for me would be far more useful and would extend the sound palette from being more about metallic and clanging sound effects to more of a sound design tool.

FREQUENCY SHIFTER

A frequency shifter does exactly what its name suggests: it shifts frequencies. The uniqueness of the sound of a frequency shifter comes from the fact that it shifts all of the different harmonics in a sound by the same numerical amount, rather than by the same ratio. As you probably remember, harmonic frequencies are

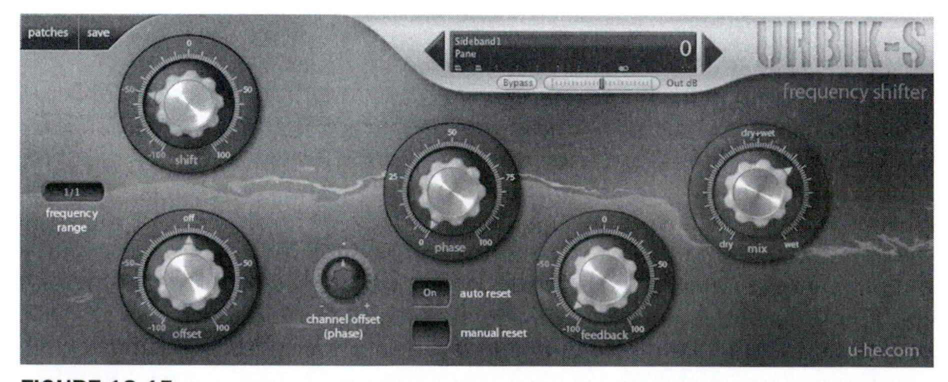

FIGURE 18.15
Under some circumstances a frequency shifter such as this u-he Uhbik-S can sound very much like a flanger or phaser. At other times the sound can be radically different.

worked out as multiples of a root or fundamental frequency. A fundamental frequency of 100 Hz would have harmonics at 200 Hz, 300 Hz, 400 Hz, 500 Hz, and so on, and if we were to increase that fundamental frequency to 200 Hz, the harmonic series would then change to 400 Hz, 600 Hz, 800 Hz, 1000 Hz, and so on. But if we took our first sequence and ran that through a frequency shifter with a shift value of 100 Hz, the resulting series would end up as 200 Hz, 300 Hz, 400 Hz, 500 Hz, 600 Hz, and so on. On its own, this would produce a very inharmonic sound, which could possibly be what you are looking for. On the other hand, there are various ways in which we can use this generated sound.

If you look back to what we said about phasers, you might remember that a phaser works using comb filtering and a series of evenly spaced (numerically speaking) notches in the frequency spectrum. What we described as a possible output from a frequency shifter is a series of evenly spaced (numerically speaking) "spikes" we could use, if we subtracted them from the original sound, to create a similar effect to comb filtering. Perhaps not unexpectedly, the output from a frequency shifter *can* sound very much like that from a phaser. In many ways, though, frequency shifters are, like ring modulation, especially adept at creating metallic, resonating, and inharmonic sounds and effects.

SUMMARY

As I stated at the beginning of this chapter, time domain and modulation effects are much more like fairy dust we sprinkle over the sounds in our mix. Hopefully, after reading this chapter, you have a good idea of what they all do and why we might want to use them. More importantly, I hope that you have an appreciation of why they should be used in moderation. In general, I like to use reverb effects to help give a sense of front-to-back depth in a mix, as well as providing specific, more dramatic effects on certain sounds. I use delay effects mostly to provide additional rhythmic interest and occasionally as *resonators* for special effects.

When I use modulation effects, I try to make sure I use them subtly and don't use the more dramatic effects on too many different sounds at once. If I do, they actually tend to lose their effect as they become lost in a swooshing and swirling fog. Chorus and ensemble effects are the least obtrusive in this respect, and flanger and phaser effects are the most obtrusive. I have no objection to any of them, and, in fact, I really like the effect a well-placed flanger or phaser can have. I often use flanger or phaser effects on the whole mix (or at least substantial parts of it) as a means of building tension coming out of a drop, and they can be really effective as a final means of adding that final lift. Like so many things, however, if you use them too much or too often they start to lose a little bit of that magic.

A surprising thing about these effects is the difference when you remove them all. As an experiment, when you have completed a mix, try bypassing all of the reverb, delay, and modulation effects and then listen to how much smaller, narrower, and shallower (front to back) the mix sounds. Even in some of the current, quite dry-sounding production styles, there is still a lot you can do with subtle reverb and delay effects that give depth and separation while still maintaining a very upfront and dry sound.

In the next chapter we look at some more common effects you might need or want to use in your remixes, before moving on to the last chapter in this section, where we look at the mastering process.

CHAPTER 19

Mixing: Other Effects

INTRODUCTION

A full look at all the different effects plug-ins (and hardware units) available would take up this whole book, given that effects developers have an ever-increasing range of effect ideas to empty our bank accounts on. Instead, I will focus on just the more common and useful ones. In this chapter, to conclude our look at effects, we consider distortion and amp-simulation effects, saturation type effects, and filter effects.

DISTORTION AND AMP SIMULATION

There are a surprising number of variations within this category, ranging from guitar pedal-like distortion and overdrive, waveshaping and bitcrushing effects, and full-blown amp simulation. Each has its place and its own merits, and they all have contexts in which they work best. For example, while an amp model might be the most accurate representation of a real guitar amp distortion sound, there are times when a slightly more harsh and abrasive effect might work better in your mix. In nearly every aspect of music production there isn't a *best* distortion effect, just like there isn't a *best* reverb or a *best* synth. What is best in each of those categories is the piece of equipment (hardware or software) that enables you to get the sound you can hear in your head out into the real world. That could be a top-of-the range Lexicon reverb or a crappy sounding cheap reverb box from the 1980s. It all depends on context.

In terms of the different kinds of distortion available, I will make a small departure from my theme up until now of mainly software-based effects and instead start by talking about hardware. With distortion effects I definitely recommend you at least take a look at a few cheap and easily available guitar pedals. There are a large number available, mostly at reasonable prices (even when new), and they have been developed over at least four decades, so there is a huge amount of experience in the designs. The effects range from subtle overdrives through heavier distortions and on to "metal" and "fuzz" effects. Of course they were designed for guitars, but they can do wonders for beefing up the sounds coming from a synth, and let's not forget their use for vocals as well. A copy of a lead vocal fed through some kind of distortion effect and then layered subtly underneath the main (clean) vocal can give a lot of attitude, and if balanced right, an increased sense of cut and clarity, because the distorted version, by its very nature, will contain significantly more high-frequency content. You simply can't get the same effect by boosting the high frequencies on an EQ because you are only boosting what is already there. When you distort a sound you are *creating* new harmonic information, which means you have more there to work with in the first place.

There is also a lot to be said for the hands-on approach of using these pedals. There are generally only a few controls available on them, so noting down the settings for future recall of the track shouldn't prove too much of an arduous task. Of course, it's not the 100% immediate total recall we get with a purely software based system. I am actually starting to incorporate more and more hardware into my studio, just because the whole experience of creating music *feels* different. Don't worry if that's really not your thing, because there are a large number of distortion effects available in plug-in form. They range from generic overdrive and distortion effects to ones modeled on specific classic guitar pedals. And the range of tones available from the plug-in effects is just as wide as those available in the hardware ones.

In general, I would describe distortion and overdrive effects as being quite raw and brash sounding. Even the more subtle overdrive effects still have an edge to them, which is very different from the sound of a mildly overdriven guitar amp. There is a very good reason for this. When you overdrive or distort a sound, you are creating a lot of additional harmonic information, and with the exception of some kind of tone control, there is little you can do within the plug-in to control this. In a guitar amp, however, the distortion created by the amp (or, indeed, any distortion pedals used on the way into the amp) is controlled a little by the frequency response of the speaker cabinet that is connected to the amp. This combination of amp and speaker does a lot to minimize the harshness of the distortion and provide some warmth instead. As I stated earlier, it's all about context, so the brashness and raw edge of a pure distortion effect might be just what you need.

On the other hand you might be looking for something with *exactly* that warmth, depth, and smoothness. That's where amp modeling comes in. There have been great advances in amp modeling technology in recent years, bringing it to the point where the better amp modeling effects provide a sound that is pretty indistinguishable from the real thing. Some will argue that the experience of playing a guitar through an amp modeling plug-in or hardware box doesn't give the response of a real amp, and there may be some truth in that. But you really do need to be a pretty good player to be able to appreciate, and make use of, those kind of subtleties. If this book were written for gigging and recording guitarists, I might look at these plug-ins in a different light. However, it is written for the remixer, who is, I would imagine, less likely to be a very experienced guitarist, so I have tried to look at these amp modeling plug-ins in the context a remixer might use them. As such, the increasingly minor differences between real amps and plug-ins are of less interest to us. What *is* important is that they can give us 90% of the vibe of a real guitar amp sitting in the corner of a room and running at outrageous volumes.

The process of modeling a guitar amp successfully is very complex. The signal goes through various stages in the amplifier (preamp stage, tone controls, and power amp stage), then there's the effect of the speaker cabinet, the effect of the sound of the room on the overall sound, and finally, the contribution of any microphones used to record that sound. Each of these aspects is an elaborate process, which, if done properly, involves modeling of some complicated *nonlinearities* (where the amount or even type of change to the sound varies depending on the level of the input signal) as well as dealing with physics and acoustics. It is no surprise that top-quality amp modeling has only really become a possibility in recent years. Early attempts sounded pretty good considering the technology available at the time, but they lacked these nonlinearities, and as a result, seemed to sound "flat" in comparison to a real guitar amp that felt alive and responsive. The knowledge of how to model these unique properties probably came along at least a little while before the technology was readily available to do it, at least in a real-time way. Fortunately we have reached that point, and now we have some fantastic software available to us, which gives the modern producer the ability to try out many different amp and speaker combinations and even use different *virtual* microphones.

There are three manufacturers' amp modeling plug-ins I use on a regular basis all capable of very impressive results. These are Amp Designer and Pedalboard, included with Logic; Native Instruments' Guitar Rig (I have version 4); and IK Multimedia's AmpliTube 3 and X-GEAR. All offer a variety of different amp models, some officially and some unofficially modeled on well-known real-world amps. If you're not really a guitarist or don't have any experience using or recording different guitar amps then you might not instantly know your JCM-800 from your AC-30, but that won't affect the end result. Each amp model has its own tone and characteristics, but even if you don't already know what these are, a quick flick through the different models available will give you some idea, and it won't be too long before you get a feel for which one to use for any given sound you want to achieve.

There will normally be different speaker cabinets you can choose, as well as different amplifiers. Most of the time the "matching" speaker cabinet will be chosen whenever you select a different amp, but there is always the option to change to a different one. To illustrate what I said earlier about the speaker cabinet itself having a significant impact on the final sound, try loading up a particular amp, finding a sound you like, and then switching through the different speaker models available. It really does make a huge difference in the final sound. If there is an option for "no cabinet" this would effectively simulate the sound of recording the output from the amp directly into a mixing

FIGURE 19.1
IK Multimedia AmpliTube is a very convincing sounding guitar amp simulator that has the added bonus of a number of classic guitar pedal effects to help rough things up!

desk. You can't actually do this because the power coming out of the back of a guitar amp is designed to drive a speaker and would probably fry any kind of mixing desk or soundcard you connected it to! There are ways of doing this indirectly, but it is quite a specialized thing to do. However, software sometimes allows us to do things very easily that hardware doesn't, so definitely try it if it is an option. What you will most likely hear is a very harsh and raw sound, much like the distortion of overdrive pedals that we mentioned earlier.

There may also be the option to choose from a number of different types of virtual microphone and, to a greater or lesser extent, change the virtual position of these microphones. The differences between microphones can be subtle and certainly won't have as much impact on the overall sound as changing the amp itself or the speaker cabinet, but it is certainly another way you can fine-tune the sound. Likewise the choice of microphone position will have some effect on the sound, although to my ears, this ends up sounding more like an EQ change than anything else. For those of you who have some experience recording guitar amps, it provides a familiar way of effecting these tonal changes.

Another very useful addition to many amp modeling plug-ins is the inclusion of a pedal board into which you can load emulations of different types of guitar pedal effects. I mentioned earlier in this chapter how useful these effects can be on nonguitar sounds, so having them available within a plug-in is great if you don't have the ability (or desire) to use the hardware versions. In fact, there have been times when I have loaded up an amp modeling plug-in purely to use the guitar pedal effects. It can be a little cumbersome to do things this way, which is why I prefer the approach of Logic's built-in Amp Designer/ Pedalboard combination as it allows the guitar pedals to be loaded as a plug-in in their own right. Of course, while this can be more convenient for me personally if I want to use the pedal effects on their own on a synth sound, the all-in-one approach of having the pedals within the amp modeling plug-in makes much more sense if you also want the sound of the amp and speakers, as it saves you having to load two separate plug-ins. You will soon find out which method works best for you in which situation.

In terms of the actual sounds of the guitar pedal effects included, software designers only tend to model classic hardware guitar pedals so don't expect to find the sheer number of different makes and models that are available as hardware. What you *will* find is a good selection that covers a lot of ground including overdrive and distortion effects; chorus, flanging, and phasing effects; auto-filters and wah-wah effects; and many different reverbs and delays. Much of the circuitry in the hardware pedals, and therefore the programming in the software recreations, is focused on making them sound good for a guitar. As such, you might not find them as flexible as more general versions of these effects, but that isn't necessarily a bad thing. With guitar pedal effects and amp modeling it is more possible than ever before to create a convincing lead guitar sound from a synth or sampler, or to just add an element of a guitar sound and feel to a synth lead sound. We become accustomed to hearing certain things in certain ways, so if a nonguitar sound has guitar effects applied and it is then run through a guitar amp, it will, to our ears, take on some of the character of a lead guitar anyway, which is certainly a very handy ability to have.

Some amp modeling plug-ins also have studio effects available. These are things like compressor, EQ, delay, and reverb, and usually they're set up like typical rackmounted effects, rather than the pedal type. As such, they are generally more adjustable and perhaps even higher quality than the included pedal effects. Studio effects are normally the last part of the chain within the plug-in, so they won't necessarily do anything you couldn't do yourself with other plug-ins placed after the amp modeling, but it can be convenient to have everything you need for your favorite guitar sound stored within one plug-in so you don't have to load multiple plug-ins to get the effect right.

Even though most amp modeling plug-ins are stereo, they normally work with just one chain of:

Pedal board > amp > speaker > microphone > studio effects

Native Instruments' Guitar Rig, however, is a little more flexible because it allows you to run parallel configurations as well. You could, for example, split the input signal into two and then run separate pedals, amps, and speakers on each of these before combining them back together at the output. Or you could do the same thing but have one chain feeding the left output channel and one chain feeding the right output channel. Or perhaps you want to have all the pedals first and then split the signal into two to get the sound of two different amp and speaker combinations. The way Guitar Rig does this is really flexible, and when used in such a way that each chain feeds a different output channel, the results can range from subtle widening of the stereo image to more extreme effects where there are totally different sounds in the left and right channels. The possibilities, while not limitless, are truly great if you take the time to explore them. In fact, if you just use the pedal effects in Guitar Rig and use this technique to apply separate effects chains to left and right channels, it can do wonders to liven up otherwise quite static and flat synth sounds. Try it!

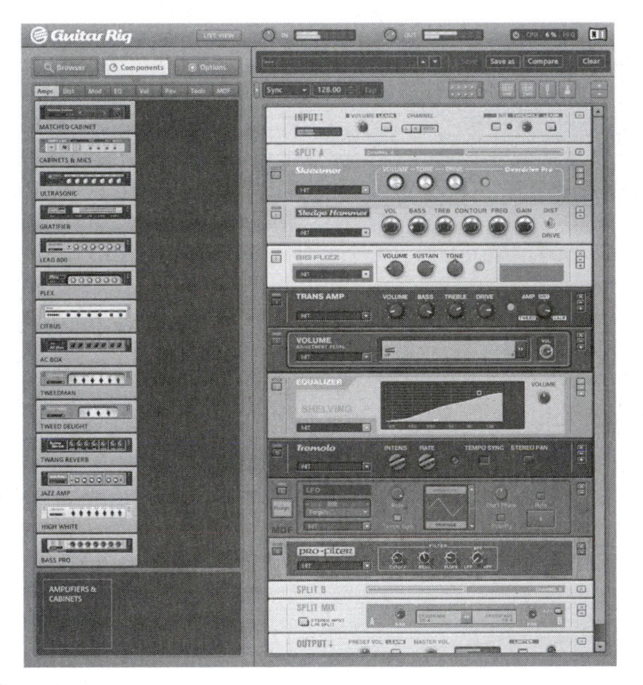

FIGURE 19.2
Native Instruments' Guitar Rig allows for the creation of set-ups that feature independent amps and effects for left and right channels.

The last types of distortion effect worth looking at are the waveshaping and bitcrushing effects. Of the two, waveshaping is closest to traditional distortion effects because of the way it works to add additional harmonic content to the sound. Bitcrushing, on the other end, should (in theory at least) remove harmonic information, because a part of the bitcrushing process involves reducing the sample rate and, hence, the maximum frequency that will come out of the plug-in. However, because of something called *aliasing* there are occasions when additional high frequencies are created, so under some circumstances a bitcrusher can actually add harmonic information. If you don't agree with me, take a look where the bitcrusher plug-in is located in Logic. It's in the distortion folder because, quite often, a bitcrushed sound will have a lot of additional high-frequency overtones that weren't present in the original sound.

SATURATION

In many ways, saturation effects are closely related to distortion effects. In fact, if you consider distortion to be the strongest effect, and then move down in strength to overdrive, saturation effects are really just at the bottom of that same scale. However, because they are subtle, the results are not designed to be heard as a distinct fuzziness but rather just a warming up of the sound. Terms like *warm* are subjective though, and saturation effects generally so subtle, that the precise effects are quite hard to put into a coherent description. What they aim to do is make digital recordings sound more like analog ones in one of a few different ways.

There is much debate about whether digital audio is better than analog audio or vice versa. Most people tend to come down on the side of analog being better in the sense that it *sounds* better. In the early days of digital audio, the results were often criticized for being "harsh" and "brittle," and ever since there has been a quest to recapture some of that analog "vibe" in a way that makes it easy to apply to a fully digital system. You might be wondering, "If analog is so much better, why not just use analog?" The truth is that digital audio systems offer many advantages over analog ones. First, there is that fact that most modern DAWs offer a practically unlimited number of audio tracks and a total recording time limited only by the size of your hard drive. With a relatively standard 1TB hard drive you could record approximately 3000 hours of 16-bit, 44.1-KHz audio (mono only) and roughly 1000 hours of the higher quality 24-bit, 96-KHz audio (mono only). All of that audio could be stored on a hard drive that costs (at current prices as I write this) about $110. A typical reel of tape for a 24-track tape machine can costs upwards of $80 and will only last for about 30 minutes or so (on each track of the tape) if the tape machine is running at 15 inches per second. It'll last only half that time if it is running at 30 inches per second. Given a total recording time of 12 hours (or 6 if you use 30 inches per second) on a cost-per-minute basis, digital has many advantages. There is also the advantage that digital recordings are much easier to copy, back up, and archive, and hard drives have a useful life that far exceeds that of tape.

Then you have to consider the convenience of being able to take a whole project with you (sequence data, effects, settings, audio files, etc.) on a portable hard drive, which can easily fit in your pocket. The fact that digital audio files can be sent around the world via the Internet in minutes or hours instead of days or weeks, and for no cost, is also significant in these days of especially tight budgets. Digital systems also don't suffer from the need for regular recalibration, servicing, and tuning, and the whole world of plug-ins has opened up the possibility for even a relatively modest studio to have access to dozens of plug-in versions of classic EQs and compressors for less than the price of a single hardware unit.

And yet, something often sounds "missing" from fully digital set-ups, so a number of plug-ins have appeared that aim to inject some of that analog character back into the digital systems. Virtually all analog hardware will do *something* to the sound being passed through it in terms of harmonic distortion, even when they aren't actually compressing or EQing or anything else. Typically though, the most common sources of analog warmth are vacuum tube (sometimes known as "valve") pieces of equipment and tape machines; as a result, these are the most common targets for emulating to give the desired effect. Like almost any kind of digital modeling of an analog system, saturation effects are relatively easy to do on a basic level but incredibly hard to do on a thorough and truly convincing level. Obviously with the ever-increasing power available to software designers, these kinds of plug-ins are now getting much closer to their ultimate goal. A couple of examples I can recommend personally are URS' Saturation and Wave Arts' Tube Saturator. Both of these can add a subtle (or not so subtle should you choose) degree of warmth to individual sounds or submix buses, or, indeed, a whole mix.

FIGURE 19.3
Saturation plug-ins such as URS' Saturation can be fantastic for adding a bit of bite and character to any recordings or sounds you might have that are a little on the feeble side.

Parameters on any kind of saturation plug-in tend to be minimal, and will be largely centered around the "drive" control and possibly a "saturation type" control if the plug-in offers different flavors of saturation. Other than this, there may be basic tone controls for the saturation effect and a wet/dry mix control. This mix control is useful because it allows you to mix a more driven sound in at a lower level, rather than using a lower level of drive. The effect is quite different, so again it's a useful, if nonessential, option to have.

I tend to find that plug-ins of this nature are best used very mildly on a number of individual sounds (possibly even every sound/channel if your computer is up to it) to build up the warmth gently, rather than simply adding the plug-in to your master output chain and winding up the drive parameter to give the warmth. The one thing I feel duty-bound to mention is that effects like these are generally subtle, and if you turn up the effect on each individual channel to the point where you can clearly hear it working, the chances are that when all of the tracks are combined, the overall effect will be far too overwhelming. To this end it is, perhaps, unhelpful to read postings on Internet forums where people talk about the difference between fully digital "in the box" mixes and mixes that have been done on an analog mixing desk as being like "night and day." Going by what you read, you'd be forgiven for thinking that the sound an SSL or Neve desk imparts on a finished track accounts for a large percentage of the sound, whereas, in truth, this "night and day" difference is really much more subtle than you might think. Subtle, but quite powerful.

There is one plug-in, VCC (Virtual Console Collection) made by Slate Digital, that is just about to be released as I write this that goes a step further than just being a saturation effect. It models the different behaviors (and, importantly, nonlinearities) of a number of different and well-known mixing desks, and is designed to be used on every channel of your mix and on the buses and master outputs. Controls are limited to "input" and "drive" on the channel version and just "drive" on the bus version, and this is really all you need, because the effects are designed to be gentle but cumulative. Early feedback on this product seems very positive, with few people able to decisively tell the difference between the same mix recorded through a real SSL desk and one processed using only this plug-in. Again, there is a massive difference of opinion. Some people say that it sounds nothing like a real desk, while others say that it is indistinguishable. Some say that the effects are really audible; others complain that they can't hear any difference. Even the developers themselves admit that the effectiveness of the plug-in really does depend on the style of music you are recording and producing. Sadly, for us, it seems to have the least clearly audible benefits on "heavy dance music."

I have tried a demo version of the VCC software and opened up an existing track and then applied the plug-in, as recommended, to all of the individual channels and buses and the results were as I expected from reading the forum posts. It didn't really sound all that much different. But then I tried something else. I tried starting working on a track afresh and inserting the VCC plug-in on every channel as soon as I created it. And, as I was working, I started to notice something. There was definitely a perceptible difference in the way the track sounded as I was working on it. Not anything I can clearly define, but it was certainly there. At the end of the track I noticed something else as well. There was far more subtle EQ and most of the compression settings on the sounds were also gentler, so I think that, for dance music at least, it definitely works better if you apply it as you go along rather than at the end, which is, of course, how it is meant to be used. I couldn't make up my mind, in the limited time I spent using it, if it gives me something I couldn't achieve any other way, but it certainly helped me get there quicker.

I think one of the reasons why it didn't have that smack-you-in-the-face quality that some people have mentioned is because dance music, more than most other kinds of music, often has that depth, solidness, and warmth built into it because of the sounds we use to make it in the first place. As such the warming qualities of saturation type plug-ins *might* not be as necessary. I am not for a moment saying that they're not worth looking at, because any nonelectronic sounds (such as vocals) will benefit from the added warmth that saturation effects can bring, but I do think that, perhaps, the effects are less critical and noticeable than in other forms of music. I will experiment with it more, and, who knows, I could change my opinion in the future!

FILTER EFFECTS

Filters are an integral part of many synthesizer designs, but they also have a place on the audio (rather than instrument) side of music production. Back in the heyday of analog synths, many of the better specified models of synth offered an "audio in" facility, which allowed their filters to be used to process audio from another source instead of the oscillators in the synth. At the time these audio inputs were just mono, and detailed automation of the filter was just a dream, but still it was a very useful feature. Jump forward a few decades to the time of plug-ins, and these filters now exist as independent plug-ins with all the automation you could ever wish for. In some respects these synth filters are just EQ bands, but with the better ones at least, they exhibit much more character than a standard EQ low-pass or high-pass filter does.

I am actually quite surprised that there aren't more filters available as plug-ins in their own right. There are quite a few plug-in synths that include audio inputs (stereo this time) and allow you to use the filters, but it seems to me to be a case of overkill. After all, why would I want to load up a fully fledged synth (with the additional CPU load) just to be able to process audio with its filter? But still, the situation is currently what it is, so we have a choice between using the relatively few dedicated plug-in filters or loading up a synth and using the filter input facilities on there. The end result, and usage, is pretty similar but does have differences, which I will come to shortly.

As we talked about in Chapters 10 through 12, the two most important parameters that are always present in a filter are cutoff and resonance. In addition, there will probably be options for filter slope (12 dB/octave, 24 dB/octave, etc.) and, in many cases, filter type selection (low-pass, high-pass, and band-pass). In an effort to capture that analog warmth (once again) there are many filters, both in standalone form and in synth form, that include some kind of "drive" control. When this is pushed up, it adds an increasing amount of distortion, as you would expect, and this distortion can further serve to emphasize the resonance control, as the increased harmonic content the drive creates gives more information for the resonant peak to emphasize.

I urge you to use this in moderation. Too much of anything isn't good in audio production. Too much saturation makes everything blend and fuzz together.

Too much reverb makes every wash around together. You will often hear professional engineers and producers talking about width and depth of recordings, and the only way to achieve that is by making sure all of the sounds have *separation*. That separation can take the form of spectral separation (when sounds occupy their own frequency range) and spatial separation (controlled by panning and things like reverb depth and type to move things forward or backward in the mix). Too much saturation and drive (including that on filters) can bring homogeneity to all of the different sounds in the mix and make it hard for things to stand out. Filters don't always have to have that gritty, heavy driven Moog filter sound to them. It *does* sound good, but there are times when a thinner, more digital sounding filter is the right decision because the slight thinning it brings to the sound will help it stand out from other similar sounds.

In use, filters can add a great deal of vibe and movement to otherwise quite static sounds, but can also be used in a more static role for tonal shaping, much like a more characterful EQ. Something I do quite a lot with layered sounds is to use different EQs or filters on each sound, but with similar settings. That way there are small differences to the tone of the filter or EQ on each sound, which further highlight the individual characteristics of the sound. This can work in two ways, both equally valid. One is to listen to the type of sound you are working on, and if it is quite a warm sound already then you use a warm sounding EQ or filter; if it is a more thin and digital sound, you use a more thin and digital sounding EQ. The other technique is to do completely the opposite and use thinner sounding EQs and filters on warmer sounds and warmer EQs and filters on thinner sounds. There really isn't a better way; it's very much a personal choice and one that can change from track to track. In my opinion and experience, "fatter" isn't always better and sometimes the best tool for the job can be the cheapest one as well.

SUMMARY

We have covered the main types of effects you will use on a day-to-day basis. But any look at effects like this runs the risk of becoming outdated very quickly because the technology and the ideas of software developers progress at an alarming rate. Totally new categories of effects don't crop up all that often, but I am sure that, in the near future, somebody will develop a plug-in effect that doesn't fit neatly into any of the categories listed in this or the preceding chapters. Such is progress! Even taking that into account, the effects noted here will probably always form the main part of your production and mixing process, so it's a good idea to get a solid grounding in the "how" and "why" of these main effects and then learn the new ideas and techniques as they come along.

Mixing tracks is something that I, and many other people, tend to do as they are working on the production. This is partly because of the deadlines we often have to work to, but in addition, it is partly to do with the fact that the songs and productions are often written as they go along. As such, it is really useful to have a very clear vibe to work to, which is close to how the final product will sound. If you have the kick drum and bass line mixed, EQed, and effected to pretty much the state they will be in at the end of the process, it makes it much easier to choose other sounds and create other riffs and arpeggios based on how much room you know there will be. If, on the other hand, the track wasn't mixed as you went along, you might come up with a riff or choose a sound that didn't fit properly at the final mix stage because there simply wasn't room for it.

I rarely work on the final mix of a track from a "zero" position, with all faders down and no compression, EQ, or effects. I have done it in the past and I do actually like working like this because you can focus all of your attention on the sonic detail, rather than having your head in a place where you are already thinking ahead to the next musical part you have to work on and trying to figure out a melody or chord pattern. That is generally considered the "correct" way of mixing tracks, but I don't know if there is really a huge benefit in working that way.

Whichever way you work, knowing a little about how the different effects work, what they do to the sound, and what the individual parameters do to the effect can certainly help to make the mixing part of the process a little quicker. The less trial and error you have, the better results you'll get and the quicker you'll get to that point, because you will probably already have some idea of the settings you want before you even open the plug-ins.

That said, you might think you would like someone else to work on the final mix of your tracks. There is much to be said for this way of working, if time and budgets allow, and I am a firm believer, under ideal circumstances, in everybody doing what he or she is best at. Dance music producers and remixers have to wear an awful lot of hats during the process of creating and mixing music, and we can become very adept at switching roles quickly. But there will always be things each individual is better at than others. If we have the chance to concentrate on what we are best at and let someone else who has a complementary skill set do the same thing, we could end up with better results. A dedicated mix engineer brings "fresh ears" and a fresh perspective to a track. He or she won't be as familiar with the track (or in some cases as bored with it) as you and will bring new ideas to the table that you might not have had. For this to work, you have to find someone who is familiar with the style of music you make and genuinely understands the subtleties and actually *likes* it. Even if it isn't something you imagine yourself doing all the time, it could be fun to give it a try on a track one day. Perhaps do your mix and then pass all of the dry stems over to a mix engineer and let that person do his or her thing with it. It could be an eye-opener when you get the mix back, or, it might not be that different. In which case, congratulations! You are well on your way to making great sounding mixes!

CHAPTER 20

Mastering

WHAT IS MASTERING?

Many years ago the mastering process was necessary when recordings were transferred from master tapes to the vinyl records the consumer purchased. Because there were large differences in the frequency response and characteristics between the two media, the process wasn't as simple as taking the tape master and making a vinyl record from it. Changes to the EQ of the recordings were necessary and additional dynamics control were often applied because of the physical limitations of "cutting" a record to vinyl. These same processes are required, at least to some extent, with preparing files for CD or digital download, and the process of mastering has remained an integral part of the overall music creation process.

The ultimate goal of mastering is not *just* to make the track as loud as possible, although in recent years that aspect of it has certainly taken center stage in the so-called loudness wars. The *real* purpose of the mastering process is to make the track sound as good as it could possibly sound for the intended delivery format. There is a valid argument that this should be done at the mixing stage, and most people do aim to do this but might not have an acoustically perfect room or have monitors that have a flat frequency response. As such, a good mastering engineer at a good mastering facility with good equipment and a good room will be in the best environment to *accurately* hear what is happening in your mix and make small tonal and dynamic adjustments to help bring out the best of what you have in the track.

Sometimes just having the final stereo mix won't allow the mastering engineer enough flexibility to make changes if something more drastic is required to correct a problem with the mix. It isn't unheard of to take stems to the mastering session as well as the final stereo master. At this stage, though, you are starting to blur the line between mixing and mastering and if I found that I was regularly having to use the stems rather than the stereo master I would have to start questioning my mixing technique. Still, it's a useful option to have.

HOMEBREW MASTERING

A dedicated mastering facility was, until relatively recently, the only option for mastering. Now, as with so many other aspects of the music business from recording to consuming, computer technology has revolutionized the way we do things, and there are now very capable mastering tools available in a plug-in format that, ultimately, makes them available to anybody with a computer-based studio. Although the quality of this mastering software is constantly increasing, it's fair to say that most experienced mastering engineers would probably prefer to work on the top-class analog equipment for mastering purposes rather than solely using plug-ins, especially at this final stage of the process.

Many major releases are still mastered in the traditional way, but what about those of us who don't have the budget to include professional mastering on all of our tracks? Can a home or project-studio owner get even *close* to the results possible from a dedicated mastering facility?

This is a hard question to answer, because the mastering process is a combination of a number of different factors: the original recording, the equipment available, the room itself, and, most importantly, the person doing the mastering. Within that "chain," a reduction in quality of any one component will reduce the quality of the final product. So while a project studio owner may have access to tools that can compete with a professional mastering facility in terms of quality, he might not have the acoustics in his room to compete with the mastering facility and won't have the advantages that using a separate mastering engineer will have.

When you are working on a track from concept through recording and programming and on to mixing, you can become very close to it or very sick of it. Either way, your perception of the track isn't going to be as objective as that of the mastering engineer who doesn't have the blood, sweat, tears, and possible frustration and swearing invested in the track. That simple fact could make a world of difference to the result even if the rooms and the equipment are identical. On the other hand, you do know your style of music intimately (or at least you should!) and you also know the eventual sound you want to achieve, so this does count in your favor as a "self-masterer."

Finding a mastering engineer who is experienced in, and also likes, your style of music is very important (much like employing a mix engineer, mentioned at the end of the last chapter) because she will be more likely to have a natural empathy with you and what you want to achieve, and it will probably be easier for you to describe the final result in terms she understands.

Perhaps for reasons of economy or time, more and more producers and remixers (especially dance music producers) are mastering tracks themselves now that there are quality tools available for them to work with. I do doubt how many some of them truly understand the mastering process; when I listen to the tracks, and even more so when I pull them into Logic or a wave editor, it becomes overwhelmingly clear that there is some immensely heavy limiting and clipping going on. In some cases a zoomed out view of the waveform overview for the track looks like someone has taken an unmastered track and then cut off the top and bottom 25% with a pair of scissors. Not only does limiting of this magnitude cause irreparable damage to the sound of the track, it also makes it *extremely* fatiguing to listen to. Nonetheless, these bad practices continue and look set to do so for a while yet as long as there is a continual striving for more and more "loudness" on final tracks.

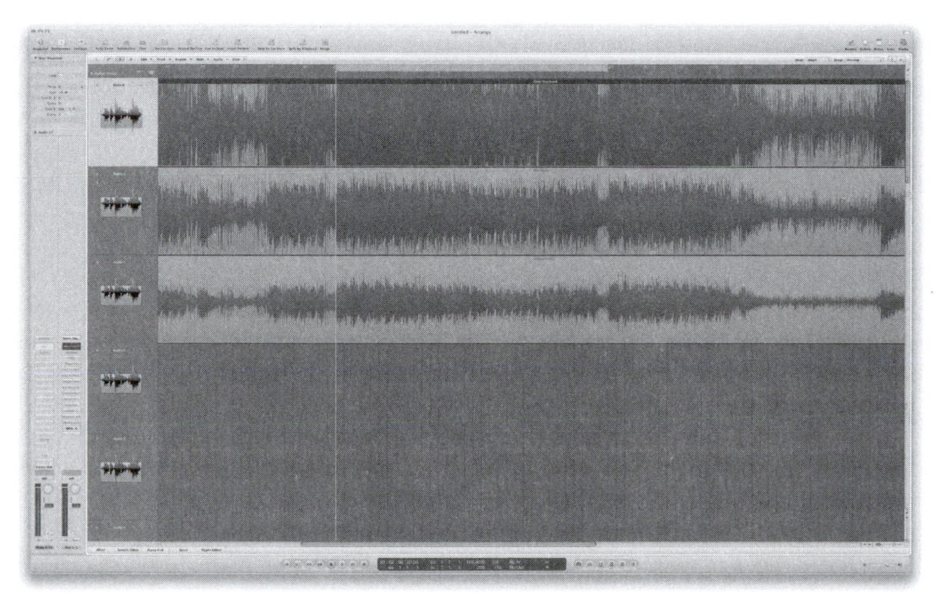

FIGURE 20.1
Here you can see a comparison between three versions of the same track. The over-mastered version at the top has clearly been pushed too far, while the mastered version in the middle has an optimum level that still retains dynamics. The unmastered version is shown at the bottom as a point of reference.

LOUDNESS VERSUS LEVEL

A distinction needs to be made here about the difference between the terms *loudness* and *level*. Levels can be measured in dB (or one of the many variations on the dB scale such as dBFS or dBu), but loudness really is a perceptive scale that is personal to an individual. Of course there is a more general perception of loudness, but what one person finds "comfortably loud" might be way too much for someone else. In addition, you could take two tracks that both had peak levels of 0 dB (as loud as they can physically be in a digital medium such as CD or MP3) but gave the impression of being at completely different loudness levels. The actual frequency content of the track can play a part in this because the human ear is naturally most sensitive to sound at around the 3.5 KHz area, so any songs with lots of energy in those areas will have a tendency to sound louder even if peak levels are the same. In addition, the actual music content of the track will have an effect, as will the dynamics of the original track. If we take a recording of a very gently played piano chord that has been normalized and now peaks at 0 dB, and we compare that to a recording of the same piano playing the same chord but has been struck much harder and is also peaking at 0 dB, there would be a difference in perception between the two that makes it seem like the more forcefully struck chord was actually louder.

One factor that is very important in perceived loudness is the average, or RMS, level of the audio. While there can be a relationship between average and peak levels, there is often little correlation between the two. For example, you could have a very subdued piece of music that had just one loud cymbal crash at the beginning but, other than that, was very gentle. If this whole mix were normalized to 0 dB there would still be a low average level throughout the whole track. On the other hand, a heavy rock or dance track also normalized to 0 dB would probably have a much higher average level. And, of course, everybody would say that the dance or rock track was louder even though the peak levels are the same. What we normally consider the loudness of a track is more often than not the "average" level of the music rather than the peak level because our ears generally perceive sound loudness based on the average level rather than the instantaneous (or peak) level.

Given that virtually all modern records peak very close to 0 dB, there is little room for differentiation in terms of peak level, so mastering engineers work toward increasing the perceived loudness of a track to make sure it holds its own against all of the other ridiculously loud records being released. The most commonly used mastering tools and processes are, therefore, EQ, compression, limiting, and clipping and, of course, a great room, accurate speakers, and well-trained ears.

MASTERING TOOLS

EQ

There is no need to reiterate the purpose of EQ, as everything remains the same with EQ in mastering as it is in mixing, with the possible exception of specialized "mastering" EQ units sometimes having *stepped* controls so that the boosts or cuts can be easily repeated. Often the steps are as little as 0.5 dB or 1 dB, so that a fine level of control is available to the mastering engineer. The biggest difference lies in the actual usage of EQ in the mastering process. While mixing EQ can be used to radically reshape a sound or remove harsh resonances, and while those techniques are still possible in mastering, there is much less scope to use them because, in the context of a final mix, any radical EQ changes will not change one particular instrument or sound within the mix but the *whole* mix. As a result the EQ used in mastering tends to be much gentler, because anything drastic could change many things at once and, as well as removing harshness from a particular sound, might also change the effective balance between other instruments.

FIGURE 20.2
The Universal Audio Manley Massive Passive EQ plug-in is widely regarded as one of the best plug-in EQs available. The hardware version is used by a great number of mastering facilities worldwide.

The overall purpose of EQ in the mastering process is to give the track smoothness and fullness while removing harshness and emphasizing clarity (where appropriate, of course, as there are times when a *lo-fi* or harsh sound is actually desired). It can also provide a tonal balance that will sound good on as many different speakers or headphones as possible. There is no such thing as the perfect mix that will sound equally good on any playback system, so all you can do is to aim for something that will sound the best it can on the largest number of systems. With this in mind, at the mastering stage sometimes there will be alternate versions produced to serve different purposes. For example, it is entirely possible that a separate version of the track might be created especially for club plays that has slight differences to the low frequency EQ of the track to accommodate the fact that most clubs have very bass-heavy sound systems. There are possibilities for similar alternate versions using different amounts of compression and limiting as well, which we will discuss shortly.

Using EQ in mastering comes under two distinct headings, which I consider to be *corrective* and *sweetening*. Corrective EQ in this sense is EQ that removes a problem that often takes the form of a particularly harsh or unwanted frequency. This would normally be achieved using a very narrow bandwidth EQ cut. Sweetening EQ, on the other hand, is for more gentle and subtle changes to the tonal balance of the track, and this would normally be achieved using low boost and wide bandwidth parametric EQ, or possibly even low- or high-shelving type EQ. There are EQ units (hardware and software) that are capable of both types of use, but often a mastering engineer will have one EQ that is used for the more "surgical" EQ correction and a different, more characterful EQ used for the sweetening. Also, when thinking about the sweetening EQ, it is always worthwhile remembering that there are two complementary ways to achieve the same effect. If the mastering engineer felt that the track was a little light on bass, this could be addressed by boosting low frequencies *or* by cutting high frequencies. Cutting high frequencies wouldn't, obviously, actually *increase* the level of the bass, but in terms of our perception of the overall sound, the bass would sound louder *relative* to the treble, which would have a very similar effect.

Speaking of increasing the levels of bass, there is a little trick we can use when mastering, in this regard. The human auditory system has the remarkable ability to artificially recreate a "missing" fundamental frequency, and this can work well for us in the mastering stage because it allows us to actually *reduce* the level of the lowest frequencies while still retaining much of the perceived loudness of the bass. Often this is achieved by using a very steep high-pass filter set to around 60 Hz or so, because our auditory system will then artificially recreate everything in the octave below (down to 30 Hz) with a good degree of accuracy. This helps us in mastering because the lower the frequency is, the more energy is required to give it a given loudness. Consequently, low bass frequencies have to be at a much higher peak level than midrange or treble frequencies to sound balanced. By removing these lowest frequencies we might actually give ourselves a few dB or more of extra headroom for the mastering process to work with. However, in an ideal world we wouldn't want to do this because it does, after all, represent a deviation from the original state of the mix, and any "artificial" low frequencies we hear as a result of our auditory process will only ever be an approximation of sound information that was actually there to start with before we removed it.

Ideally a finished mix that is presented for mastering shouldn't require too much EQ, and, with dance music especially, more often than not any EQ applied will be in the form of low frequency focused changes to smooth out the mix in the lower octave, which most smaller studio monitors can't reproduce accurately. There are many things that influence how close the mastered version is (tonally speaking) to the original artistic vision of the mix engineer and producer: the more experienced you are at working with your monitors, the better your monitors and room are, the more knowledgeable you are about how to use EQ and compression while you are mixing, and the more work you have put into getting everything right in the mix. If all these things are in place, there shouldn't be much need for EQ at the mastering stage.

Compression

Compression, as we know from Chapter 17, is all about reducing dynamic range, that is, the difference between the loudest and quietest parts of a sound (or track in this case). The reason we would want to do this at the mastering stage is to increase the average level of the track to make it seem louder. To explain how reducing the dynamic range increases the average level let's use a simple numeric example.

Let's take a sequence of 20 random numbers between 0 and 100:

47, 24, 61, 73, 7, 40, 82, 3, 50, 1, 57, 67, 6, 36, 20, 84, 98, 49, 30, 72

If we now work out the average (mean) of those numbers we get 45.35.

Now if we take any of those numbers that are over 50 and "compress" them in a ratio of 2:1 we end up with this sequence:

47, 24, 55.5, 61.5, 7, 40, 66, 3, 50, 1, 53.5, 58.5, 6, 36, 20, 67, 74, 49, 30, 61

The average of these new numbers is now 40.5, which is significantly *lower* than before. Now we can apply *makeup gain* to increase the value of the highest number back to where it was before (in this case 98) and then we get:

71, 48, 79.5, 85.5, 31, 64, 90, 27, 74, 25, 77.5, 82.5, 30, 60, 44, 91, 98, 73, 54, 85

This gives us a new, final, average value of 64.5, which is much higher than before, so we have achieved our purpose. By decreasing the dynamic range and then applying some gain to bring the peak levels back to where they were before, we can significantly increase the average level and, therefore, the perceived loudness of the track.

As for the actual type and strength of compression used in the mastering process, generally a very gentle compression is used to increase the density of the mix as a whole without making it feel too squashed. A good starting point for mastering compression is to use a ratio as low as 1.1:1 and starting with a threshold of about −30 to −40 dB and a reasonable release time of around 200 ms, as this will provide a subtle compression effect that builds up as the louder parts of the track are reached. You can experiment with changing the threshold, ratio, attack, and release controls, but the overall effect should be that, at the points of the greatest gain reduction, no more than a few dB should be taken off at this stage. The aim of the compression part of the mastering process isn't to squeeze every last drop of volume out of the track, but rather to just boost the feeling of "energy" within the track.

Parallel compression can work wonders in mastering as well because it allows the mastering engineer to add density and fullness to the quieter parts of the track without overly affecting the dynamics of the louder parts where its contribution to the overall sound is much less. Of course, the effects of the parallel compression can be made to sound more or less subtle as required, but like most of the other parts of the mastering process, a subtle usage would normally be what was called for instead of the (often) quite heavy and dramatic way in which parallel compression can be used during the mixing process. As with "normal" compression though, a good starting point is with a low threshold of around −40 to −50 dB, the shortest attack time possible to minimize the contribution of transients from the parallel compressed sound, peak detection rather than RMS to make sure the transients are captured and dealt with cleanly, a slightly stronger ratio of 2:1 or 3:1 to give a heavier compression effect, and a release time of around 300 ms to prevent audible pumping in the compressed signal. The level of the parallel compressed signal should be set to zero and then gradually and slowly increased until the desired effect is achieved. The actual level needed will vary enormously from track to track, and it is impossible to give a ballpark figure without knowing the track. As a general rule, the more energetic the track is to start with, the higher the level of the parallel compressed signal necessary to have any noticeable effect.

FIGURE 20.3
The Universal Audio Precision Buss Compressor has a "Mix" control that allows for parallel compression within the plug-in itself.

Parallel compression could well be used in conjunction with normal compression to attack the problem from both ends. In this instance you could probably go for more gentle normal compression to work on the peaks, perhaps only taking off 1 or 1.5 dB and then applying parallel compression to effectively bring up the levels of the quieter parts by another 1 or 1.5 dB to give a total effective dynamic range reduction of 3 dB. In general, it is far better and far more transparent to have a mastering process that is made up of a number of stages that each contribute a little to increasing the volume, rather than just slapping a brickwall limiter across the master outs of a DAW and getting 6 or 7 dB (or more!) of gain reduction out of a single processor. These kinds of levels of gain reduction (and therefore volume increase) are possible with the more conventional mastering process, and while they still won't sound particularly "clean," the end result will probably sound much better and less effected than the single brickwall limiter approach.

Of course, as a result of the newly decreased dynamic range (and increased loudness) there may now be less of an apparent impact going into a loud chorus from a quieter verse. This is understandable, as we have intentionally reduced the dynamics of the track. As a result, the mastering engineer may choose to include some level automation (very subtle of course, perhaps as little as 1 dB or so) to slightly lower the level of certain sections of the track to emphasize certain transitions between sections, but in a more controlled way now that the dynamics have been leveled a little.

Limiting

Limiting is another form of dynamic range control and is very closely related to compression. In fact, technically you *could* think of limiting as simply compression with an infinite compression ratio whereby no matter how much the signal exceeds the threshold level there is no increase in output level. As such, a limiter (or peak limiter as it may also be called) is really the last line of defense against clipping. If the limiter threshold was set to 0 dB it would serve to catch any stray signals from the output of the processors before it. These days, limiters are often used to squeeze even more loudness out of tracks on top of what the compression has already achieved.

FIGURE 20.4
A limiter is often the final stage in the mastering process, but sadly, the one most abused.

Once again this process isn't too bad or too damaging to the track if it is used with care and subtlety. If a limiter is used to take another 1 or 1.5 dB off of the top of a track (post compression) then it might still be fairly clean sounding. Beyond that we start to get into squashed sounding results, as the effects of the limiter start becoming audible as pumping artifacts, but in an even more obvious and unflattering way than overly aggressive compression causes. The effects can be minimized to some extent if a multiband limiter is used, because the different bands can be set up so that the frequencies most likely to push the level "over" in a single-band limiter can be clamped down a bit harder while allowing the other frequencies to breathe more.

Clipping

Traditionally clipping (especially digital clipping) is seen as a very bad thing, because it introduces distortion, but as we have already seen, distortion in the right context (and of the right kind) can actually be used to enhance sounds in a desirable way. More and more mastering engineers (and mix engineers to a lesser extent) are starting to use very small amounts of clipping in the mastering process. I am not sure if this is a conscious decision, but I somehow doubt it. I am much more inclined to believe that this technique is simply a result of them being asked to make final mixes ever louder, and the clipping technique is simply a result of them seeing just how much they could get away with.

As I mentioned previously, digital clipping doesn't sound anywhere near as "nice" to the ear as analog clipping does, so if you were working inside a DAW for mastering you would have to pass the audio out through a D/A converter and then pass it through a (preferably high quality and low noise) piece of analog outboard equipment and adjust the input gain or "drive" until the desired effect was achieved. You'd then pass that through an A/D convertor to get it back into the DAW. Depending on the analog hardware used, this might not be an exact science, but brings me to my final point.

What Is Too Loud?

If you were to compare the average (RMS) levels of a typical CD now and one from 15 years ago, the differences would be shocking! There is an ever increasing need these days (at least from the clients' point of view) to get recordings mastered louder and louder, even at the expense of a little bit of quality or fidelity to the original recording, to the point where it has become hard to define what is "too loud." From my perspective as an artist and a remixer, I would say something is too loud when you can start to hear the mastering process (or the dynamics part of it, at least) doing anything other than raising the overall level a bit while energizing the track a little on the whole. If I can hear pumping, breathing, and squashing going on that wasn't a part of the original *design* of the track, that, to me, is too loud. The trouble is that, on more than one occasion, I have had a track mastered to what I thought were the limits of what could be achieved without degrading the original track, and then, when I put it up against other commercial releases in a similar style, my track sounded *noticeably* quieter in comparison. In the end it needed another 2 to 3 dB of gain reduction to bring it into line with the loudness of the other tracks, by which time it had lost quite a bit of punch and snap, and just felt quite brash and forward in comparison to the original version.

This is, of course, all highly subjective and what I may hear as overdone might be totally acceptable to someone else. Sadly this whole process has been going on for quite a while now and each time someone raises the bar for loudness (and simultaneously lowers the bar for audio quality) everybody else is obliged to catch up. This is, in many ways, a road to nowhere because with each successive step up in terms of loudness we get a corresponding decrease in audio quality, and, more worryingly, the tracks become more difficult to listen to as our ears simply find it hard work to listen to things that have such high average levels. There is a definite backlash within the music industry against this practice, and many high-profile artists have signed up to the cause to restore some dynamics to recorded music. Whether they will be successful remains to be seen, but we certainly have to hope so.

PRACTICAL MASTERING

So all of this is fine in theory, but can someone who doesn't have a perfectly balanced room, full-range speakers, and some very tasty outboard equipment still get anywhere near the results a professional mastering facility would get? I would definitely say that yes, with patience, practice, and the right (reasonably priced) equipment, you can certainly get a good way toward what a professional mastering engineer could achieve. I think that the biggest difficulty in home mastering is the lack of perspective you can give a track that you have already worked quite extensively on. For me, this is the fundamental problem of home mastering. You will probably have already spent quite a lot of time listening to the track in often microscopic detail, so it can be very hard to suddenly switch your head from the microscopic to the macroscopic and start viewing what you are hearing not as a collection of individual sounds but as a "song."

You also have to deal with the fact that the acoustics of your home or project studio will almost certainly not be anywhere close to a match of the acoustics in a dedicated facility, and, as we have already discussed, acoustics (bad ones) can play a major role (a negative one) in the final sound that reaches your ears. Therefore anything we can do to improve the acoustics in our rooms will help hugely if we want to try mastering our own material. Many people will strongly disagree with my methods, but before you make any negative judgment, it has to be put into context. I have gotten some pretty good mastering results (good by my own standards, of course, but I think I wasn't being biased) by using headphones for mastering. I am talking about quite expensive headphones, of course (Sennheiser HD-600), and a fairly good headphone amp (a Creek OBH21). The reason why I think this has worked for me is because the headphones take room acoustics out of the equation. Even the worst of rooms won't affect the mixing or mastering process if you are working with headphones, so they are a great way to bypass typical room problems.

They aren't without issues of their own, however, and the main ones that seem to come up for headphones in general (rather than any specific make or model) are that real deep bass levels are hard to judge accurately and panning positions can be very hard to determine correctly. Out of the two, I think that the panning position problems are the least significant for mastering purposes because there is very little, if anything, you can do to change the panning positions of individual elements within a finished mix. You can, of course, change the overall balance between the channels, or with M+S (middle and side) processing you can potentially change certain aspects of the stereo width of the mix, but if you are mastering tracks yourself it would be easy enough to change things in the original track, particularly if you are mastering "live" on the main outputs of the DAW itself rather than a bounced audio file of the mix.

What might be more important is the problem with accurately judging low-frequency bass levels, because this is an essential part of the mastering process, especially for dance music. Most of us are (or at least become) quite familiar with the sound of our monitors (and headphones if we use them) and we get very good at compensating for any flaws in the frequency response or overall sound of them. In an ideal world, none of us would want to have to compensate, we would just want something that sounded perfect anyway, but that is rarely the case for all but the most professional and well-equipped studios. For the vast majority of us, we make the best mix decisions we can on the monitors we have and then take the tracks to a mastering facility to make use of their room and speakers as much as anything else. But it is the very fact of this compensation that I believe makes it possible to get reasonably good mastering results on good quality headphones. Here is how it worked for me.

Over the years I have had quite a few of my tracks professionally mastered, but unlike other music I listen to that has been mastered, I have a before and after version to compare to each other. And this helps us recalibrate our own hearing to some degree. By comparing the mix I do in my room to the final mastered version, I can get a pretty good idea of the changes the mastering engineer made. Of course, if I attended the mastering session, which I don't always do, then I would be making notes as the session progressed about what was used and what settings were chosen, and I would probably be asking a few questions along the way. But even without knowing specific equipment or settings, I can get a feel (on a "vibe" level) of what changes have been made as well as having a number of audio analysis tools at my disposal (a spectrum analyzer, displays on EQ plug-ins, full FFT analysis for greater detail, peak and RMS level meters on various dynamics plug-ins, and a few others as well) to get a more technical view of what has happened in the mastering process. Of course, I am not suggesting that *just* by looking at a few spectrum analyzer displays and a level meter I would be able to get the same results, but it does give me a hint as to what I should be aiming for, which, combined with my experience and technical knowledge, helps me out quite a lot.

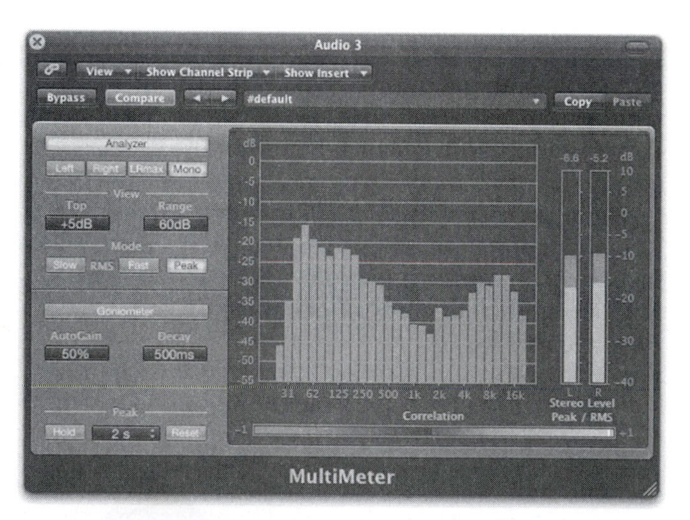

FIGURE 20.5
A spectrum analyzer such as this gives a good overview of what is going on in a mix, but you shouldn't rely on one too much when mixing or mastering; your ears are normally a far better judge than any kind of visual display.

In fact, an interesting experiment and *very* good way to learn is to take an unmastered track of your material, get it mastered professionally, and then, back in your own room, have both the mastered and unmastered versions in your DAW, and through listening and using these kinds of analysis tools, try to understand the changes that have been made and then try to get the same effect. At the same time, try to *know* how to get it rather than just moving knobs and sliders around randomly until it sounds somewhere close. Make sure you do it by design. Have a guess at what is needed, try it, and if it gets you closer, great. If it doesn't, go back to where you were before and have another think and then try something else. In this way, even with monitors (or headphones) that aren't perfect, you can get some pretty good results.

The bottom line is this: While you can certainly improve your own tracks and remixes by adding some mastering at the end of the mixdown, it is unlikely that you will get your track sounding quite as good if you master it yourself as if it were mastered by a good mastering engineer, unless you have a very well-treated room, very expensive and well set-up speakers, some great outboard equipment, and a lot of experience mastering your own tracks. Oh wait, that would pretty much make you a mastering engineer! It's certainly a great thing to try to do, and I always do a mastered version of any remix I send to clients for approval, simply because they are used to hearing things that are already balls-to-the-wall loud, so if I send over a mix that is 10 dB or so down from that level it would sound lifeless in comparison to whatever they were listening to just before they put my remix on. So the "listening copy" is always a home-mastered version, but when it comes to the final WAV file masters, I always send the unmastered file first and my "home mastered" version as a "just in case you need it" option. You might be surprised at the number of times the labels went with my own homebrew mastering, although I will never know if that is because they genuinely thought the mastering was pretty good or they just wanted to save time and/or money.

As I alluded to earlier in this chapter, some record labels even get different versions of the track that are mastered differently for different purposes. They may get a version that has the bass mastered differently, meant specifically for club plays, or a version that has slightly less compression/limiting just for radio stations (which invariably have quite a heavy limiter running across the whole broadcast and, if they are FM radio stations, have to do with the inherent compression effect FM broadcast has on the audio). I have even heard of people who create a version of the song that is mixed and mastered differently to create the MP3 versions of the track from because the actual encoding process of MP3s can (subtly) change the frequency response and dynamics of the track. In this case, it is actually mixed or mastered differently to compensate for this. That seems a little excessive to me, especially as I don't really know if it makes any technical sense because I don't know if the MP3 encoding losses are predictable and will affect every track in the same way or whether they are essentially random or based on the spectral profile of each individual track. As it happens, though, I have always thought that a very useful plug-in to have would be one you could place across your master outs that actually performs the MP3 encoding on the fly. It would undoubtedly have quite a high latency and might even be quite heavy on CPU load to be able to do this at anything approaching real time, but I think it could be great to instantly hear what the mix would sound like as a 320 kbps MP3 or a 160 kbps, 128 kbps, or even a 96 kbps or 64 kbps version. I don't know if it would be necessary to create a totally different mix for MP3 creation, but it might be useful to be able to adjust the whole mix to compensate in some way for what is, arguably, the most popular format in use today.

So if you are going to try mastering yourself, a good starting point is the following signal chain:

Source > EQ > Compression > EQ > Stereo Enhancer > Tape/Saturation/Clipping > Limiter

The reason I have included two EQ stages is that changing the EQ on the way into the compressor can change the way in which the compressor works and how it reacts. In addition, the actual process of compression itself can change the tonal balance subtly, so you might find you need to add a little EQ on the way out of the compressor as well to bring things back into line. As for the compression itself, I recommend trying multiple compression stages. The first could be a conventional low ratio/low threshold compressor to add some density, or alternatively, some parallel compression to achieve a similar result. This could then be followed by a second compressor set to a higher ratio/higher threshold setting in order to tackle some of the bigger peaks before heading on to the later stages.

I am not really a fan of stereo enhancer type effects because they can, in my experience, also change the tonal balance of the track and give it a slightly artificial, slightly smeared quality. I think that, in general, it is always better to go back into the track itself and make changes to the individual sounds to make the track sound wider, rather than to use a plug-in at the mastering stage. Sometimes, however, a stereo enhancer is more convenient, sometimes it is less time-consuming, and sometimes, if used gently, it doesn't sound too bad and gives a nice little boost to the width of a track.

Tape saturation, general saturation, and clipping type effects can be very useful in the mastering process because they are useful for some extra gain and therefore loudness, and can add some nice warmth to the sound and have the effect of "gluing" everything together. In fact, there is a growing number of people who will record the final mix onto tape and then back into a DAW in order to get that glue effect. Some people go so far as to say that this single thing can turn a track from "a track" into "a record." I am not sure I agree with all of the hype surrounding these types of plug-ins because I don't really believe they are a magic wand that you can wave over your mix that will suddenly transform it. But I do agree they can add a certain something to a mix that is hard to achieve any other way. It doesn't mean you can't get a good sounding mix without them, and it will certainly be more effective on certain kinds of music than on others, but if you have the option, it won't do any harm to include it in your standard mastering chain. After all, it can always be bypassed later if you don't feel the track needs it.

And finally we have the peak limiter. This is probably the single most abused plug-in in music production today, because it has the effect of making people believe that they don't need to worry about setting levels correctly. A peak limiter is in many ways designed to be a "safety net," but many people use them far too heavily. In use, a peak limiter should rarely, if ever, actually do anything. If it were *just* being used as a safety net it might actually do something maybe a couple of times in the whole track, but in the way a peak limiter is typically used today, it is probably working to a greater or less extent from the very beginning of the track to the very end. As much as I appreciate that things have gotten out of hand and that there *is* a constant desire to make things louder and louder, I feel there is certainly room to back off a little. Of course, a track that is mastered 5 dB (RMS) quieter than similar tracks will sound less exciting, but even if it is mastered 2 dB (RMS) quieter, or even 1 dB quieter, that will still give back quite a lot of life to the track without it sounding radically quieter than other similar tracks.

My final word on mastering is that you should only use any of the individual elements in the process if you feel the track will genuinely benefit from it. Don't apply EQ to the track at the mastering stage if it doesn't need it just because you feel that it *should* have EQ. Similarly, don't apply lots of compression if only a little is needed. All things in moderation has never been more true than it is in mastering, so if you are going to try mastering yourself rather than getting the track professionally mastered, I would say to start things gently and work up from there. If you have the opportunity to master the track and then hear it on a club sound system, all the better because you can check how it sounds in context. If it sounds good then you have done a good job. If not, go back and *slowly* increase the strength of the mastering until you feel you have it right. But please, remember that mastering isn't simply a case of putting a peak limiter across the track and then having it take about 10 dB off the top of the track. That will certainly make it louder, but at what cost?

SECTION 3
The Reality of Remixing

CHAPTER 21

Remix Walkthrough: First Steps

We have gone through a lot of the technical steps needed to prepare you for remixing and many of the creative ones as well. So now it's time to put it all into practice. The best way of doing that is to actually go through a remix with you, step by step, so you can at least see the process I follow and see how I would deal with some of the issues that come up along the way.

I was in two minds as to how to present this part of the book. I considered making it all very slick and orderly, but then I realized that not all remixes go 100% according to plan. So this chapter, and the following three, will literally cover the steps I take in the order I take them, even though sometimes that will involve going back over things. It might seem somewhat chaotic, but I honestly think that this kind of real-world approach will be more beneficial than any "edited" version, as it will give you more of an insight into the kind of things that happen, the decisions you end up making, and the changes in direction that happen as you go along.

Usually when I am working on a remix, I will use a number of synths and plug-ins from various manufacturers, but for the purposes of this book, I have restricted myself to just using the built-in Logic synths and plug-ins. I have also stuck to the included Logic sample library. I did this for two reasons: first, so that anybody who has a copy of Logic 9 will be able to load up this project without having any compatibility issues, and second, so that I could show you just what is possible using only the built-in Logic plug-ins and sound library. There is a great tendency to think that you need additional software to be able to do "professional" work but this simply isn't true. Of course there are other synths that will offer you a broader sonic palette and other audio plug-ins that offer a variety of different flavors and are, arguably, better in quality, but this doesn't stop you producing things that sound very professional and that you could be rightly proud of with just the Logic Studio package.

On this book's accompanying website you will find a number of Logic song files that go with the various stages of the remix. They are included so you can see the progress and see what I was doing along the way, as well as being able to actually get in there and get your hands dirty, so to speak. With that all in mind, it's time to begin!

STAGE 1: MESSAGE IN A BOTTLE

The first thing I always do before I even start thinking about what I am going to do with a remix is to load up all of the remix parts I have been given and have a listen through. Many times all I will be given are vocal parts or an a cappella, but in this case, I was given a complete set of stems, so I could have used anything from the original song. Sadly, because of the nature of the remix that I had in mind to do, the only parts I was really sure I was going to use were the vocals. But still, I had all of the parts imported into a Logic song so I could always listen back again when I was further into the remix to see if I could make use of anything else later on.

I do generally like to make use of parts other than just the vocals if they are provided, but sometimes they end up being chopped, effected, and manipulated quite a lot so you wouldn't really know that they were from the original song. I like to incorporate as much as I can in situations like this to keep as much of the soul of the original song intact as I can while placing it in a new body!

The next step is to take time to actually *listen* to the track I am going to be working on. And by that I mean listen on two different levels. The first is more of a casual listen, something I usually do while I am doing something else; perhaps replying to emails, perhaps updating my website, anything really, as long as I am not concentrating 100% of my attention on the song. And I usually prefer to listen to just the vocals at this stage so I don't get the original version of the song too deeply ingrained into my mind. I will often listen through a good few times just to get a feel for the melodies and rhythms of the vocal. By doing this I get a good idea of the mood of the song and what it is trying to convey. Then, once I have done this, I focus on the track properly and actually listen to the words.

Every song is a story of some kind. Some are happy, some are sad, some are angry. They take many different forms, but each has a message. Our job, as remixers, is basically to take that story, that message, and translate it into a different "language." It may have been written in the "language" of ballads but we have been hired to translate it into the "language" of electro house. And herein lies the first potential pitfall.

No matter how well you know your own language (electro house in this case) you have to also be fluent in the language in which it was written to be able to translate it.

By listening to the words of the song you can start to really understand what the writers and the performer were trying to say. Once you get to that stage and you have an understanding of the song, you are then in a better position to start the translation process.

In the case of the song we are working on, there was a very clear theme to the song both in the lyrics and in the underlying melodies. A few of the lines stuck in my head in particular and it was these that really set the tone for me:

"Did you leave me for another? Was I always just a friend?"

Coupled with the chorus:

"I'm thinking about holding on to something."

The feeling and mood for me was one of questioning turning to acceptance mixed with optimism and hopefulness. In short, it tells a pretty big picture of a whole spectrum of human emotion within a couple of minutes! But the overall impression on listening to the original track is one of a positive feeling. So with that, and with the brief of turning the track into a commercial electro house track, I started to think about what kind of chords I would use. The musical content of the original track is predominantly major and, as a remixer, I often like to change that at least a little so, in this case, what I thought I would try was to give the chords a change to a minor progression to help give a little more coolness to the track, in line with the brief. This would still fit with the lyrical content, but the chorus vocal would change in context from the optimistic and positive feel the original version had to a slightly more uncertain feel. Still hopeful, just not as convinced. I think that the message of the song is the same but perhaps the tone of confidence has been subdued a little.

The first thing to do was to very quickly and roughly time-stretch the vocal to match the tempo I want to be working at. Because I wanted to make this quite energetic, I decided on a tempo of 128 bpm. The vocal was originally at 90 bpm, so I had a choice to either time-stretch it downward to 64 bpm and then have a "half tempo" feel to the vocal or stretch it up to 128 bpm. I made two copies of the vocal track and then tried both options. Having listened to them both I felt that the 64-bpm version was just *too* slow and would bring down any club mix, and just wouldn't sit at all. The 128-bpm version, on the other hand, sounded just ridiculous in places with the sped-up vibrato giving it a serious case of the "chipmunks," but I knew I could fix it with a bit more work a little later on. At this point I just wanted something I could start working against, so I didn't worry too much about the parts of the vocal that sounded problematic, and kept the 128 bpm version in there as a guide and something to work with for now.

At this stage I worry very little about the actual sound, so I simply loaded up a standard piano preset in the EXS24 and started to play around with some chord progression ideas. As a general rule, I like to try to keep the verses quite simple in musical structure to allow the choruses to feel *bigger* somehow, even without factoring in the other techniques we can use to help build the dynamics of a track. I also like to loop around a verse/bridge/chorus section to get a feel for how the sections will work together. If I am having particular problems with any of those sections, I will perhaps loop around just that one section until I have something I think will work, but then will always listen back in context.

For this particular track I wanted to keep the verses very simple and then "open up" the pattern once we reached the chorus. This is a regularly used trick in dance music production and remixing, but it isn't always possible because the melody in the verses could be such that a more complex chord pattern was necessary to fit with the vocals. In this case, however, the vocal melody was simple enough that I could, with the exception of a little turnaround every couple of bars, keep the same chord running through the whole verse. This gives the option to have quite a dramatic change going into the chorus without really having to work too hard to get it. And that's always a good thing… deadlines, remember!

For the chorus I already had a "feel" in mind and when I played the chords that I felt might work they did… kind of. What I felt was that they worked against the main vocal melody, but the vocal harmonies were clashing, which was to be expected given that the original harmonies were sung against the original chords. I wasn't overly worried because I had the options of either retuning the harmonies with Melodyne or removing them altogether. I decided I would deal with this later. For now, I just recorded everything in against a kick drum (not worrying about the kick drum sound at the moment) and then moved on to the next stage.

STAGE 2: GROOVE IS IN THE HEART

The next thing I do is to start building up some percussion parts to give something to work the other musical parts against. I will usually have some idea in my head of the specifics of the sound I am aiming for by the time I get to this stage, and that can be quite a big influence on how I proceed with building the percussion part of the track. If I think that the production needs to be quite big and bouncy, then the drums and percussion I start with will tend to have more to it to give it more energy; whereas if the production idea I have is more minimal, I will make sure the elements I put in at this stage aren't too "busy."

The idea I had was for something quite sonically energetic, but not overly complex, so I decided to work on a main groove that was quite simple, but with some other more "busy" percussive elements in the background to help push things along. Although I worked on getting all of these elements working together, it was always my intention to have the simpler groove running through the verses of the song and then introduce the other percussive elements during the choruses to help lift the song at those points.

The first thing I always look for is the kick drum. This is partly out of habit, but also because it is such an important part of the track. In Chapters 10 through 12 we looked at the sound design possibilities for drums and percussion, and true to form, I applied those principles here. While auditioning kick drum samples I found quite a few that had certain characteristics I liked, but not one single sample that had everything I wanted. So I imported a few different sounds (ten in this case) into an Ultrabeat patch and then played with the tuning and amplitude envelopes to get the sounds basically right. Then I created ten copies of the Ultrabeat instruments on ten audio instrument tracks, loaded the same patch into each, and created a simple four-on-the-floor pattern for each. Why, you might be asking, did I create ten copies when I can play all of the sounds at once from within just one copy of Ultrabeat? The truth is that it is partly out of habit of working with samples in the way I described earlier, but it is also because I am not worried about CPU power, because once I have the sound I want, I will bounce it down to a single audio file anyway and then get rid of the Ultrabeat instruments. So with each kick drum sound playing on its own instrument track, I started to work on isolating the parts of each that I really liked before combining them in different ways to see what worked best together to give me what I was looking for.

I ended up using six different kicks layered together and processed (edited with EQ and compression applied). This might seem excessive, but if you listen to what each of the individual kick sounds is doing it will hopefully make sense, as each is contributing a certain tonal quality to the whole. Of these, one had the main weight of the composite sound, four provided top end frequencies and attack, and one was a very roomy sound. All of the individual kick drum layers were routed to a submix bus, and overall EQ and some final compression applied. At this point I wanted to bounce this sound down, but instead of just bouncing a single composite sound, I bounced the four "top end" kicks as one new audio file, the "sub" kick as another audio file, and left the "roomy" kick as an Ultrabeat instrument because I wanted to manipulate the decay time (the amount of "roominess") of this sound during the track, which is easier to do as an Ultrabeat instrument than as a sample. I should note that I bypassed the EQ and compression on the kick submix bus when I bounced them, and then routed the composite kicks back into the kick submix bus, and then applied the EQ and compression again. With the bounced kicks placed in the track, I could now move on.

I started to look at snare drum and clap sounds. I had recently been working on some remixes and was really pleased with the snare drum sounds I had gotten. They were a combination of quite a real-sounding snare drum with a lot of low-frequency content combined with a snappy synthetic snare sound to give it a bit more top end and attack, and then some short room reverb applied to give the sound a nice stereo image and some depth as well as adding some tonal interest to the end of the sound. I *could* have just reused the sound from the previous track, maybe tweaked it a little. But I thought that, for the purposes of what we are doing here, I would go through the whole process and, undoubtedly, come up with another good sound to add to my sound library!

Once again I opened Ultrabeat and loaded in quite a few sounds I liked, and went through the editing process as before. Because I had only loaded five samples this time, I ended up using them all, but even after however many hundreds (if not thousands) of hours of auditioning and choosing drum sounds, I still never like to make decisions on sounds without hearing them in context. That is a general rule of mine, actually. While I may solo a sound if I am trying to figure out a problem frequency and remove or reduce it with an EQ, I always listen to changes in the context of the whole mix because the amount of EQ you are applying (or the amount of reverb or delay) can often be misleading when you are hearing the sound out of context. The final result was close to what I wanted, so I bounced the composite sound again, but only a single variation this time around because I didn't need the flexibility I did with the kick.

I did, however, want to add two more things to the snare sound this time around. There are quite a few tracks I have heard recently that have what sounds like either a reversed clap sound or perhaps just a short reversed reverb version of the main snare sound leading in to the snare. It doesn't always happen on every snare hit, and sometimes only once every two or more bars, but I do like the effect, so I found a nice clap sound in Ultrabeat, did the usual EQ and compression, bounced it down, added it to the arrangement, and then reversed the bounced audio file. The effect is usually subtle, so I pulled the level of that sound down quite a bit and played around with the exact timing relative to the snare a little bit until I was happy with the result. I only placed it on every fourth clap so that it didn't get too repetitive. Even then, it may have been a little too much, but I left it like that for now. Just to add a little sense of space, I put quite a large reverb on the reversed clap, which would help to soften the end of the reversed part and add some depth to every fourth snare/clap hit. The final thing I wanted to add was a normal clap, so I found one in Ultrabeat that I liked, did a little processing on it, bounced it down and put in a clap on all of the beats where there was a snare drum. I didn't bounce it at this stage (although I planned to eventually) because I wasn't 100% sold on the sound just yet, but wanted to move on.

In terms of the rest of the percussion elements, I had already decided that my main hi-hats would be playing quite a straight 8th-note pattern to really give the other sounds I was planning to put in some space to work in. This, then, was what I looked for first. I ended up choosing a straightforward TR-808 closed hi-hat sound, because it was short, unobtrusive, and small sounding. I really didn't want a big hi-hat sound for this part, because, as I said, I intended to add some more interest on the top-end percussion later on. I wanted to build the complexity and fullness of the sound through additional percussion parts rather than by choosing a big sound to start with. I programmed in the pattern with maximum velocity on the notes that fell on the beat and a slightly lower velocity on the ones that sat in-between the beats, to give some dynamic interest.

Once I had that part played in and the level roughly set, I started to audition a few loops as I normally would. In this case I found 16 that I liked, but with the intention of only using a few parts of those and probably discarding the majority of them. I started going through each, EQing most of the bottom end out of them so that they became only "percussion" loops, and then adjusting levels and cutting down some of the loops in places to just a smaller section of the full loop. It took a little while, but eventually I got to a point where I was happy with the groove of everything I had chosen working together. Of course, the plan was to have these different loops coming in and out throughout the track to build the dynamics, rather than keeping them all running at the same time. I had even included one quite complex loop that I didn't really plan to use, but I decided to keep in the arrangement as a "just in case" option if the percussion felt like it needed something with more movement and energy. In the end I had a total of 9 of the 16 loops in use, but one of those I just used a small part of as a fill, one I was only planning to use if I needed a little extra push on the percussion (as I mentioned earlier), and another I was planning to use only in the breakdown section of the track. Of the remaining six, there were three that were playing a very similar pattern but on slightly different sounds, so these would seem like only one extra percussion element in the mixdown. As a final thought, before moving on, one of the loops had a really nice grungy character that I thought added something quite special, but it got a bit much after a while so I added an AutoFilter and chose a setting that had some level-dependent envelope modulation applied to the cutoff frequency. The best way to hear how that sounds is to actually listen to it rather than have me attempt to describe it. It is the part called "Dirty Electro." I planned for it to start off with the cutoff filter set at minimum, at which point you would just hear a subtle part of it creeping through owing to the envelope modulation, and then, as the track progressed, I could open up the cutoff frequency to give the full sound.

At this point I was pretty happy with what I had done, but the one thing I wasn't 100% sure of was whether the drums and percussion needed a little more "swing" to them. They felt fine as they were, but I had a feeling that, later on, I would probably be changing the quantizing on them just a little. It wasn't a big deal as it would be relatively easy to do later on. Actually, I *say* that, and it really is true. But requantizing audio files, as you read earlier, is something that has only very recently become possible in a quick and painless way. Prior to that it would have been quite a time-consuming, manual editing job. Achievable yes, but not something you would want to do on a whim just to try it out. With the drums pretty much taken care of, for now at least, it was time to move on to the bass line.

STAGE 3: BASS… HOW LOW CAN YOU GO?

When I am working on a bass line, I try to separate it into two distinct processes in my head. First there is the actual rhythm of the bass line, and then there are the notes and chords to consider. What I usually try to do is just loop one or two bars of the drums and mute the vocals while I try out different patterns for the bass line. While I am doing this I just choose the first chord of the verse chord progression as a basis for root note of the bass line. With this remix, that was actually the majority of the verse pattern, so it meant that I could have a longer loop to work with while figuring things out. You might be wondering why I mute the vocals while I am doing this. After all, it would make sense to keep them in there given that the bass line will, eventually, have to work with the vocal rhythms. And my answer would simply be that, in the past, when I have worked out a bass line against the vocals, I have felt a little restricted in terms of what I could do. So now I work out the groove of the bass line just against the drums and then unmute the vocal to see how it fits. If it works, great, and if not, I try moving a few things around. In my experience, this has led to better results because I have come up with bass line patterns that I might not have tried had the vocals been present; I would have thought they were too complex or syncopated, even though, in the end, they worked really well. This is how I work, but it doesn't mean that it is the best approach for you. Try it both ways and see what works best for you.

While I was figuring out what to do, I tried a few different options for the bass line with the one I finally settled on being named "Final Main Bass." All of the different bass lines I tried worked pretty well when I unmuted the vocals, but the one I eventually chose, I think, had the most bounce to it. Each of the others felt a little more dominant than I wanted them to be. Because the original track was quite a flowing and languid ballad, I wanted to introduce a more syncopated feel to match up with the pacier feel of the 128-bpm vocal. Had the bass line been too slow or regular, it might not have worked well with the bounce of the sped-up vocal. That said, there was another one I liked quite a lot, which I named "2nd Choice Bass"—I thought I might use it later if the one I had chosen didn't work out as I had planned.

Once I had the rhythm of the bass line worked out, I then copied the pattern throughout the length of the verse and moved a few notes around to match the changes in the verse chord progression I had built earlier. I thought it worked pretty well, but it sounded a little flat to me in places, so I introduced just a few notes into my bass line that weren't actually the "root" notes. I kept this fairly subtle by using the changed notes as transitions from one chord to the next. They are subtle changes, but help give the track a little more life.

I did feel, however, that the sound wasn't quite what I was looking for. I admit that I had chosen the sound pretty much arbitrarily by just opening up a preset I knew was in the right area of what I wanted, but now that I was happy with the basic feel of the bass line, the sound didn't seem quite right. I think that it was the envelope of the sound that felt a little wrong. Because the pattern was predominantly short notes, I felt that the sound needed to be a little more spikey or snappy and have a really short release time to accent the rhythm. The current sound felt like it was somehow blurring the notes into each other a little. But I did like the warmth of the real low end, so I opened up the ES2 that was making the sound and muted a couple of the oscillators so just the sub element of the sound was playing now. I then loaded up another ES2 on a new channel and started to create a sound that would work with the sub sound

I already had, to give the snappiness I was looking for. I tried a few different sounds before I came to the one I ended up using. It had the qualities I wanted, but now it seemed to be lacking a little in definition. I decided I would come back to that a little later, and find a third sound to layer on top. I also changed the level of the sub sound relative to the new sound and shortened the decay time on the sub sound to tighten it up a little.

I also wanted to add some extra notes to the bass line, but using a different sound, a thinner sound, and in a higher octave. Not too many, but just to place them in the gaps in the main bass line. This is something I tend to do quite a lot to inject extra energy. The sound used will not be especially bassy and will be mixed quite low, but used sparingly the results can be great. To make sure that the notes sit in-between the ones already there, I just copied the pattern down onto a new instrument track and then muted all the notes that were there. I could have just started from scratch, but this gives me a clear indication of where the gaps are and where I can place the new notes. While I tend to play the main bass line in on my MIDI controller, I prefer to add in these *accent notes* by drawing them in on the *piano roll editor*. There is no hard-and-fast rule here, and this is just my preference.

I started by finding an extra bass sound, one that complemented the sound I had chosen for the main bass line. I wanted something that sounded a little more *detuned* for this sound, and as I like to listen to things in context, I inserted a high-pass filter into the instrument channel strip right from the start so I could hear the sounds I was auditioning in the way they would end up. Having a high-pass filter on a bass sound (in this case set to a cutoff frequency of about 1000 Hz) really *does* change the perception of the sound. This is why I always apply the filter before choosing the sounds on an *accent* part like this. I went through a few different options before finally settling on a nice sound that had quite a lot of *unison detuning*. In fact, had I been choosing a main bass sound, the detuning might actually have been too much on the lower frequencies, but because those were filtered out it didn't sound quite so wobbly. I also added some delay to sit the sound back in the mix a little more.

In order to make things more interesting and not too repetitive, I made sure that I had the whole 16-bar verse set as a loop and then started to draw in some notes in the gaps between the (muted) main bass line notes on this MIDI part. I didn't want to add too many, otherwise the bass line would have been too busy. In the end, I settled on having just a couple of accent notes in each bar, but I also made sure that the position and actual notes varied through the 16-bar sequence. Variation is always good, especially if it is subtle. You have to strike a balance between regularity, which helps people recognize and "settle in" to a track, and variation to keep things interesting. It seems almost contradictory until you add the qualifier that the variations should normally be subtle throughout each section, so that the changes feel more like *evolution* than *revolution*.

One trick I have used many times for things like this is to come up with a few minor variations of a part—let's call them A, B, C, and D—and then use them in a repetitive nature! What I mean is, in an 8-bar loop I would use them, for one bar at a time, in the following sequence: 1-A, 2-B, 3-A, 4-C, 5-A, 6-B, 7-A, 8-D. By doing this, you are keeping things interesting without overloading the listener with too much new musical information. This is exactly what I did here and I was pretty pleased with the result.

Now that I had done all of this, it was time to move on and do the whole process again for the chorus chord progression. There may be times when you want to, or *need* to, change the rhythm and pattern of the bass line in the chorus. I did actually think about doing that for a while, but eventually I settled on keeping the same basic pattern for the chorus (adapted to the different chord progression, of course) and then, if it felt like the bass line needed a further lift later on, I would probably add another sound layer either to the main bass line or the accent bass line. Rather than replay the main bass line in with the new root notes for the chorus, I more often than not simply copy the MIDI part and move the notes around. This is the best way to ensure a consistent groove. If I wanted to change the rhythm or feel of the bass line in the chorus, I would, of course, have replayed it. It didn't take long to get the verse bass line(s) working with the chorus chord progression, and I was pretty much finished. However, one thing I did want to do was introduce some further small changes in the accent bass line. I just wanted to add a couple of variations to it so that, instead of using A, B, C, and D as described previously, I added in variations E and F and ended up using them in the following sequence: 1-A, 2-B, 3-A, 4-E, 5-A, 6-B, 7-C, 8-F. Although the change was subtle, I thought it did make a difference, especially as variations E and F had a little more interest to them.

One other thing to consider is that, even if you copied the MIDI part and simply changed the notes in the new section, you could always go in and make changes to individual notes by moving them around or changing the lengths of them at a later stage. At this point in time I wanted to just get something started that I could work on more later if necessary. The groove I had worked perfectly for what I wanted to achieve, so I felt that enough had been done on the bass line for now, and it was time to work on the next stage.

STAGE 4: THE SPACE BETWEEN US

Up until this point I had been using the very rough piano chords I had initially played in, but now was the time to change this for something more in keeping with the sound I wanted to achieve. I had a very specific sound in mind, that of quite a thin and reedy 1970s string machine, probably with some light phasing on the sound as well. We had a brief look at creating *pad* sounds in Chapter 12, but fortunately for me, I had a sound I had used on an earlier project that I thought would make a good basis for the sound I was looking for, without having to start completely from scratch. Sometimes it is nice to sit and "play" with sounds, but other times you may be eager to press ahead with other ideas, so you don't always need to start from scratch on every single sound!

When I had loaded the sound and heard how it sat against the bass line, I thought it was pretty close to what I was looking for, but it was a little too full bodied. A simple high-pass filter, as I had originally intended to put on the sound, achieved much of what I wanted to do, but the overall EQ of the sound still seemed a little too dominant. So I spent some time working on the EQ and ended up with a sound that combined quite a steep high-pass filter with a little low-pass filtering and a slight dip in the midrange as well, just to help limit the sharpness of the sound and to give an element of that "telephone" EQ effect. I am also quite a big fan of sidechain compression, so I added that effect to the pad sound as well, albeit subtly. I wanted the pad sounds to be mostly in the background anyway, so I probably could have not used this effect without them being too dominant, but I felt that the sidechain compression, even though it was subtle, would still help to make a little more room in the track, either for other elements to work in later on, or just to help it not feel too busy. I added some of the Logic Ensemble effect to give the pad a bit of a swirl, and then added a reverb and set the mix control to about 55/45 wet/dry to really push the sound back in the mix and give it a nice tone.

Now that I was happy with the basic sound of the pads, I decided to listen again with the vocals playing just to check that everything sounded good. Up until now I had been working with just the dry vocals as, when I am working on drums, percussion and bass lines, I never really worry too much about the actual sound of the vocals themselves. But now that I was getting a little further into the remix, I thought it was time to do some work on the vocals to help get a better sense of how things would sound. I still had to do some work on the vocals anyway because there were places where the time-stretching had made them sound fairly odd, but I didn't want to interrupt what I was doing just yet to work on those. As I have implied throughout the book, the vocals are usually the most important part of any track you are remixing, and as such, deserve special attention. With this track, I wasn't too worried about the vocals. Sure, there were a few things I had to work on, but it was a long way from being the most complex vocal I had ever had to deal with. I also had the benefit of not having a crazy deadline to work to, which always helps. So for now I just wanted to add some compression, EQ, and some basic reverb and delay to make the vocals sit *inside* the mix rather than feel like they were floating on top of it.

The vocals in the original track had been recorded very well to start with, so didn't really need a great deal of compression. In fact, the only reason I added the compression at all was to give the vocal a more *up front* and slightly *edgy* sound. I thought this might help because the original track, being a ballad, didn't have the vocal sitting against a highly energetic backing track. I often find, when remixing ballads, that the vocals benefit from some additional compression to add texture and tone, even if the dynamics of the original recordings are very well-controlled. To some extent this can be achieved (or at least contributed to) by some EQ, but compression helps to make the vocals cut through in a different, and sometimes to me at least, preferable way.

There are probably hundreds of different compressors available, and each will have a slightly different sound to it. Most of us have access to at least a few different compressors, and I would recommend trying out different ones to help you achieve the effect you are looking for. As I wasn't really looking for level control and was certainly not looking for *transparency*, I would normally have chosen something like the excellent Universal Audio model of the Fairchild 670 or perhaps even their LA-2A, but as the aim of this section of the book is to allow you to open up the files for yourself, I chose the normal Logic compressor. There are a number of different compressor models in here, so it took a bit of experimentation and a little while to set this up to get the effect I wanted, which was just a bit of warmth and bite to the vocal, without making it sound overly distorted. In the end I settled on the settings you see in Figure 21.1.

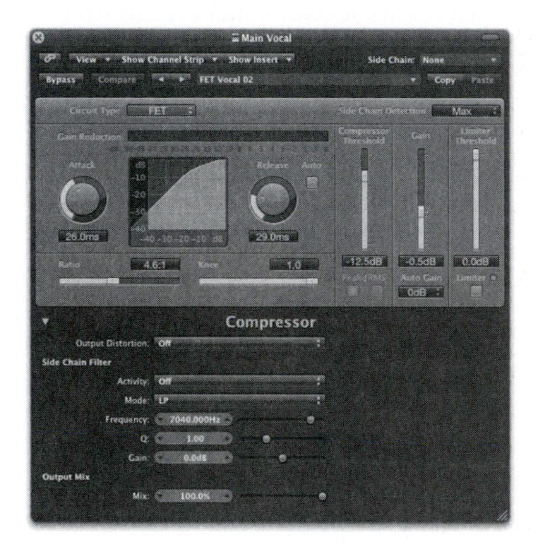

FIGURE 21.1
Here are the settings I used for the main vocal track. It was enough to give the vocal some added energy to help it blend with the increased pace of the backing track.

As it turned out, once I had applied the compressor to the vocal I didn't feel, at this stage, that it needed any EQ applied. This would possibly change throughout the whole process of working on the track, as the frequency content of the backing track changed, but again, it was fine for now. There was enough presence in the vocal to allow it to be heard over the denser backing track I had in mind, but now I had to do something about pushing that vocal back down into the mix, because at the moment, it felt like it was both on top of *and* in front of the music.

To start adding some depth to the vocals, I first added some delay courtesy of the built-in Stereo Delay plug-in in Logic 9. I tend to use a stereo delay rather than a mono one for situations like this because a stereo delay will have different delay times for the left and right channels, which can add a more interesting rhythm to the delay effect and help to increase the width of the vocal track (and to some extent the track as a whole, as a result). The actual settings I use for this kind of delay, one that I apply generally to the vocal, are subtle. In this case I chose delay times of a quarter note for one channel and a dotted quarter note for the other (to give the rhythmic interest I mentioned previously), and I kept the mix of the delay itself quite low at around 10%. I adjusted the feedback controls until I could hear about three or four repeats clearly, and then removed quite a lot of the lower frequencies and some of the higher ones from the delayed signal. I had been setting up the delay parameters with the vocal soloed, so when I listened to the vocal with the music I felt like it was a little too subtle, so I increased the mix amount ever so slightly to about 13% and increased the feedback setting a little as well.

In addition, I felt that the vocal needed just a little reverb. There are many different opinions about mixing vocals (and tracks in general, really) and about what is the "right" amount of effects—and especially reverb—to use. There is very much a general trend, at the time of writing, toward tracks and vocals that are mixed far more dry and upfront. I am sure there are reasons for this, but I like to have a little depth to my work. I also do appreciate that if everything in a mix has reverb applied to it, the results can sound flat as well, albeit flat and distant because of the reverb. I believe that to truly build up a mix with depth there needs to be a bit of both approaches. Some things need to be upfront and more dry while others need to be sitting a little further back.

In this track I didn't feel that the vocals needed to be swamped in reverb, so I had three options. The first was to have a longer and bigger reverb, but at a lower level; the second was to have a shorter reverb, but at a higher level; and the third option was to use a combination of both. This is something I have started doing more of lately, and not just on vocals. By using a combination of a shorter reverb to liven up the vocals and a longer reverb to make the vocals sound a little bigger, you can get a really good compromise. I often use a pre-delay setting of about 40 ms on the shorter reverb and sometimes as much as 100 ms on the longer reverb, to give a three-dimensional feel to the vocal. The shorter reverb, with the shorter pre-delay, gives a burst of ambience, which then quickly dissolves into the subtler but more expansive longer reverb, which occurs a fraction of a second later. This combination can work really well on vocals and on some other instruments, but again, too much of *any* kind of reverb can easily overload a mix, so use sparingly.

There have been great advances in reverb plug-in technology over recent years, and one of the developments has been the advent of *impulse response* (IR) based reverbs. As we saw in previous chapters, these reverb plug-ins use *samples* of real acoustic spaces to provide the reverb instead of relying on a mathematically created model. The advantages of this are that the reverbs sound more smooth and natural. While I am the first to agree that there are many occasions when these kinds of reverbs are preferable to artificial ones, there are also times when I believe an artificial reverb can work better. Perhaps it is a result of our ears simply getting used to hearing artificial reverb effects on recordings over the years, or perhaps it is something else, but to me, at least, sometimes an IR reverb just sounds a little *too* subtle and sublime to really cut through, especially in largely electronic instrument based dance music.

Because of this, I often choose an IR-based reverb to supply the shorter ambience reverb, and then an artificial reverb to supply the main body of the reverb sound. An interesting twist to this, however, is the fact that, lately, there have been a number of IR libraries developed, which instead of sampling real acoustic spaces, use the same principles to actually sample the sound of classic reverb units. Of course, these samples don't have the programming flexibility and adjustability of a real hardware reverb unit, but if the actual settings that have been sampled are appropriate for what you are looking for, they offer a way to get the sound of reverb units costing several thousands of dollars into a relatively inexpensive home studio!

For this vocal I chose the "0.5s Bright Vocals" setting from Logic's own Space Designer IR reverb plug-in to supply the ambience, and then, after trying quite a few different IRs, eventually settled on the "02.4s Long Snare Reverb" (yes I know it's supposed to be for drums, but it sounded good!) setting for the longer part of the reverb. I spent a little while adjusting the amount of reverb signal from both reverbs until I had something close to the effect I wanted. However, I didn't spend too much time fine-tuning these settings as I knew they would change later, once I had additional instrumentation and possibly a slightly different vocal sound to contend with.

Now that I had reached this point I would consider I had put down the basics of the remix. I had something I could use to build on by adding additional sounds, by automating certain parameters of the sounds I already had to provide some dynamic interest, and by bringing different sounds in and out of the mix entirely. The one final thing I wanted to do before getting stuck into the remix was to decide whether I was going to work on the radio edit or the club mix first.

There are two schools of thought regarding this, but the seemingly accepted wisdom is that if you are mainly aiming the remix at clubs then you should start with the club mix, as this will be the main target market of your work. Similarly if your main intended market is radio play then the Radio Edit might be the more sensible starting point. In either case you will probably need to provide a full package of mixes (Club Mix, Dub Mix and Radio Edit being the main ones). To be honest I don't think it really matters which route you take as you will be arriving at the same destination in the end. I think you will discover which route works best for you fairly quickly. We discussed in Chapter 6 the different approaches of working on a club mix and radio edit, perhaps one or other of these feels more comfortable to you and whichever you choose you can work on the other version after.

I generally work on the radio edit first. I have a few of reasons for doing this. First, for the majority of the remixes I work on, the aim is to secure radio play wherever possible. Second, and more practically, it is quicker to work on a radio edit; there is less work on the arrangement to do, less time spent getting creative with the vocals as, for the most part, the vocals come in very close to the beginning of the arrangement and then pretty much run straight through until the end. Finally, there is less time spent bouncing down (or recording to CD/DAT) mixes before being able to send them to the record label. It might seem like you are saving only minutes at a time, but as I mentioned earlier, remix deadlines tend to be almost impossibly tight, so if you are planning on making remixes an important part of your career then *anything* you can do to streamline your workflow and make things happen more quickly (without compromising the quality of what you are doing, of course!) is welcome. In an ideal world, one without deadlines, I would spend much longer on my remixes, but for my workflow I think I would still start with a radio edit anyway, in spite of not having to save time. Who knows, that may change over time, but right now I like the immediacy of working on the radio edit in the first instance.

With that in mind I decided to do some quick arrangement of the relatively few parts I had so far in the mix and map out a very quick radio edit arrangement of the track before moving forward to the next stage, where all of the hard work, fine-tuning, and real attention to detail would come. If you listen to the remix, even at this very early stage, you will start to get a feel for how things could end up sounding. In the final Logic file for this chapter I have also included two versions of the original (as muted audio files at the top of the Arrangement window). One of them is the original track at the original tempo and the other is the same track but time-stretched to match the tempo of the remix we are working on. It is interesting to listen back to the original track at certain points during the remix so you can hear how different it sounds and make sure you have come far enough away from the original version (not difficult in this case as the original was a ballad, but it's not always like that) while at the same time making sure it still sounds like the same song! It's not always an easy thing to balance those two seemingly contradictory goals, but you get used to it and become a good judge of it after a while.

CHAPTER 22

Remix Walkthrough: Working on the Details

INTRODUCTION

The next stage in the remix, and in many ways the most challenging, will actually vary quite a lot depending on which genre you work in. If you are working with a more minimal sound, there is far less detail to fill in, at least in terms of additional instrumentation. This stage in the process will be far more concentrated on making sure the sounds are right, the grooves are right, there is a good degree of movement and interest created from the subtle (or sometimes not so subtle) changes in the sounds through automation or perhaps creating some additional complexity through extra percussive and FX elements throughout the track. If you were working in a genre that had a more complex and layered production style, you would probably spend more time working out extra musical parts before working more on the details of individual sounds.

This is also the stage in the process when there seems to be a lot of back and forth work. Sometimes things will progress smoothly and as planned, in a flurry of delicious creativity. Other times you will hit what seems like a brick wall, and it will take a lot of work to overcome it. Most of the time it will probably be somewhere in the middle. One thing I have learned over the years I have been doing this is to not fight what you feel. If you had one idea for the remix, and then, midway through, you have an idea that would take you in a different direction, it is always worth at least *exploring* that idea. If you feel you don't have the time to experiment, by all means push ahead with what you are doing, but if you find yourself stuck and trying to make your initial idea work, then by exploring a different direction, and in spite of the fact that you might have to go back a few steps, you can actually end up saving time.

The only caveat to everything I have just said is to consider any brief you may have been given by the record label when you think about possible different directions to take the mix in halfway through. There have been many times I have had what I thought were *great* ideas for a remix, both at the very beginning of the process and halfway through, but I couldn't explore them because they ran contrary to a specific brief I had been given.

The main thing to think about, in my opinion, at this point is how to approach filling in this extra detail. This is something I cannot give you any hard-and-fast rules about, as this part of your workflow is very much a thing that will develop and be unique to you. The best I can do is explain what I do, and *why* I do it, in the hope that it might help guide you on to your own path. And my process basically comes down to two main points: start at the *high point* of the song and work backward, and put in *all* your ideas because you can always take them out again later. Let me explain a little more.

By "high point" I mean the part of the remix where there is the most going on in terms of musical and production/sonic complexity. This is often, although not always, the section of the remix just after the main breakdown. The reason I tend to work on this section first is simply because there have to be enough elements in the remix as a whole to fill this section up (enough that is relevant to your genre) while also making sure that there isn't too much going on. I will come back to that last point shortly. Because of this, it is usually the case that any element that appears in this section of the track will have appeared (in one form of variation or another) at some point earlier in the track. As a result, all of the musical elements you have used here can be reused (with variations and changes where necessary) earlier in the arrangement and can be gradually brought in (and later taken out) as the course of the track progresses, ultimately culminating in this section of the track we are talking about.

I mentioned just now about not having too much going on so that it is not overcrowded, but this in some ways seems to go against my second main point: put in *all* your ideas. It really doesn't, when you consider that I mean you have to make sure that *eventually* the mix isn't too crowded. I always try out any ideas I have. Maybe 70% of them never get used, but I never regret trying things. You might come up with something that you only use for 16 bars in the whole track in this high point, and nowhere else, but it is the attention to detail that will turn a good remix into a *great* remix. And variation, be it subtle or more obvious, is one of the keys to this attention to detail. It just shows that you have made an effort to make sure the track progresses and evolves over the course of its 6–8 minutes, rather than just throwing a few things together with a "that will do" attitude. The differences might be subtle but they are enough to keep an A&R exec, a DJ, or a radio listener fully engaged and interested until the end of the song, and that in itself is half the battle won!

One final thing I will mention before moving on to the details of how I approached this stage in the process with the remix on the website, is that, during this stage, I don't tend to fuss or obsess too much about the fine-tuning of the mix in terms of EQ, levels, compression, and effects. Of course, to some extent you work on this as you go along, but I prefer to spend this time being creative rather than technical, and going with the flow.

BUILDING UP THE LAYERS

I had taken a break from listening to, and working on, the remix for a little while and the first thing I felt I needed to add was some rhythmic, but musical as opposed to purely percussive, interest. The bass line was already providing a nice rhythm, but I felt it needed a nonbass musical part that played off of the bass line rhythm and emphasized and supported the pad sounds that were already there. Because I didn't want this new layer I was adding to take up too much space, I started to look for an appropriate sound with a fairly short decay time. What I actually wanted was something along the lines of a pizzicato string sound, but with a synthetic tone to it. As this is a type of sound I have used quite a lot before, I already had a good idea of how to create it, so I loaded up the ES2 and started working on the sound.

I wanted something wide and full sounding, so I started with all sawtooth oscillators, slightly detuned relative to each other, to give some fullness, and set the cutoff frequency about 25%, the resonance at 0, and the filter type to 12 dB/octave low-pass. I set the attack time on the amplitude envelope to minimum, the sustain level to maximum, the decay time to about 500 ms, and the release time to around 200 ms. I then set up similar envelope settings on the filter cutoff envelope, made sure the filter cutoff envelope control was set to maximum depth, and started to play some chords to fine tune the sound a little. The first things I adjusted were the decay/release times for both envelopes. I felt that the filter envelope times needed to be adjusted a little, and I also played around with the filter envelope depth. Then I actually changed one of the oscillators to a square wave and tuned it up an octave to give it a little more bite. In the end the overall sound seemed nice in the context of the music so far.

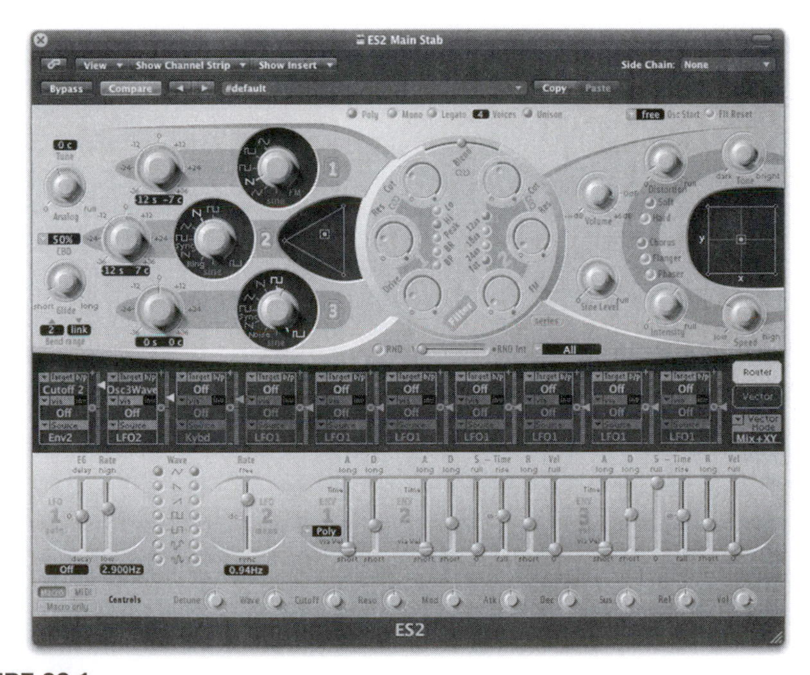

FIGURE 22.1
This is what became the "ES2 Main Stab" sound. You can see the settings used here. Note the fairly short decay and release times on the filter envelope. Those, combined with the fairly low filter cutoff frequency and filter envelope depth, are what give the sound the majority of its character.

When I started working on the actual notes and chords I was going to use for this sound, I began by putting a fair few of these stabs into the chorus, but the more I played with the sound, the more I felt that they should be quite a regular, albeit offbeat, pattern. Quite by accident (it happened while I was playing and adjusting the sound at the same time) I found that having a repeating rhythm of one chord every three 16th notes seemed to work quite well. This had the effect of making a new rhythm of 3s that ran over an existing loop. In the end I settled on repeating the pattern every two bars so that the 3s didn't become too much of a focus. I played in this rhythm, following the chord sequence of the chorus, and then sat back to have a listen. I really liked what it was doing but I felt it definitely needed some spatial effects to help fill it out.

On sounds like this that are quite sparse in their usage, I really like to use some kind of delay plug-in in Logic and set up delay times that work to create an additional rhythm with the main sound. Because the pattern for the main sound was every three 16th notes, I set up the delay with the same note values. With this stab sound soloed, I adjusted the levels of the delay signal as well as the feedback settings, and as I nearly always do with delay effects, removed some of the low and high frequencies with the controls within the delay plug-in. The reason I do this is because by removing those frequencies you help to differentiate the echo from the sound itself, so it sounds like an echo effect rather than a repeated playing of the sound. Plus it helps to sit the echo further back in the mix. When I thought it was about right, I took the sound out of solo to listen to it in context, and then, after hearing how the delay effect sounded, decided to increase the levels of the actual delays a little to make them a slightly more audible.

Next up was reverb. I didn't want this sound to be *too* distant, but I did want a little ambience to it, so I used the PlatinumVerb in Logic and used the "Bright Long Verb" preset as a starting point, and adjusted the settings until I thought they were fairly close. The reverb signal ended up being quite loud, but I was planning to use sidechain compression on the stabs, and as I will discuss shortly, if you use sidechain compression on a sound with quite a lot of reverb the compressor actually *pumps* the reverb signal as well, which can be a great effect. In the end, the dry signal was set to 67% and the wet to 33%. Once again, I wasn't assuming that this is the level it would stay at, but it was certainly fine for now.

At this point I should explain why I go against what is the generally accepted approach of using reverb and delay effects as *send effects* rather than *inserts*. Using just a few different reverb units (hardware or software) and then sending different sounds through them using an auxiliary send tends to make sounds feel as though they are in the same acoustic space. If you think about it, for "live" band type music this makes a lot of sense, as, subconsciously at least, we would expect to hear all the members of the band playing in the same physical space at the same time. Electronic-based dance music, however, is not subject to those same subliminal preconceptions. In addition, many the people who make this kind of music have not learned their trade working in a traditional studio environment, so don't necessarily subscribe to these "rules." As such, when listening to electronic music, our ears (and therefore our subconscious) are used to hearing things existing in different "virtual spaces" at the same time.

One of the other reasons why these traditional techniques evolved in the first place, aside from the recreation of a more realistic sounding recording, was the very limited resources available in older recording studios. As studios always used to rely on hardware reverb processors there were, typically, only a few of these available at any one time, so rather than use them as insert effects on a particular track, they used them as send effects so they could be used on a number of instruments/tracks at the same time. Now that the technology is available to use a virtually unlimited number of software reverbs, this limitation is no longer relevant. Of course many studios still use hardware reverbs, but they will most likely have the option to use additional software reverbs if they choose to. And because we have the option to experiment more now because of the technology we have, well, it would be rude not to!

There will always be people who prefer the older, "proper" way, and they will often be very vocal about things being done "the right way," and there will be those who want to push things forward. What I do really depends on how my mood takes me, but most often I will use a combination of insert effects on tracks I want to separate out from the mix and send effects for tracks I want to place in the same virtual space. This combination gives a sense of realism combined with some interesting spatial effects for certain sounds.

I mentioned just now the idea of using reverbs as insert effects rather than send effects to get an interesting effect with sidechain compression. If you have a sound you want to use the sidechain compression effect on, and you put quite a large and obvious reverb effect on as an insert effect, and then have the sidechain compressor *following* the reverb, then the compressor has the effect of pumping the reverb signal and the original sound. In some ways this can make the sidechain compression effect even stronger because it goes totally against what we are used to hearing. Reverb, in one form or another, is around us pretty much 24 hours a day. Most of the time it will be very small acoustic spaces that we don't actually perceive as "reverb" as such, but it is still there. Still, most of us, at some point or another, have been in a church or concert hall or even a long tunnel and have heard these larger reverb effects.

Because of this, our ears are used to hearing reverb. In fact, if you have ever been into an anechoic chamber (a specially designed room with absolutely *no* reverb at all) you will know how strange it is to hear sounds without any reverb on them! But what our ears are *not* expecting is to hear that acoustic space rhythmically pumped by a sidechain compressor. It is a very strange effect, far more impressive than sidechain compressing a dry sound, because it really does make you feel like the space you are in is being compressed! It's an effect that is certainly not appropriate (or even usable) in every situation, but you owe it to yourself to give it a try. The effect it has is awesome. So, after the short digression, back to the remix!

The new sound I had put into the track was working well, and was something I thought I would want to use in the verses of the song as well, but certainly in a more background role. I copied the MIDI part over to the verse and set about changing the notes to match the chord movement in the verse. Once I had done this, I started thinking about how I would further change this part to work in the verse. I still wanted to keep the feeling of the pattern working in 3s and crossing over the bar as it did sometimes, but I did want to vary it a little.

At first I tried varying the pattern of the part, but no matter what I tried, I didn't like the change of flow from verse to chorus. So the next step was to try changing the sound a little between verse and chorus. To do this I tried opening up the filter a little on the sound in the chorus, and this certainly gave a bit of a lift, which was what I had been looking for. But it still didn't do everything I wanted, so I chose to do what I often do, which is to look at layering sounds. What I was looking for was something that would create an additional lift going into the chorus without necessarily changing the whole feel of the sound. In fact, I knew exactly the sound that would do this but there was one small problem. The sound I wanted was one I had programmed on the Korg Legacy Cell, and given that I wanted to create this entire remix using only Logic plug-ins I had a little work to do. Fortunately the Legacy Cell works on what is, at heart, a standard analog synth type architecture, so it was fairly easy to set up a sound that was very similar in the ES2. After a bit of tweaking to the basic sound and then working on the effects, I had something that was fairly close to the original Legacy Cell sound. It wasn't 100% the same because the oscillator waveforms (although in theory the same) would be a little different, and the sound of the filters, and even the effect of the envelopes, would all have a bearing on the sound. All in all though, it was close enough even in isolation, and when I compared the two side-by-side in the context of the whole track I thought it was just about right.

I copied the MIDI part for the original sound in the chorus on to this new sound and had a listen. When the track reached the chorus, there was a definite increase in the sense of energy. At the moment the change felt a little harsh, but I could fix that by filtering in the new sound rather than just having it coming in out of nowhere. I just wanted to try one thing before moving on. Because of the key of the track and the chords I chose, the original stabs were quite low in frequency, and as a result, had quite a dark sound to them, which even opening up the filters couldn't remove completely. I decided to try transposing the new sound up one octave, so in addition to the new sound coming in (which was, incidentally, not radically different from the first but just different enough), there was an additional higher octave musical part that definitely added a nice sense of brightness to the chorus.

The next thing that would be obvious to put into the verse is the pad sound. Although this sound was meant to be in the background, I was a little reluctant to use the sound in the verse and the chorus simply because it would mean less of a dynamic build into the chorus. But at the same time,

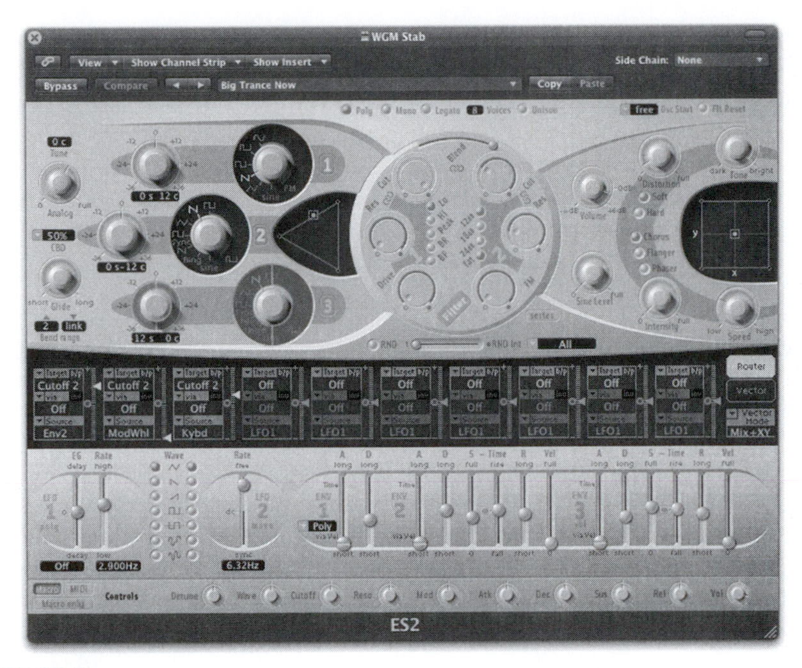

FIGURE 22.2
These are the ES2 settings for the additional stab sound that comes in in the chorus (called "WGM Stab" in the arrangement).

I generally prefer my songs and remixes to be smooth in nature and I am not a huge fan of sounds just coming out of nowhere. So I copied the MIDI part for the pad sound over to the second half of the verse, changed the chords to match the verse bass line, and then added a low-pass filter to the pad channel strip and drew in some automation that raised the cutoff frequency from 20% to fully open over the course of 8 bars, but with a quicker build-up toward the end of the pattern. If you look at the cutoff frequency automation for this sound you will see that the increase in frequency isn't linear (in other words, not a straight line), but more of an upward curve. I did it this way to make it feel like a quicker build-up leading into the chorus. It might seem like a subtle and small detail, but much like the small change to the verse pattern on the stab sound, all of these small changes might be insignificant on their own but give the whole finished track a very gentle sense of evolution without needing to be really obvious changes that *could*, potentially, almost catch people's attention *too* much.

Listening back at this stage I wasn't actually sure if this gave the effect I wanted. There was a gentle lead-in to the chorus pad sound now, but at the same time, it made this sound feel less important. Perhaps a big part of this was that the sound itself was very quiet and in the background. I wanted to make it more of a feature without it being dominant. What can be useful in situations like this is to bring in another layer of sound to make the sound feel more open than just a filter cutoff change can do. I could have perhaps changed the low-pass filter setting on the EQ to allow more of the high frequencies of the sound itself to come through, but to be honest, I liked the character of this sound as it was and didn't want to change it! Because I really didn't want to make this combined sound too full or dominant, I thought that the best approach would be to find (or create) a sound that was a bit more bright and edgy than the sound we already had.

The first thing I tried was a bell-like sound. Not a bell in the sense of a church bell, but very much a classic D-50 type synthetic bell. I tried a few sounds like that, but in the end, they all sounded a little *too* sparkly to be a main feature. I kept my favorite one in there in case I decided to use it later on! My next idea was to try something completely different. For those of you who know "California Gurls" by Katy Perry, there is a sound near the beginning of the track that plays some chords. It almost sounds like the kind of sound you would expect from an old '80s computer game (think Commodore 64 SID Chip sounds), and I had created, for another project, a sound like that. It has quite a nice edginess to it, so I wondered how that sound would work out as a sustaining pad sound rather than a shorter, more stabby sound. I loaded up the ES2 preset and changed a few of the envelope settings, and then copied the MIDI part over for the chorus chords. It sounded good! I copied over the reverb settings I used for the main pad as well, adjusted the

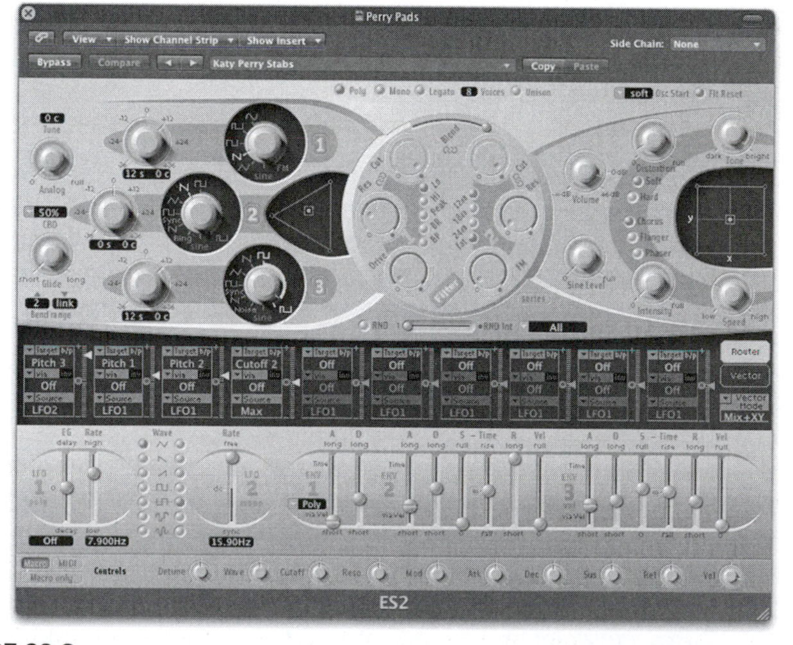

FIGURE 22.3
ES2 settings for the "Perry Pads" sound that gives a nice edgy quality to the pads even though it is mixed quite low.

wet/dry balance to be more dry, and then played with the level a little to get the balance between the two sounds right. But I was left with something that had a decidedly retro feel and was thin enough to be a background sound but wasn't harsh or tinny in any way. Just for fun I also unmuted the bell sound I had created, and even this worked well with the two other sounds. I still wasn't sure if I wanted to use it though, not because I didn't like the sound, I really did, but because it felt like it might get too dominant at that point. But it was nice to have it as an option. It could, for example, be something I only brought in during the final chorus of the track to give it one final push before the ending. With pads sounding good, I saved the file and took a quick break to refresh my ears before having another listen to what I had so far.

There was still something I felt was missing from the overall sound, but couldn't quite figure it out. I didn't want to get caught up trying to figure it out, especially as I had already decided what I wanted to do next and was keen to move forward and not lose track of my idea. In some ways that can be a hard part of remixing, or music production in general. Sometimes you get a real flow of ideas, and it can be frustrating when it takes longer than you want to finish what you are working on at the moment. I have, on quite a few occasions, stopped working on programming a sound for one particular musical part to just quickly play in an idea for another one before I forgot the rhythm or melody that was in my head. It is easily done, because you are thinking about one rhythm or melody while, probably, listening to a totally different musical part. I know many people who actually keep a Dictaphone type voice recorder handy for recording ideas when they are away from their computer, as for many—myself included—some of their best ideas come at totally awkward times such as lying in bed just about to drift off to sleep. These recorders can be had relatively cheaply these days (even the digital ones), and if your budget allows, I do recommend one, as it will, in all probability, come in very handy.

The idea I had wanted to move on to was a simple melodic idea that worked along with the stab sounds I had already created. I wanted this part to be the same rhythm as the stabs, but instead of being full chords to just be a single changing note that created a melody. It would have been possible to add additional (higher) notes to the chords that were already there but I wanted the sound to be distinct from the sound of the stabs. Although the mix was a house mix the sound I had in mind was very much a trance sound. There is very little that is *genuinely* new in any kind of music, but there are many combinations of genres that haven't yet be explored in any great depth, so because I have a number of varied influences, I often like to combine elements (in this case sounds) from one genre into another. Add to that the fact that, at the present time anyway, there is a definite movement toward a blend of house and trance, and it seemed like a good idea.

My first choice for a sound like this would normally be something like the Korg Legacy Collection or ReFX Nexus, but for the purposes of this remix walkthrough, I wanted to use synths and plug-in effects I could include on the book's accompanying website so you can read about what I am doing and listen to it, and actually get in there and get your hands dirty (so to speak) with the files themselves. With that in mind, I ended up programming a sound on the good old ES2 synth that had a good majority of the attributes I wanted.

I already had the melody idea in my head, so I just recorded the notes in and then set about adjusting a few parameters on the basic sound I had loaded to make it fit where I wanted it to. Quite by mistake I actually ended up softening the attack of the sound a little, and even though it wasn't what I was planning, I thought it worked quite well. By having a slightly slower attack than I had initially programmed (ending up at around 100 ms) it gave the stab sounds a little more room. Because this melody was following the rhythm of the stabs (for the most part anyway), it almost felt like a part of the stab in some places, and in others it felt separate, which was an effect I liked. As always, I wanted to try layering another sound to give this lead part an extra dimension, so I loaded up an ES1 and created a simple sound based around a straightforward sawtooth wave with some pitch modulation (vibrato) added to give it a little wobble—enough to separate it from the other sound. I gave this sound a sharper attack, however, and went into the MIDI part and made all of the notes legato so that it wasn't following the rhythm of the stab parts. The combination of the different attack times and different rhythms worked nicely when I pulled the new sound down in level relative to the softer sound.

When I listened to the lead sound in the mix without the vocals, it sounded really nice (after a little tweaking), but when I unmuted the vocals it got to the point of being way too much! I did want to use the melody, though, so I routed both sounds through a submix bus and applied a simple low-pass filter (Autofilter) to the bus. This meant I could keep the sound really filtered down while the vocal was there and then open the filter up during an extended chorus, probably in the club and dub mixes more than the radio edit.

I now felt that there were a good amount of layers of sound in the mix, so it was time to take a short break and then go back in and start working on each of the sounds in more detail before deciding if any other musical parts were needed. I had a feeling that, if anything, any extra parts would be minimal, at least in the chorus parts of the track. I still needed to work on the *breakdown* of the track and the final chorus of the song (after the breakdown), but I often leave that until the end of the process.

One coffee later and the first thing I wanted to double check was the balance between the kick and the bass line sounds. I had already spent some time working on the bass sound, but listening again now more critically, I wasn't totally happy with it. The problem for me was that I didn't know if it was something I needed to add, or something I needed to change. In a situation like this there is only one thing to do: try it and see. I was saving multiple versions of the Logic project anyway because of the need to illustrate all of the steps I was taking for this book, but I would normally make sure I had saved a backup of the file before making any changes like this so I had a quick and easy way to get back to where I was before I started. Another option, if you have a powerful computer, is to use Logic's "Copy Channel Strip" and "Paste Channel Strip" to create a copy of the sound you are looking to change. Copy the MIDI part to the new Instrument track and you can then bypass (or mute) the original sound while you try different sounds or different settings. Of course, this soon starts loading up the power requirements on the CPU, so be warned if you are working on an older or less powerful machine. It's fine to do it, but once you have compared the two and made a decision, stick to it and remove whichever sound you are not going to use. It will free up CPU power and memory, and assuming you have been saving regularly and saving incremental versions (that is, not just saving over and over the same version), you should still be able to load up an older song file if you really want to go back to the earlier sound.

There is mental process behind working out some of these sound changes, but it's something that's just about impossible to put into words. The creative thinking process varies so much from one person to the next, and that's something you have to develop for yourself. But I can try to explain in technical terms why I wanted to change the sound, and hopefully you will then have in your mind a more "artistic" view of the reasoning behind it in terms that make sense to you.

One of the questions I get asked most often is how long I take to work on a remix. The truthful answer is "as long as the record label gives me." Sometimes that is as much as two weeks, but more often than not it is in the region of three days or so. That may sound like a long time to you, or it may sound like an impossibly short time, but that is the reality. You need to make sure that, whatever you are doing, you are doing it in the most efficient and fluid way you can. As a result of this, I tend to avoid anything but tweaking of sounds when I am working to deadlines, because it is very easy for me to get caught up in programming sounds and for time to pass far too quickly. In other words, just because I probably *can* get a sound to give me what I want, it doesn't mean that is the best use of my time. I do have sessions where I just mess around and play with sounds, possibly creating quite a few new patches along the way and storing them for later use, but those are almost always at times when I don't have deadlines to meet and schedules to keep. I would never advocate rushing a job unnecessarily, and in some ways I guess that moving on to a different sound source could be seen as doing exactly that, but it really *does* come down to timing.

It took me a while auditioning different bass sounds before I found some I thought might work. In the end I chose another two sounds: one that gave me a nice, tight, "subby" sound (from the EFM-1) and one that gave me more of a detuned Moog-like sound (an EXS-24 sound called "Trance Bass"). This combination of sounds gave me that unison-detuned/chorus/ensemble effect I was looking for, combined with a nice warmth and weight to it and quite a solid attack. At this point I wasn't 100% convinced that the sounds might not be better with a slightly gentler attack and a slightly longer release to give it more of a '70s analog sound, but I didn't want to dwell on it too much right now. I decided to revisit it at a later stage. I added in a very gentle stereo delay with 3/16th and ¼-note delay times for the left and right channels respectively. However, I removed a great deal of low frequency from the wet signal to the point where you almost couldn't tell whether it was an actual delayed version of the bass sound or a rogue percussion element that hadn't been muted. I did this because I didn't want the (already quite bouncy) bass to overpower the other sounds. This was turning into quite a rhythmically complex mix, so I kept the delay subtle.

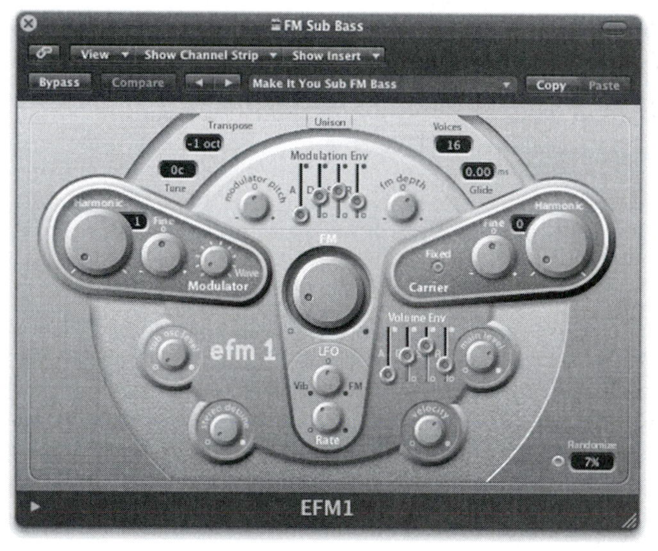

FIGURE 22.4
Settings for the EFM-1 used for the "FM Sub Bass" sound that underpins the bigger "Fat Moog Bass" sound.

At this point I had addressed the issues I had with the basic bass sound and tweaked the EQ on the bass submix bus as I had planned to, and changed the compression settings a little as well. I also spent a little time adjusting the level of the bass line overall compared to the kick drum, with just those two things playing. I listened on my main monitors (Dynaudio BM 12A) at quiet, moderate, and "pretty damn loud!" levels, and double-checked the balance between them on both sets of good headphones I own (Sony MDR-7509HD and Sennheiser HD 600) and even a little pair of ear buds. The reason for all of this effort at this stage was to make sure I had a balance between the kick drum and the bass that didn't just sound good on my relatively expensive monitors. As I have already mentioned in this book, the target audience is *so* important with any kind of music production and they will rarely have acoustically treated rooms and professional monitor speakers or headphones with price tags that run into many hundreds of dollars. Also, given that one of the hardest things to get right in club music is the balance between the kick drum and the bass line (as well as the lower frequencies in general), it is certainly worth spending a little more time on this aspect of the track.

I definitely have a preference for quite deep and warm sounding bass lines in my mixes, so after lots of listening and double-checking I decided to do just a couple more little changes to get things working just that little better. First, I added a compressor to the main bass sound and set it up for *sidechain compression* (with the sidechain being triggered by the kick drum) and an attack of around 1.5 ms, a release of about 5 ms, and a threshold and ratio combination that gave me a substantial gain reduction of about 9 dB at the most. I know that sounds like a lot but I adjusted the settings here by ear until they sounded right, rather than going by numbers. At the points in time when the kick drum hit I wanted *that* to be mainly responsible for the real low-end in the track with the subbass then becoming more of a factor in-between the kick hits.

Now that I had made those few changes, I decided to look again at the stab sound. I wanted to do a little more work on this before working on the pad sounds, because the stabs were, to me, what were now driving the overall pace and movement in the remix and I thought they deserved a little more attention. I was actually pretty happy with the basic sound for this part. Unlike the bass and pads there wasn't anything fundamental that I wanted to change with this sound. However, I still felt like the chorus part needed something extra. I tried automating the filter cutoff again, but this didn't really give me what I wanted because I lost the "staccato" nature of the sound completely when I did that, and it felt like it lost some of its rhythmic energy in doing so. Once again, I was looking at an additional layer of sound to give me what I wanted. Although it seems like there are beginning to be a lot of layers involved in this remix, you have to remember that for all of the musical parts where I am layering multiple sounds, the effect, if done right, is just one of a more complex composite sound, rather than sounding like lots of different sounds playing at once. In that respect, the overall finished mix shouldn't sound overly complex. That's the theory at least. Time will tell if it works out that way!

What I wanted for this extra sound was something that was already quite open sounding, and I had something in mind. There is a great sound I programmed on the GForce impOSCar, which would do exactly what I wanted here. I loaded that up to check how it sounded. With a few small changes and some level adjustments, this new sound started to blend nicely with the others. I had an idea: rather than simply copy the whole stab pattern, I would copy it but then mute some of the chords so that this new sound worked more as an accent to the existing sound rather than another full-on layer of the sound. After trying a few different alternatives, I settled on the rather simple solution of keeping all of the stabs that are part of the ascending chord run at the beginning of bars 1 and 3 of the 4-bar pattern and muting all of the ones that just repeat the same chord in bars 2 and 4. This had the additional benefit of giving me a rather nice pattern to use in a breakdown section, where I needed to hint at the chorus pattern while at the same time having a lot more space. This would probably prove useful a little further into the remix.

Unfortunately I had the same problem I had earlier, insomuch as having a sound that I loved but not being able to use it because I wanted everything to be Logic only. So once again the ES2 got a bit of a workout and I managed to program something that was a fairly faithful copy of the original impOS-Car sound. The result is called "80s Stabs" in the Arrange window, so named because it has that retro synth brass sound I can easily hear playing "Jump" by Van Halen!

This sound was then routed to the same stabs submix bus as the other two, so that it had the same sidechain compression across it, which actually helped pull the three sounds together a little more. I felt that the new sound had a little too much bass in, so I put a simple high-pass filter from the Channel EQ on it at about 170 Hz and then spent a little while adjusting the balance between the three sounds. The new sound was really good, and there was a strong tendency for me to make that the most prominent of the three sounds. But I pulled it back a little, as it began to feel like there was too much of a change between the verse and chorus sections as I was only planning to use the new sound in the choruses. I then saved the project before taking another short break.

Next up would be the pads.

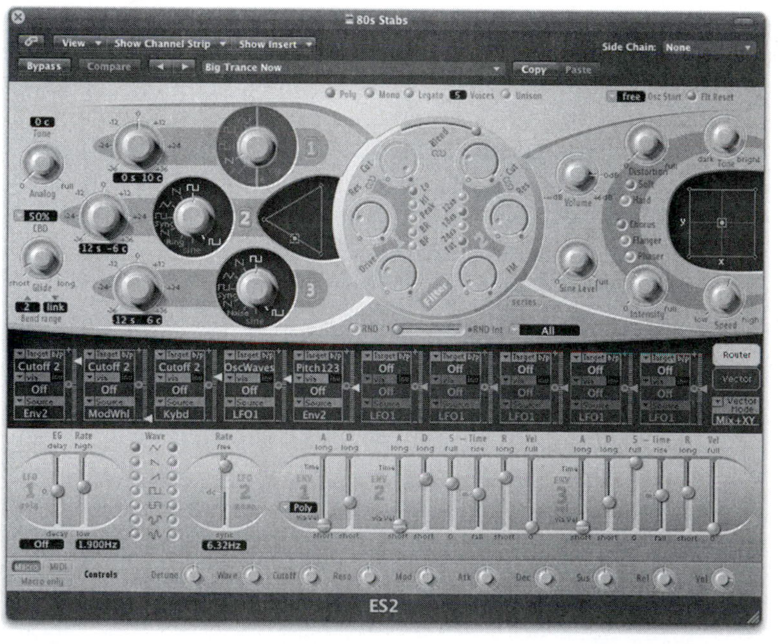

FIGURE 22.5
The "80s Stabs" sound I programmed was an ES2 "copy" of a favorite patch on the GForce impOSCar.

Once again I thought that the sound was pretty much there, and I decided that having the "Bells" in all the way through might not be a bad thing if I turned the level down a little. I thought that the "Perry Pads" sound was competing a little too much with the "String Machine" sound, so I applied a little EQ to the "Perry Pads" and increased reverb level on the sound to push it a bit further back. The biggest issue with me for the pads now was the overall level in the mix and the amount of pumping from the sidechain compressor. As it stood right now, it felt like they were being compressed a little too hard and were a little too loud in the mix, so I chose to take a look at both of those things. I started with the compression, because adjusting the amount of pumping would change the relative volume of the sound before we have even touched the volume fader. As it was, the sidechain compressor was ducking the pads by about 12 dB, which I felt was too much. It is a lovely effect and were the other sounds less dominant I might have actually kept it somewhere around here, but with the pads ducking this much I had to keep the level higher for them to be heard at all. If I reduced the amount of compression, I could probably lower the level overall, because the pads would be heard more when the other sounds were compressed and less when they weren't, which gives them their own little space in a roundabout way.

I kept the threshold on the pad sidechain compressor where it was and just dropped the ratio down to 2.0:1, which reduced the amount of ducking to about 6 dB. This was still enough to give it the sound I wanted, but it didn't feel anywhere near as drastic. As a result, I also lowered the gain on the compressor to 0 dB to bring things down to a better level, but I was quite prepared to adjust the bus level as I went along.

One thing I sometimes do to figure out the level of a sound (or group of sounds in this case) is to listen to the track playing with the sound in for a few bars, perhaps once through the verse or chorus, and then mute the sound and hear how much of a difference that sound was making. This is especially useful if you are trying to set the level for sounds that you want to be just about there but not really obvious. When I started this remix I had anticipated the pads being a much more prominent part of the overall sound, but things have a habit of changing as you go, so now I just wanted them to be there, in the background, really just supporting what the bass and stabs were doing. It took a few times around this in/out cycle of listening and quite a few adjustments before I felt I had the level of the pads set about right overall. Of course, this level wouldn't necessarily stay static throughout the entire duration of the mix when it came time to working on the finer details later on, but all I wanted for now was to set a good starting point level that I could adjust if needed.

At this point things were really starting to feel more complete, although having said that, there was still a lot of fine-tuning to do. I hadn't really done any work on tweaking the sounds I had for the lead part as yet because I needed to get the vocals knocked into shape some more before I could figure out where I would have space to use that lead part. I also had to decide what (if anything)

I would need to change about that sound to get it to fit in with everything else. And the vocals were going to take a bit of work to *de-wobble*, so I was going to leave that until the next day. I also tend to work on a lot of the detail in terms of EQ, compression, effects, levels, and "spot effects" right at the end of the process, so both of these things would be dealt with the next day (see Chapter 23). What I did feel was necessary before moving on to the last job of the day (the rough arrangement of the parts I had been working on) was to do a little work on the percussion part of the mix.

I thought the basic groove and sounds were fine and I didn't want to change them, but it was definitely a bit too repetitive. I hadn't put in all the little repeated kick drums at the end of a verse or chorus, the little *dropouts* on the kick drum for a couple of beats to really highlight a transition, the little sweeps, and FX to aid those transitions, and perhaps some additional drum fills. I wanted to introduce some small elements that provided additional interest throughout any given verse or chorus as well as thinking about which of the existing percussion elements I would use in the different sections of the song.

To tackle the second issue, I listened to different combinations of the basic elements and came to the conclusion that I wanted to keep the verses quite "stripped out," so they would just have the simple hi-hat pattern I created early on along with the "Beachball Loop" running throughout the verses. To pick things up a little in the second half of the verse, I would introduce the "Dirty Electro" loop, and then, in the chorus, I had both the "16th Note" loops (all three of them) and the "2-Step Flux Beat" loops that I could bring in. Perhaps I would bring in the "16th Note" loops in the first chorus and save the other loop until the second chorus, or perhaps even the final chorus after the breakdown. There were other little tricks I could use as well, such as gradually introducing some of the elements by using chopped-up versions of the full loops as an intro. But the fundamental idea was there.

The other thing I wanted to tackle was the issue of the semi-irregular short percussion noises throughout the track to make it feel a little less repetitive. I have, over a period of time, programmed and collected a lot of little noises (some from synths, others from recordings as mundane as opening a bottle of Pepsi!), which I can use as percussive sounds and little rhythmic accents. To do this, I opened up a track I had worked on a little while back where I had found some really good percussion sounds and had done quite a bit of work on manipulating them into interesting patterns and variations. Rather than redo all of that work, I will admit that I was a little lazy and simply bounced down each of the different percussion parts to a new track to import into this remix and see if they worked at all. Some of the parts had a very distinctive pitch to them (which I had done intentionally in the previous track), but fortunately, the two tracks were in the same key so no real effort was needed. Moreover, the tempo was the same, so that saved a bit of time as well. If that hadn't been the case, I could have just time-stretched and pitch-shifted the relevant parts to bring them into the right key, or in some cases, have gone into the EXS-24 on the original track and simply repitched the samples to be in the right key for this track (B minor).

When I listened to these sounds in the new remix I quite liked what a few of them were doing and thought that a few weren't working too well. As I didn't want to add a great deal—just a few sounds here and there that I could vary a little to keep things interesting—it might just be enough. I wanted to create a degree of planned randomness so I started moving the different sounds around to see where they might fit. To be honest there was no real logic behind what I was doing here, I was simply moving the sounds around a little and cutting them up until I had a few variations to use. Once again, this was a very small change, but do you remember what I said earlier about subtle changes? Enough of these subtle changes ended up leading to a track/remix that felt like it *evolved* as it went along, and that was exactly what I was aiming for here. The new percussion sounds hadn't made a massive difference to the overall track, but it was yet another little thing I could use in variations throughout the track when I started working on the final details, so once again, it would do for now. I could always program additional variations if I needed them, after all.

The next, and for today final, thing on my list was to work on the arrangement of the remix, and that included working out what I was going to do for the breakdown section of the track. Before working on that I needed to map out what I had over the full length of the radio edit, because at the moment, I had just a single verse and chorus worked on and I was getting to the point where I needed to start considering the *flow* of the remix as a whole. The first thing I do when I am mapping out the arrangement is to take the verse/bridge (absent in this case)/chorus section that I have been using to build up the groove and sounds and simply copy that whole "block" over the following verse/bridge/chorus in the vocal. Sometimes the second verse is shorter or longer, in which case I will obviously adapt the sections as necessary. In this case the second verse is only half the length of the first, so some adaptation was required, which I will come to in just a moment. Of course, if we simply kept both verse/bridge/chorus sections of the remix the same, it wouldn't be particularly interesting so there would obviously be differences between the two.

Sometimes I will start the first verse on a minimal level with a lot of things muted out, in which case the second verse would have some more elements in. Alternatively, if the first verse has pretty much the full complement of sounds I have worked on for that section, I might take a few of them out for the second verse, or perhaps simply low-pass filter the music in the first part of the second verse instead. Whichever of these two (or any other) options you choose, the end target is the same: to have familiarity (as in not completely different instrumentation to the first verse) but, at the same time, variation. Right now, and because of the idea I had in place in the verse already of dropping to the "roomy" kick drum at the beginning of the verse section, I was leaning toward the idea of keeping the first verse really stripped back like I had it and then having the second verse have a bit more power than the first. For now all I did was mute the roomy kick during the second verse and replace it with the full kick sound, and then strip the musical parts down to just the main bass line (keeping the "HPF Bass" muted now as well) and then bring in the stabs during the second verse. I did this by adding an Autofilter to the stabs submix bus, starting the second verse with that filter cutoff at about 30%, and opening it up fully by the end of the (now shorter) 8-bar verse. The transition from the end of the chorus to the beginning of the verse didn't feel 100% comfortable with me because of the sudden drop in the filter cutoff on the stabs submix bus, but I would work more on the finer points of the arrangement tomorrow, so now it was on to the chorus.

The chorus in this song is a little unusual in that it is 12 bars long rather than the more usual 8 or 16 bars, but that isn't too much of a problem. What I decided to try for now was to cut back the length of the first chorus to just the 12 bars where the vocal was present and then, for the second chorus, to add another 4 bars to what was already there, to make 20 in total (12 vocal and then another 8 of just instrumental) to allow a little bit of breathing space before the breakdown. I wasn't sure how it would all flow once everything was in place, but it would be easy to cut that back to just 16 or even 12 if need be, so I left it at that.

There would probably be a few bars of intro before the first verse, so taking that into account, by the time I got to the end of this slightly longer second chorus the total running time up to this point was around 2 minutes 8 seconds, which is exactly what I was hoping for. Most radio edits seem to be around the 3-minute mark these days, so that gave me just short of a minute (or around 32 bars at 128 bpm) to use for the breakdown and the final chorus. As a general rule, I like the final chorus to either be a double chorus or, at the very least, an extended version of the chorus. The reason for this is simply that the chorus is (at least it is supposed to be!) the high point of the song and the part with the most energy, so it is always nice to have that lasting just a little longer than people expect at the very end of the song. Given that the chorus as it stands is 12 bars long, a double chorus would be 24 bars, which left me 8 bars for the breakdown, which as it happens, is a nice length for a radio edit breakdown. In this case, there was a slight complication because the breakdown (or more appropriately in this case, the "middle 8") of the original song wasn't actually 8 bars long, but ran to 10 bars instead! Clearly some additional thought would have to go into this, but in order to finish mapping out the basic arrangement, I left a gap of 10 bars before copying over the first chorus (the 12-bar version) parts and then simply repeating the whole block to give that 24-bar last chorus. At this point it was time to work on the breakdown.

There is a good reason why I work on this section of a remix last, and that is that, when listening back to the finished remix, the breakdown has to drop out of the main track (typically a chorus section) and then do whatever it does before building back up, building tension, and ultimately dropping back into the track (again, usually a chorus section). To my mind, if I haven't pretty much finished working on the chorus section I don't really have a clear idea of what I am dropping out of and back into, so it can be hard to judge the dynamics properly. A good breakdown will build up and raise tension to almost fever pitch before it drops. If you misjudge this, you can almost make the end of the breakdown *too* big, so that when the full track starts again, it feels like an anticlimax. Sometimes you can have a huge build-up and then drop back into just a kick drum and a bass line to devastating effect, but this is clearly a deliberate decision. If you drop back into a musically busy section and it sounds too small in comparison, people will, perhaps quite rightly, feel that you totally misjudged the dynamics of the track. It is unlikely that they will actually think those exact words in their heads, but it just won't feel right to them. So we need to make sure that the build-up and release back into the next section is dynamically right, hence my desire to get the chorus section (or whatever comes after the breakdown) to the point of being essentially complete before I work on the breakdown.

Because this was the radio edit, the breakdown didn't need to be too long or epic, or indeed, actually drop (dynamically) too far, but having said that, I wanted it to at least be reminiscent of what I was planning for the main breakdown in the club mix. Earlier in the book we discussed many of the "standard" arrangement techniques for club tracks in general, and as you may recall, I said that there were really only three different kinds of breakdown: a quick drop and a slow build, a slow drop and a quick build or, more rarely, a slow drop *and* a slow build. What each different track or remix needs does vary, but in general, I prefer the quicker drop and the longer build because it creates (if done correctly) a greater sense of tension and anticipation followed by release when the track kicks back in.

What I had in my head for the breakdown in the main club mix was to have a total drop after the second chorus to nothing, but a sustaining bass "drone" (or perhaps a very low note on a pad sound) and have the last word of the vocal from the chorus echoing off into the distance. Over the top of this I would slowly start filtering up the stab sounds, but only using a single chord rather than the full pattern. I would also filter up one of the pad sounds playing the same chord as the bass note. I anticipated that this would last about 16 bars. At the very end of that first 16 bars of the breakdown I would have some kind of *riser* effect and maybe sweep the filter up on the stabs and the pad sound from about 50% to fully open over the last couple of beats. From there I would bring a couple of the percussion elements back in (but definitely not the kick and ideally not the snare/clap) and add in a high-pass filtered version of the bass line that would filter down over the course of the rest of the bass line but only going maybe three quarters of the way back down before the track came back in. I would also use the full chorus chord pattern here. Then maybe some additional, more effected, vocals over the course of the next 16 bars and have a much longer riser effect and possibly even some kind of snare drum roll. The final thing would be to have a high-pass filter on the whole "Instrumental" bus, which would sweep up over the last 8 bars to thin the track out a little before the whole track came crashing back in (with the "Instrumental" bus high-pass filter obviously bypassed at this point).

So with that in mind, I figured I would use an adapted version of this plan for the radio edit. However, the very first thing I had to do was go back and listen to the original version of the track (which I hadn't done since the very beginning of the remix) to take a listen to this middle 8 section and work out exactly what was going on musically. And, sure enough, there were two sections of 5 bars each. I am certainly not going to criticize anybody for doing things that are somewhat uncommon in terms of musical creativity like using 5 bar sections instead of 4 bars; on the contrary I admire such tenacity. However, as a remixer it can make your life difficult because almost all dance music tracks work on sections of 4, 8, 16, or 32 bars, so I had a choice to make here: do I stick with the original 10-bar layout or do I somehow try to manipulate things so that there are only 8 (or possibly 12) bars here? In the club mix breakdown it wouldn't be as much of an issue because I could easily insert some space between the different lines to make it fit to the more usual 16-bar section, but with the radio edit I felt that 16 bars for this breakdown section would be a little too much. It was time to try a few options.

First and foremost, and after listening to the original middle 8, I felt that my original idea of having a held bass (or pad) note and then going back to the original chord sequence might not work so well. The chord sequence in the original 10-bar pattern goes as follows:

B minor, A, G, E minor, F# minor, G, A, B minor, F# minor, E minor, F# Minor, G, A

So all of the chords matched up to those I was already using in my main chorus chord pattern. However, the sequence was very different because of the 5-bar nature of each half of the "middle 8." If I played my chorus chords over the first 4 bars of the middle 8, I thought that it worked okay, but then I had that extra bar to deal with. I actually tried cutting the vocal at the 5-bar point in the Middle 8 and then moving that a bar earlier so the sections overlapped, but that just sounded messy because you couldn't make out the last line of the first section or the first line of the second section. In the end I tried something I didn't think would work, but actually ended up sounding surprisingly good.

I kept the majority of the vocal running as it should, even though this meant the vocal melody didn't line up with the chords, as it should do. What I mean by this is that the first line of the second half of the Middle 8 *should* fall on the first chord in a sequence, but with what I had done here, it actually fell on the second chord and the rest of the lines were consequently 1-bar late. However, given the way the vocal fell over the chord sequence I was using, I thought it would work okay. I had to do some more work on the vocals tomorrow anyway, so I believed that I would be able to sort out any minor problems at that stage. This still left me with the problem of the 10 bars though. If you listen to the original version you will hear that the last bar of the 10-bar "middle 8" is actually silence, which means I could possibly get this down to 9 bars. "How are 9 bars better than 10?" I hear you ask! Well, sometimes you can get away with an extra bar at the end of a breakdown if you use the "gap" in the music to just heighten the tension. In fact, there have been many times when I have done just that, even though I didn't need to because of the vocals. The only real difference here is that I would normally use a delayed vocal to carry over that gap, and here I had an actual line of vocal (the last of the middle 8) to use to fill that gap vocally.

With that figured out I could start work on choosing which musical elements to include in the radio edit breakdown. I had a drum loop I picked out earlier specifically for the breakdown so I put that in the right place to start with and then copied over the original stabs sound just to get a feel for how things might work. I was fairly happy that the chord movement of the chorus pattern would work, but then I remembered I had wanted to use the "80s Stabs" part during the breakdown because it had a little more "space" in it, so I muted the stabs I already had and copied over that part instead. As I listened to it I liked how it sounded, but thought it would be better in a longer breakdown because using this sound dropped the dynamics of the whole track a little too much for the relatively short radio edit breakdown, so I switched back to the first stab sound I had. I also wanted to add in some pads to keep a little more of the density and energy in the breakdown, so I copied over the main "String Machine" pad and used the low-pass Autofilter I had on there and dropped the cutoff frequency down to zero for the beginning of the breakdown, and then filtered it up to fully open over the course of 8 bars. I also copied over the other two pad sounds for the second half of the breakdown and applied some gentle volume automation to bring them in a little more subtly.

The next thing I wanted was a nice long riser effect, so I opened up my usual ES2 and started working on programming one. Many typical riser effects are based on white noise, so I would obviously include that, but I also wanted to have a pitched element to it that could rise over the course of its duration and end on a really high-pitched held note. I started off using oscillators 1 and 2 for the melodic part of the sound, and quickly set up the modulation wheel to control the pitch of both oscillators as well as the ES2 "detune" parameter, which serves to detune the oscillators relative to each other. With both of these modulation routings set up with very large "depth" values, the end result is two oscillators that start off tuned 3 octaves apart but end up at the same pitch. Then I also set up the modulation wheel to control the filter cutoff, so that when the modulation wheel was at zero, so was the cutoff, and as the wheel moved to maximum so the filter opened. This was a good start, so I moved on to the white noise part of the sound.

I set up a few more modulation routings so that now there was an LFO set to control the filter cutoff. That LFO frequency was set to 31 Hz so that it was a really fast oscillation of cutoff frequency, but then I also set up the modulation wheel to control the speed of the LFO and slow it down to virtually nothing as the wheel moved up. In addition, I set up the modulation wheel to have an inverse control of the amount of modulation the LFO gave to the cutoff frequency, so that as the wheel moved up, the depth of LFO modulation of the cutoff frequency reduced. I also set up the modulation wheel to have an inverse control over the resonance amount of the low-pass filter so it reduced as the wheel moved up. The end result was a sound that started out quite dirty and with a lot of modulation, but became more static (albeit with a much higher pitch) toward the end. One final thing I wanted to add was a high-pass filter in series with the low-pass filter that started off at quite a high cutoff frequency to reduce the amount of low end rumble, and then, using the modulation wheel again, have the high-pass cutoff frequency drop as the modulation wheel moved up. A few little tweaks to the balance of the three oscillators, a bit of filter drive, and a touch of chorus, and this sound was done. It sounded a little too dry and upfront at the moment, of course, but I was going to push it back further into the mix using effects next.

It wouldn't need much to push this sound back into the mix, so some simple Stereo Delay and a nice big setting on SilverVerb did the job. I wanted to have a pumping effect on this sound, which, ordinarily, I would use a sidechain compressor for. However, as there was no kick drum running through this part of the track I had a couple of options. I could have created another kick drum track and used this as a sidechain input to a compressor with the actual audio output of the new kick track set to "No Output," or as I decided to do, I could use the Tremolo effect as a makeshift sidechain compressor type effect.

FIGURE 22.6
The Logic Tremolo plug-in can give a great sidechain compression effect when set up appropriately. These are the settings I used to give the regular, 4/4, pumping effect.

The actual settings I used are shown in Figure 22.6, and you can probably see that the shape of the tremolo waveform looks much like you would expect the output of a sidechain compressor to be (in terms of output level). When the "Rate" is set to "1/4" it has the same effect as having a sidechain compressor triggered by a kick on every beat. In actual fact I should say that it has a similar effect. The gain reduction effect when using a Tremolo follows a very regular "curve" shape set by the symmetry and smoothing controls. In a sidechain compressor the gain reduction "curve" is dependant on the signal coming in to the sidechain input and, as such, the shape of the curve can be very different from the regular shape of the tremolo effect. For something like this where it is a background sound I just want to give the pumping feel to, it does the job more than well enough.

As it stood right now, the breakdown felt pretty flat, even with the new rise effect. It needed more energy and tension. I liked the breakbeat drum loop, but I felt that it almost slowed things down, so I muted it out for now and then copied over some of the top-end percussion sounds from the chorus of the song. In fact, I copied all of the percussion elements over except the main kick, snare, clap, and 808 hi-hats. This definitely sounded better, so I kept these in but still wanted to use that breakbeat loop somehow. So I put a bit of a high-pass filter on it to take some of the weight out of it, added a bitcrusher to give it a really lo-fi and retro sound, and then loaded up Delay Designer and flicked through a few of the "Warped" presets until I came across one I liked. I then adjusted the wet/dry balance and cut the loop so it only came in during the second half of the breakdown. I pulled the level down and it sounded great!

It was time to add a few little extra FX and then I was going to call it a day. The first thing I wanted was the typical "boom" kick. This is usually made just by adding a large reverb to a kick drum sound, so I could have added an appropriate reverb to the kick I already had and then automated the "Bypass" setting so that it was only switched in when I needed it. However, I prefer the ease of having this as a separate sound, so I copied the existing "sub" kick to a new audio track and had a listen to a few large reverbs to see if I could find one that did what I wanted it to. The most obvious choice for me here was Space Designer because it has some awesome large "indoor spaces" and this was my first port of call. After trying a few I eventually chose one called "07.0s Bright Cathedral," but as the name implies, this was a little brighter than I wanted it to be, so I went into the EQ plug-in and rolled off everything over 5 KHz and this gave the final sound much less air and brightness.

I added a high-pass filter to the "Instrumental" bus and drew in an automation curve that raised the frequency from its lowest to about 50% over the last four bars of the breakdown, before dropping back down to its lowest at the very last second before the track came back in. Actually, one thing I do want to mention here is a little tip regarding automation on submix buses. In general, on an audio or audio instrument track you will have automation during places where there is something going on in the track. As a result there will usually be a region (audio or MIDI) where the automation is happening. By default, when you move or copy a region it will ask you if you want to move or copy the automation data because it is "attached" to that region. However, if you create automation on a submix bus, you will get a "track" in the Arrange window where you can view, draw, or edit automation, but because there is no region that it is attached to, if you insert or cut out sections of the track this automation data won't move or follow the arrangement in the same way it does for audio or audio instrument tracks. To allow for this, I will draw in an empty region on the submix bus track in the Arrange window and make sure it starts prior to the first automation I want to deal with and ends after the last of it. This ensures that the automation will move should you cut/insert sections in the arrangement, and gives you a quick and simple way to copy automation from one section of the song to another should you wish to do so.

As I listened back I did feel that there needed to be just a little bit more tension and a greater sense of anticipation, so I loaded up a standard Roland TR-909 "kit" into the EXS-24 and put in a snare drum roll of continuous 16th notes for the whole of the 8 bars. Had the volume level of this roll been consistent it would have sounded terrible! First, there wouldn't have been any kind of build, and second, it would have dominated things way too much. So I drew in some volume automation starting with the volume at zero and then gradually fading up over the course of the 8 bars so that the maximum volume was reached by about the 7th bar.

This sounded much better, but I still wasn't totally happy, so I added a low-pass filter and did a similar automation *ramp* with this, which adjusted the cutoff frequency from about 100 Hz up to fully open over the 8 bars. Because I now had the volume *and* the filter automation, I had to adjust the automation on both of them slightly. This gave a nice background sound to the snare drum at the very beginning of the roll.

The second problem was that the low-pass filter cutoff automation was a linear increase, but once the cutoff frequency got beyond about halfway, there was very little change in the sound. So I made the automation a curved line in order to sweep the frequency up more slowly at the beginning, only opening up fully (and more quickly) right at the end of the roll.

The final change was to add some accents to the snare drum roll so it didn't sound like continuous 16th notes all the way through. To add a little bit of variation, I put in some additional notes between some of the 16th notes (which make those particular points feel like little snare drum *flams*) and then, in the very last bar of the roll, added these extra notes in more consistently. The final effect is, in my opinion, one of adding to the feeling of something building up in a dramatic way. If you listen to the breakdown section with the snare drum roll in, and then again with it muted, I think you can hear the difference quite clearly.

At this point I decided to take a short break to give my ears a rest and allow me to be a little more objective when I listened to the track again. I was close to finishing up for the day and I wanted to have a final check on things before delving in deeper with fine-tuning of the mix tomorrow. Sometimes when I take a break like this I don't listen to any music at all, but other times I do, as long as it is *nothing* like whatever I am making at the time. After about 15 minutes I was ready to listen to how things were sounding. I wasn't going to make any major changes at this point because I wanted to leave that until I was refreshed and focused again in the morning, but if anything jumped out at me as being obviously in need of fixing I would do it now.

To get the best idea of how the track might sound to different people listening to it on various devices (iPod, hi-fi, car radio, etc.), I listened to it at multiple volume levels on my monitors and my main headphones again to get a good overall impression of the sound. I may have stated this before, but you should never underestimate how different tracks can sound on different systems, and while you will never be able to create a mix that sounds perfect on all different systems, I try to create something that will translate reasonably well, at least.

There were a few things I knew I would be working on the next day, but nothing that really warranted spending any longer on today. I always try to not work stupidly long hours. There will always be times when you are given a crazy deadline that involves late nights or early mornings (or both!), but I find that after a certain amount of time, the decisions I make about sounds or levels or effects aren't always the best ones, and time then has to be spent the next day correcting those decisions before moving forward from the place where I should have called it a day. I decided that my work for the day was done and then started enjoying my evening!

As you can see, I took care of quite a few of the things that needed doing in this session, and the mix was sounding much closer to a finished product. From this point on, things would start getting more detailed and would involve working on the finer points of EQ, compression, levels, and effects in order to get the radio edit finished up. I would also work on some basic mastering of the track to get it sounding a little bigger and more impressive for the client to listen to. Once that was done I would move on to the other versions of the track before finishing up and sending (or uploading) masters.

CHAPTER 23

Remix Walkthrough: Fine Tuning

After a good night's sleep and plenty of rest for my ears, I loaded up the remix and the first thing I did was put on my headphones and take a listen at a medium volume. The reason I chose my headphones was because I find them very useful for working on the finer points of the mix in *most* cases. The only areas where my headphones, and headphones in general, tend to be a little misleading are in judging the bass levels accurately and in judging panning positions and stereo width. One of the problems with judging bass levels accurately is simply that smaller "speakers" like the ones used in headphones struggle to reproduce lower frequencies at the same (relative) levels as monitor speakers, which can, potentially at least, have tuned ports to help with bass reproduction. There have certainly been many advances in headphone technology over the years and many of the better headphones you can buy today have exceptionally good frequency response. I have actually mixed a few tracks purely on my headphones, and they ended up sounding just as good as the tracks I mixed on my monitors, but the job was a little harder and it took me quite a bit longer (and a lot more A/B-ing against other tracks) to feel that I had the bass levels set correctly.

Earlier in the book we also saw that panning positions can cause problems on headphones, so what I tend to do is position sounds roughly where I want them to be on my headphones and then listen on the speakers to fine-tune them. If I then go back to working on my headphones, I just ignore the fact that the sound feels like it's in the wrong place because I know that isn't the case. If that doesn't sound like something you would be comfortable doing, there are a couple of options in the form of hardware or software units that introduce a certain degree of *crosstalk* between the headphone channels. What this means, in simplistic terms, is that, even if you panned a sound fully to the left channel, there would still be a small amount of it audible in the right channel of your headphones as well. While this doesn't simulate the effect of being in a room completely it does allow your panning decisions to be a little more accurate when working on headphones alone. With both of these disadvantages you might be wondering why I choose to use headphones to work on the details of the levels and EQ. The disadvantage of taking the effects of the room "reflections" into account when panning becomes an advantage if your room has less than perfect acoustics.

The mix didn't sound too bad but I knew I had a lot to do (including some big changes to levels and EQ that I heard more clearly after a break from the track), and the first thing really putting me off was the vocals, so I made that my first task of the day. As I said at the beginning of this remix walkthrough, the vocals are, for me, more often than not the most important part of the remix, so I tend to put a lot of time and effort into making them sound the best they can be. Sometimes that is an easy job. When the vocal parts I am given have been well recorded, well EQed, and compressed (but hopefully not *too* much!), and are at a tempo that isn't that far away from the tempo I will be working at, things can be quite easy. Often (apart from time-stretching) a simple EQ tweak, perhaps some additional spatial effects, and some "riding" of the levels will be all that is needed for them to work for the remix. However, on other occasions it can be a whole day's work just getting the vocals to sound right!

I have already discussed many of the vocal techniques I use in Chapter 15, so I won't describe the process I used in great detail here but will simply say that, because of the quite large time-stretch (from 90 bpm to 128 bpm) there were quite a few places where the vocal sounded excessively wobbly, so I needed to fix those and then see what I could do about the few places where the vocal harmonies didn't match up 100% with the new chords. I had been working up until now with just a bounced a cappella, which had all of the vocals mixed together, but seeing as I had been given separate lead vocals, backing vocals, and a few ad-libs, I wanted to work on them separately to get the best results. As it was by far the most important, I decided to work on the lead vocal first.

When I listened to the lead vocal in isolation, the magnitude of the task before me became more apparent. Clearly I would need to correct the sped-up vibrato on at least one part of every line in the lead vocal. When I was listening in context with the track some of the lines sounded okay, but given that I wanted to get the best possible result, I wasn't going to just settle for "Okay," and wanted to correct every line I could. So I rolled up my sleeves and got stuck in!

Given that you have read the chapter on how I fix up vocals in this situation, I won't detail every single step, but will just give an overview of what I did. To not get distracted, I created a new Logic song and just imported the three original vocal tracks I had. The first thing was to obviously stretch them all from 90 bpm to 128 bpm. Whereas I had just used one of the (quicker) built-in Logic time-stretching algorithms for the rough stretch I did earlier, this time I chose to use the iZotope Radius time-stretching algorithms as, to my ears, they are the best I have found. Because I wanted to use the original vocals I had as well, I created a copy of each of the original vocal parts, imported those into Logic and did the time-stretching on these files. Once I had the original 90 bpm and stretched 128 bpm versions of each vocal part in the arrangement, it was time for the hard work to really begin. I had a listen through the lead vocal (stretched version) and cut the file roughly at all the points I thought would need the vibrato fixing. I wasn't worried about being too exact with these cuts because I would take time to make sure the pasted slower vocals had good edits anyway. For now all I needed was an indication of where the parts needed to be replaced.

Once I had done all of these, and there were quite a few, I cut the original slower vocals in the corresponding places. I then created a new, empty audio track beneath the stretched vocal and then copied over all of the cut regions from the slower vocal to the relevant places on the stretched vocal track. I wanted to focus on each edit, so I would set up a cycle in the Arrange window so that I was just looping around one line of the vocal at a time, and then, once all of the edits had been done on that line, I would move the cycle forward to cover the next line, and so on, until I reached the end. So to begin, I set up the cycle around the two bars containing the first line of the vocal and zoomed the waveform view in so I could see better where I might be able to make the first edit. One of the key things I look for is a point in the file where the waveform crosses the *zero line*. The reason I do this is so the edits sound smoother and reduce the chances of clicks and pops. The bigger issue in determining exactly where and how to make the edit is actually looking at the shape of the waveform overview and trying to find the same point in that shape in both waveforms (original and stretched), and using that exact point as the place to do the edit.

Of course there are a number of places where you can do the edit, but I always like to keep it as close as possible to the point where the vibrato becomes a problem. The main reason for this is so the transition between the two is as unnoticeable as possible, and to make sure I use as little of the original as possible purely for practical reasons. In the case of a vocal like this, that has quite a substantial stretch, the original regions I paste in are always much longer than the stretched ones, and as a result, you can run into problems with them not finishing before the next line of the song starts. That actually happened on several occasions in this vocal, and I will explain what I did shortly, but for now, I will just finish explaining the general process I used.

Once I determined where the edit should happen, I zoomed right in to make sure that both the end of the faster vocal and the beginning of the slower vocal were both happening at zero-crossing points, and then nudged the faster vocal around while listening to make sure it was in the best place for a smooth sounding edit. What might look in theory as the most obvious place doesn't always give the best sounding edit. But having said that, it's always a good place to start! If I can get things sounding nice and smooth then great, but there are times when a short crossfade between the two regions can help to even things out, but I always make sure I keep it as short as possible as longer crossfades can sometimes lead to the vocal sounding like it is phasing or flanging in places. This would take care of the front end of the edits.

The back end of the edits proved, in part at least, to be just as challenging, if not more so. If you look through the Logic file with all the edits, you will see that in some cases I needed to cut a section out of the sustained note with vibrato to make it fit into the length I had available. That can be difficult sometimes because if the note is naturally dying away in volume then no matter where you do the edit, there will be a drop in volume (even if only a small one) at the edit point. Fortunately that wasn't too much of an issue for me with this vocal, but it can be at times, so watch out for that one. On a couple of occasions I faded out the end of some of the edited regions because, even though they finished before the next line of the song, it felt like there wasn't a natural gap between the last word and the first word, so I applied a fade out to try to give a bigger sense of space.

Eventually, after quite a bit of work, some head-scratching, and even a little bit of profanity in a few places, I got the lead vocal finished. I would estimate that, in total, I spent about four hours just on the edits for this one audio track, but if you listen to the result, I think you will agree it was worth it! There are still a couple of places where you can hear the wobbly vibrato, but the vibrato happens for such a short period of time that it would have been almost impossible to paste any meaningful amount of the original vocal in such a short space, so I decided to leave those and hope they would be inaudible in the context of the whole track. If it turns out that there were still any issues, I could always load this project up again and see what I could do. While we are on the subject of saving projects, I feel I should mention that I saved this project literally after every edit was done. I know it might sound a little paranoid saving every five or so minutes, but some of the edits were a bit of a nightmare and I didn't want to risk losing what I had done and then not being able to find the good edit point again quickly. So the few seconds it took to save the file each time was more than worth it in my book.

Next I decided to tackle the "Adlibs" track, simply because it was very short and I needed something quick and easy after the edit job I had just worked through! The process was exactly the same; it just took much less time. And I was a little bit naughty and took a shortcut here. Both lines of the adlibs sounded identical to me, so once I had fixed the first line I just copied the whole line into the place of the repeated line and that was that! Now it was time to move on to the "BV" track. This presented the same problems I had with the lead vocal, and some additional ones. The lead vocal was always just one voice/note (monophonic), but multiple voices/notes (harmonies) don't really present too much of a problem when doing edits like these. What *did* present a problem was that there were overlapping lines of lyrics in the one file. This could be a potential minefield for me because it was more than possible that there might be a sustaining note with vibrato that needed cleaning up being overlapped by another melody line. Where I would normally take the unstretched vocal and paste it in the place of the stretched one because it was just a held note, I couldn't do that here because I had the extra melody to think about.

To make matters worse, there were quite a few lines that needed fixing in the backing vocals. On the flip side, the chorus was repeated throughout the track so

I only had to make the changes once and then use that fixed chorus during each of the choruses, so that was some small consolation. As I set about working on these vocals I soon realized that I was in a very different situation to when I was working on the lead vocal, because with the chorus vocals the lines with vibrato tended to flow into the next word directly rather than having some space after them. The more I listened, the more I realized that this was going to be impossible to do in the way I had just worked on the lead vocal. There were a few places in each chorus where all of the voices were singing the same lyric at the same time and those I could probably do in the way I had just been working, but there were still quite a few that overlapped, which I would need to approach differently. So I worked through all of the parts I could fix using the cut/paste technique and then loaded up Melodyne to tackle the other rogue parts.

All I did here was select all of the parts of the track I hadn't been able to fix with the other technique and use the control for adjusting pitch modulation on the selected notes. By pulling this down to around 15% it got rid of a good amount of the wobble that was there before, but at the expense of the vocal now sounding a little phasey. I decided to import both the "pre" and "post" Melodyne parts back into the main Logic song and see how they sounded in there. In all honesty, I probably could have contacted the clients and asked for them to rebounce the backing vocals as individual tracks, and this would have given a better result in addition to giving me an almost infinite amount more flexibility in the next stage of the process. But I wanted to keep this as "real world" as possible, so I figured it was best to continue with what I had been given to demonstrate how to make the best of a less-than-ideal situation.

When I heard the fixed vocals in the song I was really quite happy and felt that many the worries I had about the backing vocals weren't justified. Now that they had even basic effects on them and the detail was covered by the music, the wobble on the vocals was pretty much inaudible. I was hoping to be able to use Melodyne again on the backing vocals to make changes to some of the harmonies that were, when I had been working on the music, feeling a little out of place. I wasn't 100% convinced that I would be able to change them that well, but I thought it was worth a try. The worst-case scenario was for me to remove the backing vocals altogether, which I wasn't keen on doing because I did like what they were doing. Alternatively, I could simply run the whole backing vocal through a vocoder and use that to effectively set the pitch and harmonies of the backing vocals. Even that would mean a compromise on sound, and I wasn't particularly looking to have that robotic tone to the vocals in this remix. With that in mind I loaded up the Melodyne plug-in on the backing vocal track and just had a quick look to see if it would do what the marketing blurb states and allow you to manipulate individual notes within a polyphonic audio recording.

I have used this feature of Melodyne to great effect in the past, but that was on a very clean and clear recording of a funky electric guitar part. Even then there were some noticeable artifacts in the final version but they were just

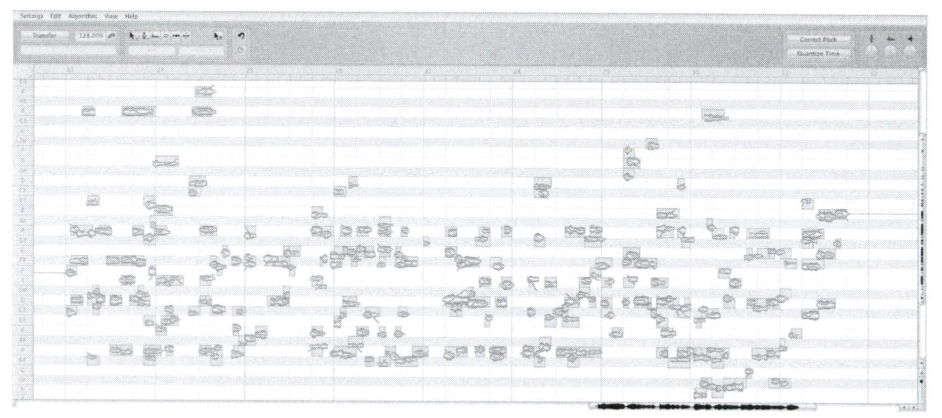

FIGURE 23.1
The polyphonic harmonies of the backing vocals as analyzed by Melodyne. If you listen to the backing vocals alone while looking at this image, you will see that it looks as though it hasn't quite got it right. In a way that's understandable as there are lots of overlapping melodies.

about acceptable in the track I was working on. I was more concerned here, because we were dealing with multiple notes at the same time (as you get in a guitar chord) and multiple different melodies, sometimes hitting the same note at the same time but with a different word, and I had my doubts if even the wonderful Melodyne could get me out of this tight spot. When I recorded the first few lines of the chorus into Melodyne I started to worry a little. You could see clearly that there were separate notes in there but the notes and patterns showing on screen were not matching up to what I thought I was hearing so it was time to get in there and try moving some things around to see how it sounded. After just a few minutes I could tell that it wasn't going to be easy, and I wasn't even sure if I would get away with it, so I decided to leave the backing vocals as they were for now and work more on the mix of the track. Now that I had spent a good amount of time away from working on it, I heard a number of things I wanted to change about the levels and EQ and overall feel of the track.

We all strive for perfection but when you are on a tight deadline you have to decide whether the ever-increasingly minute changes that take an ever-increasing amount of time to do are worth it. And given that I wanted to make this a realistic remix, I had set myself a goal of getting everything (including alternate versions and "stems") finished in just four days. That might not sound like much, but if anything that is a little longer than I usually have! By now I was a good way into the second day so I really wanted to get a move on. The first thing I did, seeing as I had my new and improved vocals in the track now, was to improve on the basic vocal effects. I had put a pretty standard delay and reverb effect on the vocals when I started working on the mix just to give them some depth, but rather than try to change these to sound better, I decided to start from scratch.

The vocal parts I had been given were very clean and clear, and sounded like they had been compressed a little already, but as the remix I was working on was much more energetic than the original track, I decided to push them quite a bit with compression to make them sound more "in your face" and to give them some bite. It took a little amount of experimenting with the Logic Compressor to get the desired effect, but in the end I think the result sounded good. It was still a little too clean for what I had in mind, which is one of the reasons I would ordinarily use a compressor with a bit more character. Modeled "vintage" compressors like this will often add a certain amount of dirt to the sound. While modern recording systems are capable of exceptional clarity and punch, a great number of people yearn for some of that "vintage" character, and huge amounts of time and money have been invested by companies in modeling some of the more well-known classic units from the past.

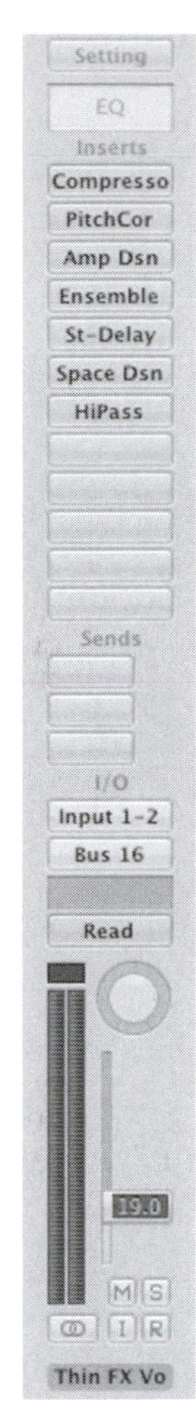

FIGURE 23.2 Channel strip for "Thin FX Vocal," showing all of the plug-ins used to create that one vocal sound.

I did still want to add some more *drive* to the vocal, but I wanted to be more in control of the amount that was added and the level, so I created a new track and copied the compression settings over before adding a few other plug-ins to give this version of the lead vocal a really heavily effected Pitch Correction plug-in set to a very fast correction time for that typical Cher/T-Pain vocal sound. The reason I did this was because you get an almost *double-tracked* sound, because the main vocal pitch isn't exactly in tune, while this version is. That means there is a tuning difference between the two, which varies with every note. Next up was the excellent Amp Designer to give the vocal some edge. I didn't want too much distortion, and ended up using one of the clean sounds because it added just enough character while still retaining the intelligibility of the vocal. Then it was on to the Ensemble plug-in, to add a lovely modulation to the tone, and then, finally, some delay, reverb, and a high-pass filter to remove a lot of the body from the sound. All of these together gave a really modern, edgy vocal sound, which I could mix in underneath the main vocal. I set the level roughly so it could be heard but wasn't the dominant sound, and then went back to working on the main vocal again.

Sometimes, probably more often than not, I will use delay and reverb effects as *inserts* on a vocal, but I wanted to try something a little different with this track. I thought that the vocal sounded pretty good being "upfront" as it was now (without any reverb or delay), but I did feel that it needed something to put it in a physical sounding space. So instead of just adding reverb and delay in the way I normally would, I set up two more auxiliary sends: one for reverb and one for delay.

For the delay, I chose a similar setting to the one I used for the stab sounds and then added a bit of stereo delay to give it some width. I figured that this would give echoes that sat in the same rhythmic pattern as the echoes on the stab sound, but with a little more width. I also added a high-pass filter after the delay effects so I could make sure there wasn't too much "body" in the delayed sound. I wanted the delay to be subtle, but the two delays in sequence gave it a lovely spacey sound with a slight bit of distortion from the tape delay, which I thought would work really well.

However, when I was trying to set the level of the delay signal, I ran into a problem. The issue was that if I set the delay level so the echoes felt like they were at the right volume, they seemed to crowd the vocal part, but if I set the levels so they felt like they weren't in the way of the vocal, when the vocal stopped and just the echoes were left, they felt too quiet! I could have automated the level of the echoes, but that would have meant an awful lot of work throughout the track. I decided to use a very handy little shortcut: sidechain compression. This is something I use an awful lot, but normally, I use it to provide regular and rhythmic *ducking* of pad (and bass and other) sounds. On this occasion I wanted to use it to duck the level of the echoes, and because I wanted them to be ducked when the vocals were present and come back up in level when they weren't. Instead of using a kick drum as the sidechain trigger input I used the vocal channel itself.

To make this sound fairly inconspicuous, I set up (relatively speaking) long attack and release times of 80 ms and 750 ms respectively, and adjusted the threshold and ratio until I was getting about 6–8 dB of gain reduction on the loudest parts of the vocal. This felt like just about enough to push the echo level down so it was audible, but not overpowering when the vocal was actually present. I think the results are very effective and certainly a timesaver over the alternative method of automating the volume of the echoes. And I thought it gave a good compromise between a dry sounding vocal and a sense of space. So good, in fact, that I wanted to try something similar with the reverb.

On the other auxiliary send I had set up, I loaded up Space Designer and tried some different presets to get the basic reverb sounding right, before I added in the compression. What I wanted was not something super-realistic but something that was actually quite shimmering in tone. I needed a reverb with far more high-frequency content than low frequency, and that had quite a long decay time, which I would then keep quite low in the mix as well because I wanted the space and the vocal to sound slightly separated, as if the vocal was positioned in front of the space it was occupying. I know that might sound quite strange when you read it, but it does make more sense when you hear the final effect.

I found a really nice-sounding plate reverb setting on Space Designer, but the decay time wasn't as long as I wanted so I used the "1/2 Sample Rate" control to the left of the plug-in window to double the decay time. Unfortunately, because the sample rate is halved, that also means the frequency response of the resulting reverb is halved, and in doing so much of the sparkle is lost. I also adjusted the Pre-delay setting to around 100 ms so there was a very small delay before the reverb signal actually started. This is a very useful technique for adding a sense of distance between the dry signal and the reverb. There was also too much low-frequency energy in the reverb now, so I used the built-in EQ features to cut a lot of the low frequencies out using a low shelf, and boost a lot of the high frequencies in an attempt to get some of that sparkle back. The high shelf EQ did help to restore a little of the top end, but it wasn't what I was originally looking for. However, it did have quite a nice pseudoretro feel to it, so I went with it for now to see how it would sound in the mix. I set the reverb signal level to where I thought it sounded good as a tail on the final words of a line or section, but as with the delay, as it stood now, the reverb was washing over all of the other words in the line.

I copied over the exact same settings as I had on the compressor on the delay auxiliary, applied them to a compressor on the reverb auxiliary, set the sidechain trigger to the vocal channel once again, and had a listen. I felt that the 6 dB or so of gain reduction was a little much, so I lowered the compression ratio just a little and increased the release time to around 1000 ms, because at 750 ms, it felt like there was an obvious "pump" to the sound, which made the reverb feel a little out of place.

I wasn't totally convinced of the reverb sound at this point, so I tried a few other settings and eventually came to a good combination of a Space Designer preset called "03.3s Concrete Basement" and adjusted the compression parameters. Last of all, I lowered the actual level of the reverb send channel and backed off a little on the amount of compression on the delay send. I was really happy with the sound at this point, so I decided to spend a little time setting up some effects for the "Adlibs" and "BVs" as well, before getting to work on the levels and EQ of the musical parts of the track.

For the "BVs" I just set up another compressor using similar settings to the one I had used for the lead vocal, but adjusting the threshold and ratio so it was a much more gentle compression. I also increased the attack time so the compression wasn't so fierce. After that, I applied a very simple low-shelf EQ set to 164 Hz using the Channel EQ, and set up sends to the same delay and reverb I had on the lead vocal. I wanted both sets of vocals to sound like they were in the same space, so it made sense to use the same effects. But I set the send levels lower for the "BVs" because I thought that with the increased density of the sound on the "BV" track it might be a bit much to have too much reverb and delay on it. The "Adlibs" were a pretty simple affair, because I already knew exactly what I wanted to do with these. There is a Channel Strip preset called "Filtered Delay Tail," which gives a really nice sweeping filter effect. I simply loaded this up and reduced the resonance on the Autofilter plug-in, added a high-pass filter set to 770 Hz, and then used the reverb send I had used for the other vocals to give a good amount of distance to the sound. And that was it! I would probably automate some of the send levels or other parameters when I got further into the mixdown, but for now that was enough.

Listening back to the whole track, I felt that the vocal still sounded very punchy, clear, and "upfront," but that added sense of depth the delay and reverb gave it helped to make it feel like a part of the track rather than being something that was floating on top of it. The final things I did to the vocal sound before finishing up was to add another compressor to the reverb auxiliary, and again, use sidechain compression, but this time using the more familiar kick drum trigger. I didn't want this to be a really obvious effect so I set the threshold and compression ratio so that, at most, there was only about 3 dB or so of gain reduction. Once I had done this I raised the overall level of the reverb signal by about

1 dB and then listened back. The dry vocal and reverb were working together *really* nicely now, with the ducking happening while the vocal was present and the rhythmic sucking of the kick-drum–triggered sidechain working together to create a space that seemed to move around in three dimensions, back and forward, and up and down.

Now that I had the new vocals sounding *much* better, I wanted to get stuck in and do some more work on the mix. Rather than list out every detail, I will summarize the main changes I made. In truth, there was a lot of minor tweaking, listening, changing something else, listening again, going back to the first version, comparing, and generally messing around. The process of working on the actual mix of a track is rarely a *linear* process, because a small adjustment in one place might necessitate another corresponding change elsewhere, and it takes an enormous number of minor adjustments to get to the final mix. So if, for example, I made 10 different changes to the EQ of a particular sound, I won't describe each in detail, but will, instead, explain the final settings and why I chose them. At least this will give an overview of the final changes I wanted to make while avoiding the laborious (and I am sure very uninteresting) minutiae of the actual mixing process.

I wanted to start off by getting the drums right and building everything from there, so I muted everything except the kick to check that the sound was right, and once I had confirmed this, I unmuted the "Loops" bus and had a listen. I felt that the kick needed to be a little more dominant in relation to the other drums, so I took a few minutes to listen to several commercially released tracks that I thought were in roughly the same area sonically, so I could compare. Listening to these confirmed what I had thought, so I dropped the level of the "Loops" bus until I thought the balance was about right. There is no hard-and-fast rule for the balance between kick drum and other drum and percussion sounds; it varies quite a lot between genres, producers/remixes, and even track to track from the same team. Given the sound I was going for, I needed the drums overall to be mainly driven by the incessant and big-sounding kick, so that was my main consideration.

Next I wanted to make a few small changes to the individual drum and percussion parts, but it was mostly rebalancing and a few EQ changes, not major changes to the sounds, as I was already pretty happy. I turned up the level of the "808 Hats" part just a little and then pushed the higher frequencies on the "Room Kick" sound up a little. I wanted this sound to cut through a bit more than it was, and the EQ certainly helped, but I still felt like it could do with a bit more "snap." So I loaded up the Enveloper plug-in and used that to emphasize the attack part of the sound a smidge more. I also pushed up the level of the "Shaker" part that came in only during the chorus, because I felt that it really picked up the pace and just wanted to bring that out a little more. Listening through a bit further into the track, I now felt that the "Soft Break" part was a tad too loud in the breakdown, along with the "909 Snare" roll, so I pulled the level down on both of those parts. I did actually think about backing off on the bitcrusher effect on the "Soft Break" part, but I really liked the tone of it so just settled on a volume change in the end. Obviously with all of these changes to the individual levels of the drum parts, I had to go back and change the overall level of the "Loops" bus relative to the kick drum.

The next thing I wanted to do was take a look at all the different FX sounds I had already and look at creating a few more. I liked the sound of the little pitched FX sounds I had, so I didn't want to change anything about those other than dropping the level just a touch, but I was really missing the obvious crashes and bigger "drop FX" sounds I use regularly. I hadn't bothered spending time looking for those so far because I had been more interested in getting a good vibe happening with all of the main musical parts, but the time had now come. Once again, I would normally look on one of the many excellent sample libraries I have for a really nice crash cymbal and some drop FX sounds, but I was sure I could find (or make) something just as good using only Logic plug-ins and sound library.

FIGURE 23.3 The settings used on the Enveloper plug-in help give the "Room Kick" sound just a little more "snap."

I looked for the crash cymbal first and found a nice one, to which I then added some delay (3/16th note delay time) and reverb (SilverVerb), to push it quite a ways back in the mix. After hearing it in context and adjusting the level so it wasn't so "in your face," I added a Channel EQ and made a fairly gentle EQ cut (490 Hz, −4.5 dB, 0.27 bandwidth) as well as applying a 12 dB/octave high-pass filter at 250 Hz to clean up the bottom end of the crash cymbal and get rid of any unnecessary weight and clutter that was there. It sounded quite nice after a tiny adjustment of reverb and delay levels but I still wasn't totally happy with the overall sound because I felt it sounded a bit too "real" for the feel of the rest of the production. I decided to create a little white-noise–based crash to double it up with, to give it more of a synthetic sound, which might work better.

The most commonly accepted definition of bandwidth (or Q) in an EQ is that it is calculated by dividing the center frequency (the frequency chosen to boost or cut) by the difference between the two frequencies at which the actual boost or cut is half of that specified.

I loaded up an ES2 on a new audio instrument channel and quickly opened up the "808 Ride" preset as a base to work from. I muted the two pitched oscillators and turned down the sine wave level control to zero, so that only the white noise oscillator was left. I then increased the release time on the amplitude envelope to make the sound longer and more spacey, and I lowered the cutoff frequency of the low-pass filter and lengthened the release time on the filter cutoff envelope (envelope 2) so there was just a touch of a natural decay and high-frequency rolloff as the sound died away. I copied

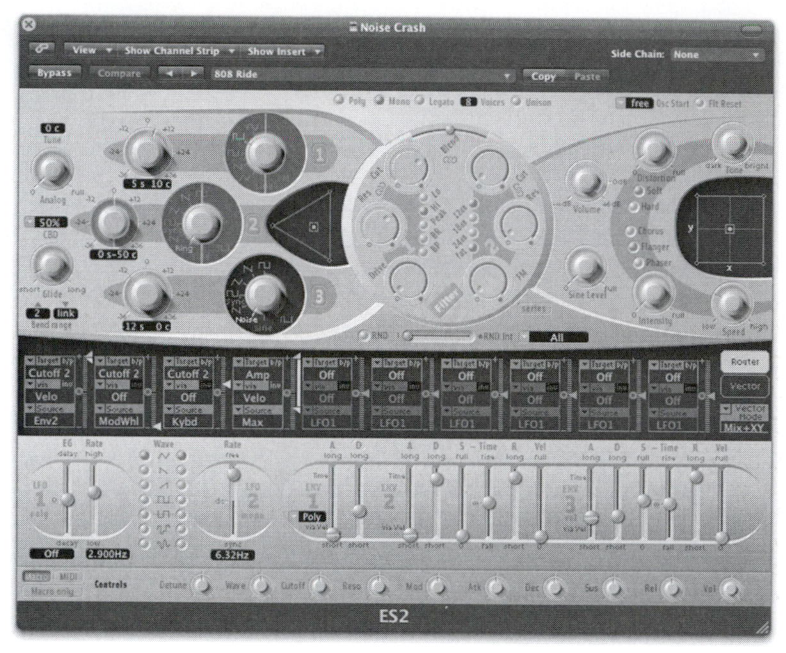

FIGURE 23.4
ES2 settings used to create the "Noise Crash" sound. When you solo the sound, you will hear that it doesn't sound much like a real crash cymbal, but it does a similar job in creating a burst of high-frequency energy.

over the delay and reverb settings I had on the crash cymbal and adjusted the balance of the two sounds against each other. In the end I think the "Noise Crash" sound was better being the slightly dominant one. I had one quick idea I wanted to try before moving on to the next stage. I am a big fan of mixes with a wide stereo image, so I added a Tremolo plug-in to the "Noise Crash" and set the rate to "1/8D" (which is equivalent to the 3/16th note delay times I had been using), and this gave quite a nice stereo panning effect especially when layered with the static crash cymbal.

Next on my list was creating a nice "drop FX" sound, and in light of the fact that I had set myself a deadline for this mix, I took a bit of a shortcut and loaded up an ES2 patch I created for another project, because it was quite a big and fairly complex sounding "drop FX" sound I thought would fit in perfectly here. Once the patch was loaded, I routed it to the "FX" bus and then had to pull the level *way* down, because it was a much bigger sound than I remembered and totally dominated everything when it came in. I tweaked a few of the parameters to help it fit just a bit better, but the sound you hear is about 90% the sound I originally loaded. Listening back through the mix once again, I was happy with the FX sound now, but there was one more thing I wanted to add: a "riser FX" sound I could use in the transition from the verse to the chorus.

Ordinarily I would use a sample or programmed sound that was predominantly of the standard white-noise type, but as I was already using a noise crash and the drop FX sound was white-noise–based, I wanted to try to avoid that if possible. Instead, I tried something I have heard on a few tracks over the years and always sounds great to me: a reversed piano note. To create this sound I added an EXS24, loaded the "Bosendorfer Studio" patch, and recorded two low "B" notes an octave apart, which I held for 8 bars. I edited the velocity of these recorded notes to be full velocity, because I wanted the piano sound to be as bright as possible. Then I bounced the part down to an audio file and pulled it into the Arrange window. Once it was there I simply reversed it and then had to nudge the position of it slightly so that the very end of the file (what was the beginning originally, of course) now lined up exactly to the beat and then moved it to where I wanted to place it. I really wanted this sound to be sitting in a similar position to the other sounds I had used as FX, so it had a 3/16th note delay applied, only this time I rolled off quite a lot of low frequencies from the delayed signal (setting the high-pass filter on the Tape Delay to about 580 Hz) and then added quite a long reverb setting on PlatinumVerb with a reverb time of around 7 seconds. Finally I added a Channel EQ and used this to add a high shelf boost of 7.5 dB at 3800 Hz to really bring out the ringy, metallic nature of the sound. As an afterthought I trimmed the audio file back so it was only four bars long (the last four of course!) and applied a fade-in to the first half or so of the resulting file to make sure it was a nice, smooth build-up into the end of the sound. With this copied through into all the places I wanted it and with the level set so it was clearly audible but not dominant, it was time to move on to the bass and, as a result, the part of the whole mix that would give me the most problems.

In simple terms, the problem I had with mixing the bass was that I could get the balance between the kick and the bass right in the verse (where it predominantly stays on the same note) but then, when moving to the chorus, there were some notes that sounded too loud and some that seemed to drop substantially in volume. After checking that there weren't any major discrepancies with the actual recorded notes, I spent a good 10 or 15 minutes looping around the last 4 bars of a verse and the first 4 bars of the chorus and adjusting the volume level of the bass bus up and down by tiny amounts, along with very small adjustments in low-frequency EQ. In the end, I couldn't find a single combination of settings where things felt "even," but I am convinced that is actually a physical problem, because when you start getting into the area of these really low frequencies you have inconsistencies in the frequency response of both your room and your monitors to deal with. Indeed, when looking at the level of the bass notes on the channel level meter, there was very little difference in level between the notes that seemed loudest and those that seemed quietest, so I put these jumps in apparent bass level down to the acoustics of my room and the frequency response of my monitors. I would ideally have liked to have had a really big set of main monitors to test this theory out on (or at least a system with a subwoofer), but as I didn't, I decided to spend a bit of time working on headphones. Normally I wouldn't even think about this, because it can be really difficult to accurately judge bass levels and bass/kick drum balance correctly on headphones, but seeing as I was having problems anyway, I thought it was worth a shot because at least I would be taking the room acoustics out of the equation.

A bit of critical listening on headphones later and, in spite of the inherent difficulties of mixing on headphones, I thought I had found a combination of level, sidechain compression settings, and EQ that sounded reasonably consistent in my headphones, more so than when I started, at least. I had reduced the strength of the sidechain compression effect and removed the wide bandwidth mid-frequency cut I had on the EQ, and added a 24 dB/octave high-pass filter at 35 Hz (to remove the really lowest frequencies from the bass). Now for the real test. I took the headphones off and listened quietly on my monitors. It sounded okay, but then my monitors weren't especially bassy at low volumes, so I probably wouldn't hear those really big bass notes until I turned it up quite a lot. And as I did, I got a bit of a surprise, because contrary to what I was expecting, the time I had spent working on my headphones hadn't resulted in a completely unbalanced bass end to the track. It didn't really sound any better than before, but more importantly, it didn't sound any worse! The actual notes that were standing out had changed now, and the emphasis seemed to be more on the upper and lower midrange rather than the subbass, but the change seemed to help, so I was happy it sounded no worse on speakers and better on headphones. It wasn't perfect in my eyes (or ears) but it was a significant improvement.

Moving on from there, I added in the pads and made a couple of very small changes. I wanted them to be a little more in the background than they currently were, because they were taking up a surprisingly large amount of space in the mix and were interfering with the punchiness of the stabs a little. I pulled the level of the "Pads" bus down a reasonable amount and brightened the sound with a Channel EQ inserted and a fairly gentle boost of 2.5 dB applied at 6600 Hz with 0.71 bandwidth. At this point I did go back to the "Loops" bus and made a minor adjustment to the level there to compensate for the additional space that was now being taken up by the pads.

After the relatively easy job of sorting out the minor issue I had with the pad sounds, I set about tackling the stab sounds. I needed to spend a little more time with these because they formed the main part of the music of the track. I thought that the combination of the three sounds was right, and that the "80s Stab" sound that came in only in the chorus was enough of a "lift" in that section, but I just wanted to make sure the combined sound was at the right level and taking up enough (but not too much) space—tonally speaking—in the final mix. To do this I listened to each sound individually and made any adjustments before listening to the combination and making any final tweaks there. For the "ES2 Main Stab" sound I left the levels and EQ alone but went into the ES2 and lengthened the release time on the amplitude envelope just a touch to make the stabs themselves ring a little more. I felt that the sound needed to bounce a little more to make it less punctuated and abrupt. After that I moved on to the "WGM Stab" sound. On this one I bypassed one of the bands on the Channel EQ to bring back a little of the *body* of the sound. The band that I bypassed had previously been applying a little bit of EQ cut at around 260 Hz. Finally I moved on to the "80s Stab" sound and on this I lowered the low-pass filter cutoff frequency so the stabs weren't quite as open and bright. I thought this was needed because, when they came in during the chorus, the sound almost felt like it was slightly separated from the others, whereas what I had wanted was for this to feel like a part of a combined sound instead. After I had made these changes, I spent a little while setting the level for the overall "Stabs" bus. When I had got what I felt was the right level I saved the Logic project and took a quick break before having a full listen through the track.

I actually listened through the track twice: once with the vocals playing and once just as an instrumental, and I thought that things were really starting to come together now, but there was an overall feeling that the sound of the mix felt a little bit *scooped*. If you're not familiar with that term, it means that it sounds like the bass and the treble are all really deep and crisp respectively but there is a big dip in the level of the midrange frequencies. It's something that is quite hard to describe, but if you load a finished (and well-mixed) track that you like into Logic and then apply a Channel EQ, set the frequency to about 1500 Hz, and the bandwidth as wide as it will go, then gradually reduce the level of that band, you will soon hear the effect I am talking about. In this mix it wasn't sounding horribly scooped, but I did want to see if I could do anything to take that out of the equation. I suppose the best way to do this is to go back into the individual sounds and do some EQ changes and some level changes, but for the sake of speediness, I wanted to see if I could do anything about it by just applying some EQ to the overall mix.

Because I use a lot of submix buses when I am mixing a track, all of the music ends up running through an "Instrumental" bus before passing to the master outputs. I tried adding a Channel EQ to this "Instrumental" bus rather than the master outputs because I didn't want any EQ changes I made to the music to change the sound of the vocals. After a bit of experimentation I finally settled on a relatively gentle EQ boost of 2.5 dB at 4850 Hz, and this had the effect of opening up the midrange a little and taking away that scooped sound it had before. I was actually quite shocked that the frequency was so high, because I had started boosting at around the 1500 Hz mark but it wasn't doing what I wanted to the sound so I just closed my eyes and moved the frequency control around until it sounded like I was in the right area. When I opened my eyes and saw that it was up near to the 5 KHz area I was definitely surprised. If you are wondering why I closed my eyes to do this, I may already have mentioned that it is very easy to get caught in the trap, when adjusting levels and EQ settings, of having a preconceived idea of how much you think you need to add or take away, and this can influence how you hear things by a surprisingly large amount. As such, if you adjust things *purely* by ear and don't even look at what you are doing, you are going to get a more natural and probably accurate result because those preconceptions don't come into play. As long, of course, as you resist the temptation to change the settings that you chose by ear once you have opened your eyes and seen that they aren't what you expected!

With that adjusted there was only one more thing to do before I started working out the details of the arrangement. When I had listened back to the track with the vocals in they sounded just a little bit bass heavy and felt like they needed a little bit more "cut" and clarity. Before doing anything more involved with the vocals (like more detailed level automation) I added a Channel EQ to the "Vocals" bus and thinned out the bass a little using a cut of 2 dB at 265 Hz (bandwidth of 0.98) and then added a 1.5 dB boost at 4100 Hz (0.71 bandwidth) to give them a little more bite, and a cut of 1.5 dB at 10,600 Hz (with a bandwidth of 0.71) to just take out a tiny bit of harshness that the vocals had. With that adjustment done I felt that the overall vocal felt like it had come forward in the mix a little, which was what I had hoped to do. The reverb and delay I had added earlier were still giving it a lot of width and depth, but now that depth was just starting a little further forward in the 3D space.

I now felt in a position where I could work on the finer details of the arrangement knowing that everything I was hearing was at least very close to a finished product. Some people like to work a different way, but for me, I like to get the mix mostly done before finalizing the arrangement, because having a better mix means that I have a better idea of how much space I have to occupy and how I should occupy it. In truth, I had actually been eager to get to this point because there were a few things I wanted to try out: a few little tricks and one, more substantial, change. The little tricks were largely just different edits—dropping out a kick drum for a bar or so, drum fills, little turnarounds on some of the musical parts, filtering the music up or down for a bar or two to emphasize a transition, that kind of thing—but the more substantial change was actually a bit of a strange idea. I could hear a really nice synth lead melody over the instrumental version of the track when I played it, but I didn't know if there would be space for it when the vocal was in. What I had in mind was coming up with a melody, hopefully one that *would* work with the vocal melody, but even if it didn't I wasn't overly bothered because I was anticipating using it only (or at least a *lot* more) in the club and dub mixes. There might be a couple of short spaces I could use it in during the radio edit, or perhaps I could just have it filtered right back so it is "there" but not fighting with the lead vocal for attention, but, as I said, I was thinking about this mainly for the club and dub mixes. And the reason I say this is a bit of a strange idea for me is that, generally, I like my radio edit and club mix to essentially be the same mix, just with a different arrangement. It is rare that I create musical parts that *only* appear in one or the other. In this case, though, I felt it might work. Plus, if the client *did* ask for an instrumental mix, it would give me something that could take the place of the vocal in terms of offering a memorable melody.

After muting the vocal parts, I loaded up an ES1 and just played around with a few melody ideas using the default sound. I didn't want anything too complicated because the idea wasn't for this to be an indulgent keyboard solo, but rather something that could easily be a vocal-type melody. It took a few attempts, but I soon came up with something I liked the sound of. I hadn't yet heard it with the vocal, so I had no idea if there was even a *chance* that I could use the melody alongside it. It was time for the moment of truth. As it happens it worked quite well but it highlighted my initial worry that the combination of the vocal melody and the new synth melody might get a bit confusing. The sound I was playing the melody on, being quite sharp and cutting, was certainly not helping matters, so I decided to record the melody in, tidy up the timing a little, and then work on the sound for a while because I was fairly confident that if I could find the right kind of sound, I could get this melody to work alongside the vocals, and, of course, in sections where the vocal was muted.

Quite a few of the sounds I had used so far had a softer edge to them, which reminded me, in a way, of typical '80s synth sounds, so I looked for sounds on the basis that I wanted something that wouldn't stray too far out of the sonic world it was already inhabiting. I thought that anything too brash or harsh just wouldn't blend at all. Don't get me wrong, there are times when a sound that is completely different from anything you have is *exactly* what you need but, for the kind of sound I was going for in this remix, I needed something that would draw a little less attention to itself.

Once again the ES2 was my first port of call. I looked for a sound that had a lot in common with the "80s Stab" sound I was already using but also had a more sustained character. I didn't want the sound to be exactly the same, otherwise it would have been hard to separate this new sound from the one I was already using (even though the rhythm and melody of this new part was different to the stab part). I also wanted a slightly softer attack to the sound to make it sound gentler. I spent a little while working on this sound, and after adding a few effects, I had a listen. I had certainly created the sound I was hoping to, but now that I listened to it in the track, I thought it was a little *too* gentle, so I had a couple of options. The first was to brighten the sound a little, decrease the attack time, and possibly even push up the volume a little, but I was reluctant to do so because I did actually like the sound that was in there.

You have probably already seen that I am big fan of layering sounds, so I took a moment to have a think about what sound I might be able to layer up with this one to give me the extra tone I was looking for. Then, the lead synth sound from "Blue Fear" by Armin van Buuren popped into my head. If you don't know what sound I am talking about then you can have a listen to the original sound on YouTube (*http://www.youtube.com/watch?v=_cHwb8tOEHM*) and then compare that to the sound I eventually created using the ES1 called "Blue Lead." The basic character of the two sounds is similar, as they are both based on a sawtooth wave and both have a very obvious pitch modulation (vibrato) on them. I didn't want to recreate this sound exactly, but I liked the combination of the slightly sharper edge to this sound and the original, softer lead sound. Also, the vibrato on this new sound worked well against the constant pitch of the original "Soft Lead" lead. I spent a little while looping around the chorus section of the song and adjusting the balance between the two sounds and the level of the overall "Leads" bus, which I had routed these two sounds to.

Once I was happy with the balance, I wanted to listen to the transition from the verse section into the chorus section now that I had this new sound in the chorus to help lift it. I only listened to that transition once before deciding that the extra sound(s) that I had just put in felt like they were coming out of nowhere, so I quickly copied over the chorus melody pattern to the last four bars of the verse and checked to see if it worked melodically. Once I had established this, I loaded up an Autofilter on the "Leads" bus and drew in some automation, which gradually opened the filter from fully closed to fully open over the four bars leading up to the chorus. I had a listen back and then changed this from a linear increase to a slight curve to bring up the brightness of the sound more quickly toward the end of this small part. Then I saved the file and took a quick break.

I had one more quick listen through with the vocals muted, to focus on any changes I would need to make on the arrangement, and made a list of a few points I wanted to look at. For the most part, though, I was very happy with the structure, and as I mentioned earlier, the changes I wanted to make were more about little fills and edits rather than anything major. I decided that the very first thing I had to work on was a way of getting into the track.

As it stood right now there was silence and then a big burst of sound as the track started. I never like tracks that start like that. In a club mix you can generally get away with coming from nowhere and into a drum or percussion part because people will never really hear that part as it will be used by the DJ to mix in from the last record. But on a radio edit there is a much greater likelihood that people might hear the track in its entirety, so I wanted to make sure it all worked as well as it could. I thought about running the main drum/bass line groove for a few bars before dropping to the point I currently had, but it didn't really make sense to have quite a full sound for only a few bars before dropping a lot of the weight out of the track for the next 8 bars.

In situations like this another potential lifesaver is the use of a reversed crash cymbal or some other kind of "riser FX" type sound to give a short(ish) build-up to that first impact. In this case I already had both of those types of sounds in the remix, so I just copied them over, put them into the appropriate place within the arrangement, and changed a few levels here and there. When I listened back, things already sounded better, but there was still a bit more I wanted to do. The next thing was to copy over the "Human Beatbox" loop that I had chopped up to use as a fill and drop that in at the appropriate point along with the reversed piano rise I had used later in the track. Naturally, the more of these kinds of effects I put in at this point in the track, the more I had to create some automation data to temporarily drop the levels of each because the combined sound was getting much too big for that point in the track. I shortened the length of the "Reverse Piano Rise" effect to just 2 bars and ensured that both this file and the "Reverse FX 01" file had fade-ins placed on them to make sure the intro was nice and smooth.

Up until now I hadn't really noticed it, but the reverse piano effect sounded *very* dry at this point and I thought it might be a good idea to add some delay effect to carry the tail of the sound over into the first verse. This wasn't going to be as much of an issue later in the track because there would be other sounds and instrumentation in there that would cover any abrupt ending like that, but for the intro at the very least, I wanted the delay effect. I chose a tape delay for this sound and set up the (by now) standard 3/16th delay time, and immediately this was another great improvement. I had to lower the level of the reversed piano now, and after a couple more times of listening through that section I realized that I couldn't really hear a damn thing because the "PB MIP FX" sound was totally washing over everything, so I lowered the level of that sound quite a bit and then had yet another listen.

The transitions from the verses into the chorus felt like they needed a tiny bit of work in order to create a little more drama so I used the Autofilter I had already added to the "Instrumental" bus and drew in an automation curve that swept up over the last bar of the verse before dropping abruptly back down in time for the first beat of the chorus. It is a technique I use a lot (and sometimes its counterpart of sweeping down a low-pass filter) for adding a little extra emphasis to a transition like this. I took some time to get the automation exactly right because getting the timing of the sweep back down exactly right can be a little difficult. You don't want to sweep back down too early, otherwise you lose the impact, but if you do it too late you risk the first kick drum and bass note sounding odd because they still have their lowest frequencies filtered out. Once I had this right for the transition from the first verse into the first chorus, I copied it over into the same place in the second verse/chorus transition.

While I was going over this section I noticed that the lead melody I had played didn't feel like it was coming in strong enough or soon enough, so I looked at the low-pass filter automation and decided to bring it up quicker. But even that didn't feel like it was quite enough, so I added in the extra sound ("Soft Lead") that had been muted. This proved to be a little too much too soon, so I drew in some volume automation on the "Soft Lead" sound only to bring the additional weight of that sound in a bit more gently. After a bit of experimenting with the automation curve, I ended up with a result I was pleased with.

One quick thing I noticed and wanted to take care of before looking at the vocals was the fact that the "Boom Kick" seemed to be masking everything else when it hit, so as much as I loved the weight and effect of having it that loud, I decided to reduce the level a tad so it wasn't quite so dominant. I had assumed that a fairly small reduction in level would do it, but in the end, I lowered the volume by much more than I had expected I would have to. Situations like this are a prime example of why you should set levels using your ears and not your eyes, because I was reluctant to keep turning it down *thinking* that I had already reduced it by more than I expected, but when I just listened and adjusted the fader level without looking at it I soon got it sitting nicely.

Now it was time to work on the vocals and this is the part of any (vocal) remix I enjoy the least. It's not that I don't like the process of working on them from a technical perspective, on the contrary I really do, but I always have problems with the vocal levels. In spite of having worked on over 250 remixes, I always have this internal struggle when it comes to setting the overall vocal level. As a producer/remixer who is often very excited about the sounds and music I have created, I have a natural tendency to be proud of that and want people to hear what I have done. However, from my experience working on quite a few remixes for pop artists, more often than not the client wants the vocal to be mixed very high. Of course, I understand why because, as I've already stated, remixing isn't all about showing off your talents as a producer. With more commercial remixes, at least, it's more about showcasing the song (and hence the vocalist) in a different market sector.

To deal with this dilemma I have come up with a system that works well for me. What I do is mix the vocals and work on the automation until I have them sounding good (in the sense of the levels I would like them at), and then I add a Gainer plug-in across the "Vocals" bus and add 1.5 dB or even 2 dB to the overall vocal level! It might sound crazy, but more often than not that does the trick and the labels come back happy. You might be wondering why I don't just mix the vocals louder in the first place, but having them at the volume the labels often request them to be at just doesn't feel natural to me, so it's a much easier process for me to mix them in a natural way and then add an overall volume increase afterwards. Please note that this is just what works for me, and I am not suggesting that everybody work this way.

It probably took me at least half an hour of constant listening and adjusting to get the vocals to a point where I felt that the overall level throughout the track was right, and even then I wasn't 100% decided. It was getting toward the end of the day and I really didn't want to make any final decisions on levels (vocal or otherwise) until the next morning, but for now everything felt good. I should add that if you look at the vocal level automation for the vocal track, you will see that I am not one of those people who will put in vocal level changes for every single word. I know that many people do it, but I have rarely felt the need to. There are two big reasons for this, and the first, and biggest, is that in almost all cases, when you receive the vocal parts from a client you will receive the "final" version, which has been compressed, automated, EQed, and otherwise balanced to give a nicely consistent vocal performance without any words jumping out. As such there is often very little for me to do in that regard. The second reason, and this is more of a personal preference, is that I am really not a fan of vocals that have had every tiny bit of expression and subtlety squashed out of them. I understand that sometimes you want things to be really upfront and "in your face" and that you might not want a lot of expression and dynamics, but that doesn't mean that you have to level everything out!

Even in the most well-prepared vocals there is still some work to do, because the levels and automation on the original version will have been worked out against the original backing track, which means that there may well be subtle (or not so subtle) overall volume rises or falls in the original vocals that follow or react to dynamic changes in the original song, and it is highly unlikely that your remix will match those dynamic changes completely. This is where I *do* work on level automation, but it tends to be in broad strokes rather than fine detail because I will put gradual increases or reduction on vocal levels on a section-by-section or line-by-line basis so that the vocal *seems* to stay at the same level relative to a backing track with changing average volumes.

There is actually a plug-in that does, or claims to do, exactly this, but automatically. It is called Vocal Rider by Waves, and it does two things. The first thing it aims to do is provide a constant vocal level by adjusting a virtual fader within the plug-in to keep the vocal level within a desired range. In many ways this part works very much like a compressor, by reducing the peaks of a signal and bringing up (relatively speaking) the levels of the lower volume parts. The second thing is that it will take this "leveled" vocal part and, through the use of a sidechain input to the plug-in, will adjust the overall level of the vocal part to react to the level of the music (which you feed into the plug-in through the sidechain input). So, in theory, once you have set your vocal level for the chorus of the song (often the busiest part) using your DAW channel fader, Vocal Rider will aim to keep the same vocal level *relative to the music* throughout the track. What this means in real terms is that if you had a part of your track where you filtered all of the music down using a low-pass filter, the vocal would, unless adjusted, *seem* much louder relative to the music, even though the level may have stayed the same. Vocal Rider should, in theory, see the reduced average level of the music being fed into the sidechain input and adjust the overall gain of the vocal (while still leveling the individual words or lines using the other process) down to the same relative level.

FIGURE 23.5
Waves' Vocal Rider plug-in could possibly take a lot of the work out of automating vocal level "rides," and offers the option to go in and edit the automation data created automatically afterwards.

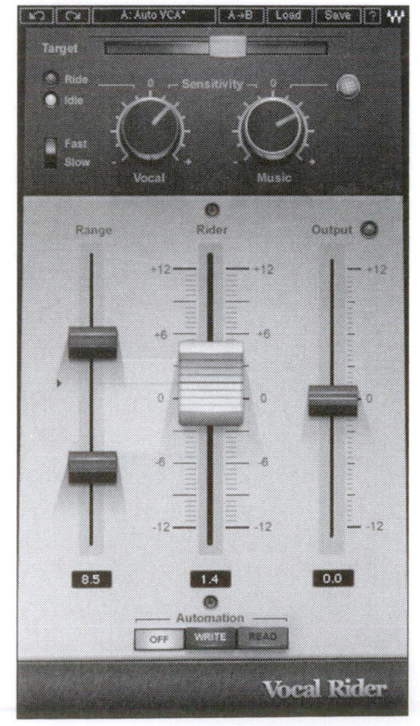

I don't actually own this plug-in but I have tried a demo version and I could never get it to do what it promises to do quite as effectively as I would have hoped. It is, I am sure, one of those plug-ins that seems very easy superficially, but requires quite a bit of tweaking and a bit of time spent experimenting with to get the best out of it. Perhaps I will try it out more in the future, as a few producers I know do have it and are getting great results, but for me, right now, it's on my "That's Interesting" list.

Anyway, to get back to the remix… I had automated the levels in the different sections of the song and they didn't need a huge amount of variation in my opinion, because the natural dynamics of the vocal overall lifted where I needed it to, so the chorus vocal level only needed a slight push in level to make it clear and understandable. I did want to hear the backing vocals a little more in the chorus, so I increased the level of the "New BVs" track by just over 1 dB and I actually muted the "Thin Vocal FX" track in the verses because I thought it made the vocal there sound just a touch harsh and edgy and, because muting it in the verses meant it felt like there was another element lifting the chorus. I quickly opened up the EQ on the "Vocals" bus and added a bit of a boost at around 6900 Hz (3 dB or so) to bring out the presence of the vocals.

I wasn't sure if it was just my ears getting tired toward the end of the day, but I felt that the overall tonal balance of the music and vocals still wasn't 100% right. I wasn't even convinced that the high frequency boost I had just added to the vocals was necessary, but I figured I was going to be finishing up for the day very soon so I would make the changes anyway and then listen first thing in the morning to hear if I had gotten it totally wrong. It wouldn't be the first time! I opened up the Channel EQ that was on the "Instrumental" bus and increased the low frequency shelf boost to 2.5 dB and added 0.5 dB of boost at 50 Hz with bandwidth of 2.80 to bring out the weight in the kick a little more. I also felt that the "Leads" bus needed turning up a little, so I added about 1 dB to the level and reduced the level of the "Break Rise" effect, as it was a little too powerful right at the end.

I should probably have called it a day at this point, but, as so often happens, I wanted to do one more thing before finishing up. I had been thinking for awhile about a couple of *spot effects* on certain words, and I had in mind a kind of lo-fi delay effect I could use on certain words, just to give some echoes that would trail off into the distance and fill in a few gaps. Because I have used this kind of effect quite a lot, it was pretty quick to set up, so I created a new audio channel and then added a Channel EQ with a high-pass filter set to 450 Hz, a bitcrusher using the "AM Radio" preset, a tape delay set to a 3/16th note delay time, and finally a tremolo set up to give a 3-bar autopan effect. I had already picked out the words I wanted to use while I had been working on other things, so I cut those words from the ends of the lines of vocals and then copied them over on to this new track so the delayed effect would fill up some "space" in the vocal.

At this point, I really did finish for the day. I was quite happy with the progress that had been made and was generally pleased about the fine-tuning of the mix, subject, of course, to giving it another listen in the morning with fresh ears and making any necessary adjustments then. Once I had done that, I would bounce down the radio edit and then move on to the club and dub mixes. Once those were done, all that would be left was the mastering stage and any jobs I consider administrative, such as uploading the files and preparing stems if asked to. It looked like I would meet the deadline and have a great-sounding mix at the end of it too.

CHAPTER 24

Remix Walkthrough: Finishing Up

As I sat down to listen to how the remix sounded the next morning I was very aware that my deadline was to have everything finished, mastered, and ready for upload to the client by the end of the day. I was also fairly confident that it wouldn't be too much of a problem, because when I had left the studio last night things were sounding pretty good, and, after all, I just had the club mix to check and then, once any last-minute adjustments had been made, I just had to work on the dub mix and then master all three versions. So without further ado, I fired up Logic, loaded in the club mix and had a listen.

The first thing I felt, and it was almost immediate, was that the whole track needed a little more level in the midrange, so I added a new Channel EQ on the "Instrumental" bus and worked on getting a good setting to rebalance things a little. In the end, I used a 1.5 dB boost at 1620 Hz and 0.64 bandwidth. There was already a Channel EQ on this bus, but I added another one so I could quickly bypass this EQ to compare to how the track sounded yesterday. If I had boosted a couple of frequencies in this second EQ, I would have only been able to bypass the bands in sequence and couldn't have directly switched from "Yesterday" to "Today." Adding in an additional EQ allows me to do just that, and once I have settled on the settings I could always go back into the first EQ and add the bands in there. Also, as a result of this extra EQ boost, I noticed that the level on the "Instrumental" bus was getting close to clipping, so I pulled the fader down by 2 dB to give just a bit of headroom.

With that done, I had a few things to quickly change on the vocals. I lowered the level of the "Vocals" bus by 1 dB to take into account the lowered level of the "Instrumental" bus, but I only lowered this by 1 dB instead of 2 dB because I did feel that the vocals overall needed to be a touch louder, as this was intended to be a very commercial (and therefore vocal-led) remix. The last line of the first chorus on the "Thin Vocal" channel ("Baby make it you") felt a little loud now, so I drew in some automation to lower just that one line by 2 dB. It felt like there was a bit too much empty space in the vocal in a couple of places, so I created a new audio channel (called "Delay FX") and added a tape delay (with a delay time of 2 beats), a high-pass filter in a Channel EQ (set to 450 Hz), a bitcrusher, and tremolo (giving an autopan effect with a 3-bar cycle), and then copied over certain words at the end of the chorus to fill in some of the gaps. A little adjustment of the levels and that was done.

Now that I was listening to the remix, I was actually happy to keep the "lead" parts in all the way through as they blended quite nicely with the vocal. It had originally been my idea to take these sounds out in the radio edit and only use them in the club mix, but after a small adjustment to the level, I felt that they worked here too. What I might do in the club mix is actually raise the level slightly during a chorus section with no vocals to bring out that melody a little, but otherwise I was happy to keep them as they were.

So that was it… the radio edit was done! At this point, I bounced a full vocal mix, an instrumental, and an a cappella. My reason for bouncing the last two is that I have, in the past, been asked for both an instrumental and an a cappella quite some time after finishing the remix, so I now do it as a matter of course when bouncing down the masters, in case I get asked for one later. Also if the client asks for a "vocal up" mix, I have the instrumental and a cappella versions so I can either load them into Logic and adjust the balance between them, or alternatively, just send both to the client and let them make any fine-tuning adjustments themselves. I was also planning to bounce down full stems for the radio edit, club mix, and dub mix, but that is something I would do once the remix had been approved by the client.

With all of that done, it was time to move on to the club mix, so I spent a little while cleaning up the arrangement and making sure everything was colored

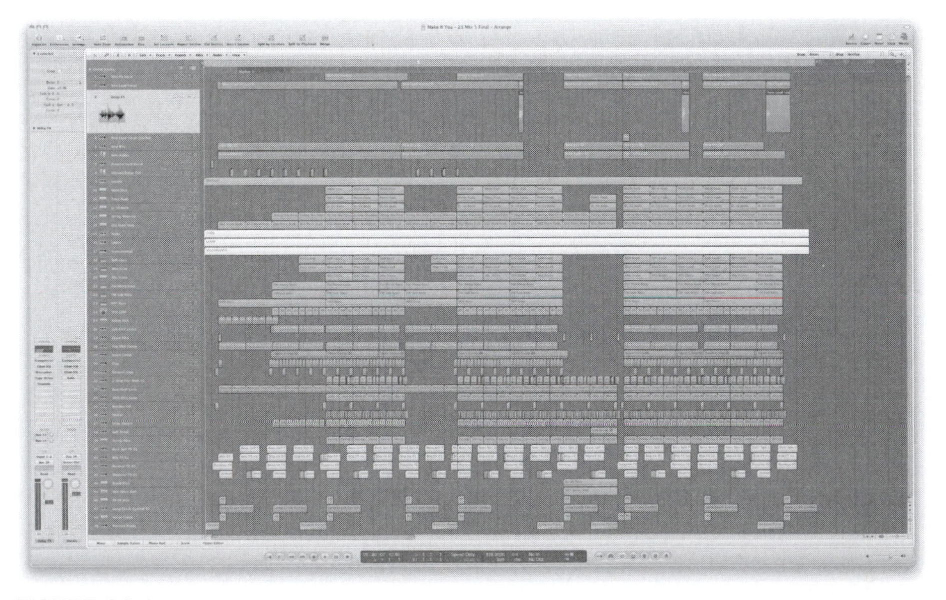

FIGURE 24.1
The Arrange window view of the final arrangement of the radio edit.

and named as I normally do, and then saved a final version of the sequence. I then saved a copy of the radio edit sequence to work the club mix from and had a think about what I would do to extend this version out to a 6+-minute version. I decided first off to move all of the parts over so that the first verse would come in at around 1 minute 30 seconds (which worked out to bar 49 at 128 bpm), and this gave me a good amount of time to have a nice DJ intro. I then inserted 8 bars between the end of the first chorus and the second verse to give some breathing space from the vocal, and as an afterthought, I inserted 4 extra bars at the end of the first chorus that I would use as an instrumental section before dropping to the second verse. This would mean that the "Lead" riff could be heard a little better and that the chorus then ended up at the (for me at least) more familiar length of 16 bars.

For the breakdown section I decided to add 8 bars at the beginning, so I had some space to drop the dynamics down quite low before building back up again. I have made breakdowns longer than 16 bars in the past, but that was a good place to start. I could always add more if I needed it later. With the extra additions I had made so far, it meant that the end of the breakdown was around the 4-minute mark. The "double" chorus after the breakdown added 45 seconds to the length, which brought the total to about 4 minutes 45 seconds, and I was assuming a 1 minute(ish) "outro," giving a total running time of 5 minutes 45 seconds, or so. This meant I still had a little bit of time to play with, so I decided to add 8 bars to the beginning of the outro chorus so I could have a short instrumental section there before kicking in to the final vocal chorus.

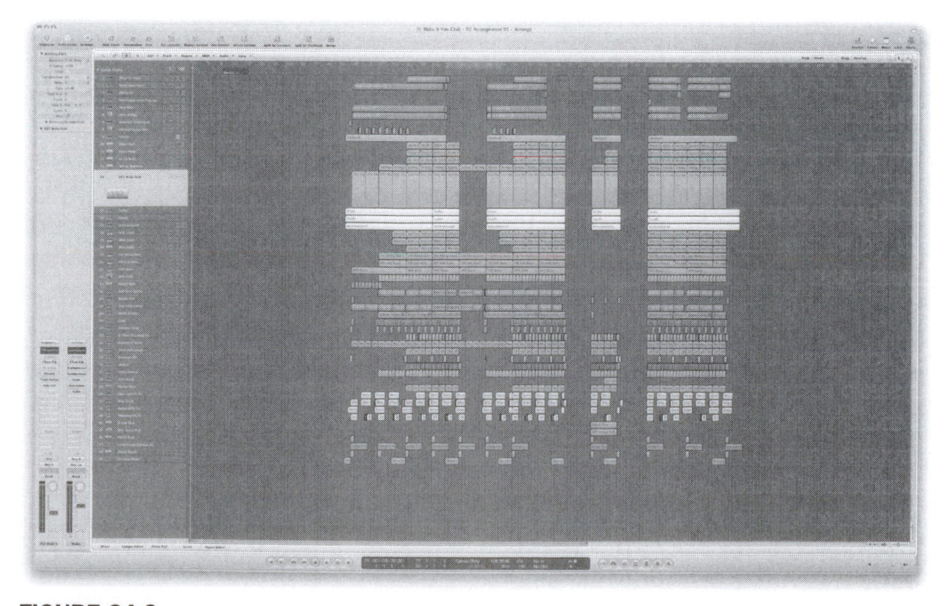

FIGURE 24.2
The club mix arrangement starts to take shape as I add in the spaces for the various sections I want to insert or lengthen.

While I was working on the arrangement changes, I muted the vocals because I wanted to make sure the music was flowing well before figuring out the vocal changes. I decided to work on the section between the first chorus and the second verse first. In order to fill in the 8 bars I had inserted, I copied across the kick, bass sounds, and the main pad sound from the second verse. This kind of dynamic drop in the middle of a track always works really well and sets up the second verse and chorus to sound bigger than the first, even if they are virtually identical in structure and sound. At the start of the newly inserted section I made sure that the pad sound was filtered down, and I adjusted the filter cutoff automation so the increase up to "fully open" happened over the full 16-bar section that was now available to me.

At the point where the second verse vocal would begin, I added in the verse pattern for the "ES2 Main Stab," but made sure the filter was set quite low, and filtered that up over the course of the 8 bars that remained before the second chorus. As in the first verse, I used a curve on the filter cutoff automation so it seemed to rise up quicker just before the second chorus. I added a couple of the drum elements back in at this midway point between the first and second chorus to help pick things up a little leading back into the chorus section, and as a way of introducing the second verse vocal at this midway point, I created a small drum fill using just the kick drum and the snare sound in a little offbeat pattern. Not only does it serve as an introduction for the vocal, but it also, just for a moment, breaks up the relentless 4/4 feel that the whole of the rest of the track has. It's just a little touch, but I think it works really well. I only like to use things like that once or twice in a track, because I think it's very easy to overdo it with those kind of arrangement tricks, and at that point, they become less special. Less is definitely more.

The last thing I wanted to do with this section was to copy over the "Reverse Piano FX" to the end of the preceding chorus to help smooth over the transition. With that done, I had a listen through this newly created section from halfway through the first chorus until halfway through the second chorus to make sure everything flowed as I had hoped it would, and once I was happy with that, I moved on to the next section.

I already knew pretty much exactly what I wanted to do in the longer breakdown that I would have in the club mix. I did mention a while back that I had an idea, but as so often happens, that plan had been amended as things went on and what I had in mind now was really quite simple. The extra 8 bars I had inserted would allow me to drop down to just the "ES2 Main Stab" with the filter closed down quite a bit, and then let it run on its own for those 8 bars while slowly opening up the filter and leading back into the beginning of the breakdown from the radio edit. From that point on, I intended to keep things as they were in the radio edit version.

I copied the main stab sound over from the breakdown section that was already there and adjusted the filter cutoff frequency down to a point (about 37%) where I felt it was closed enough, but still with a sense of being able to hear the chords clearly. I then automated that filter back up to where it was in the next section over those 8 bars. I also copied over all of the kick drums sounds as well as the "PB FX STAB," "Crash," and "Noise Crash" parts to the first beat of the new breakdown section to give a real *impact* going into the newly thinned-out section. I started playback about halfway through the second chorus and listened to the drop into the breakdown, and I loved the build-up at the end of the chorus and the big impact followed by a second or two of not really being able to hear anything before the stabs crept into (sonic) focus. I carried on listening and was quite surprised that the "old" breakdown section came in so quickly, and with quite a "bang."

It wasn't really what I wanted, because I thought that the section I had just created with only the stabs would feel like it went on for longer, but those 8 bars went by really quickly so I decided to insert another 8 bars and allow those stabs to run and filter up on their own over 16 bars instead of 8, to see if that felt more natural. It most certainly did and it felt like the right length now, but I still had to deal with the issue of building up the breakdown dynamics more gradually, because right now, it felt like there was a gradual build-up while it was just the stabs, then a huge jump into what was the old breakdown, and then a gradual rise toward the end of that section. What I wanted was something to help ease into the original breakdown section.

To do that, I copied over the "Beachball Loop" and "2-Step Flux Beat 01" into this next section, but they felt a bit harsh coming in as they did, so I added a low-pass filter (Autofilter) to the "Loops" bus and set fatness to 100%, cutoff to 14%, resonance to 22%, and then drew in an automation curve that increased cutoff to 100% with a slight curve, while simultaneously reducing resonance down to 0%. I also copied over all of the FX and kick sounds from the very beginning of the breakdown to 8 bars into the breakdown, but because all of the music had dropped massively in dynamics, I had to lower the volume on all of the parts by anything between 7 dB (kick drum parts) and 14 dB (FX sounds) to get them to sit nicely against the stabs.

I copied over the "WGM Stab" part to come in 8 bars into the new breakdown, but reduced the volume to zero and gradually raised it back up to −9 dB (where it was) over the course of 16 bars, again with my preferred curve, so that the increase in level was more marked toward the end, helping to ramp up the level of tension. This extra "layer" of melodic sound helped to smooth out the build of dynamics during the breakdown as well as adding a little extra "body," which I thought I would get away with in the club mix.

I had another listen back and this sounded good for now, but I decided I might need to make some adjustments later when I unmuted the vocals. Any changes that I did make would probably be levels/automation, as I thought that the right sounds and musical parts came in at the right time and in the right way, so it would just be a case of making small adjustments in a few places, if anything. I had a feeling that I wouldn't need to adjust much because the arrangement of the last 8 bars of the breakdown, which is the part that would have the most vocals in, wasn't really any different from the radio edit arrangement, and I had been happy with that. But time would tell. For now though, on to the next section.

Next up was the extra section after the breakdown and before the final chorus. This was going to be fairly easy as I just wanted it to be an instrumental version of the chorus but with a few little tweaks, so I started by copying over the full chorus parts from the outro chorus. At the end of this inserted section of eight bars I copied over the high-pass filter automation from the "Instrumental" bus I had used going in to each of the choruses so far, as this would help create that little bit of extra impact one more time leading into the final chorus of the track. To give this final small edit some extra impact, I also dropped out a couple of the percussion parts for the very last bar before the chorus vocal returns. All minor details I know, but they all add up to make a bigger whole.

Having this 8-bar instrumental section also allowed me to raise the level of the "Leads" bus by 3 dB to give a brief (but welcome) opportunity for the lead melody to come to the front a bit more. I did try pushing the level up even more than 3 dB here and, while it sounded okay all the way up to 5 dB in terms of being balanced with the other sounds, at anything above 3 dB higher you really started to feel that it was "missing" when the level was dropped again and the vocal returned, so I settled on 3 dB as a compromise. I had yet another listen through before deciding that it was okay to work on the next section. At this point what I was left with was, basically, an extended version of the radio edit.

For the DJ outro I dropped back into the straight verse pattern but left the pads running and kept the stabs filter open. I wanted to gradually fade or filter things

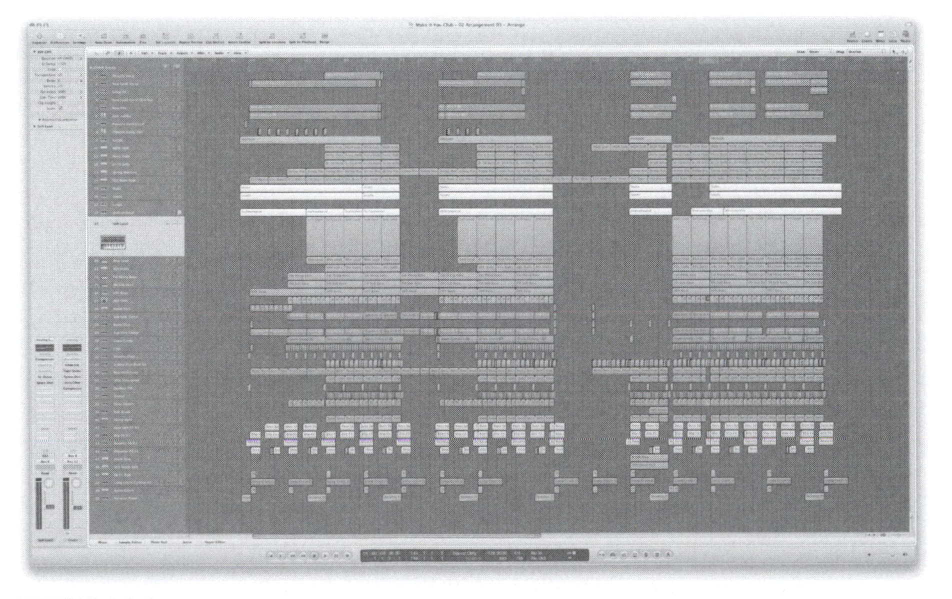

FIGURE 24.3
As I start to fill in the spaces, the arrangement takes shape even more. At this stage it is largely an "extended" radio edit that forms the main "body" of the club mix.

out so there was a nice gentle transition into the end of the track. I know a number of producers/remixers who will just stop parts playing without necessarily fading them or filtering them out, but my personal preference is for a smoother change. I can't say that one or the other is "right" because it really comes down to personal preference and, to some extent, the standard practice of the genre you are working in. So to get this gentle fade-out in the outro, I started off by closing the pad filter down to 0 over the next 16 bars and closed the cutoff on the ES2 stab down to about 14% over the same period. I then closed down the stabs bus filter from 100% to 38% over the next 8 bars. At that point, I stopped the stabs from playing. Meanwhile, 16 bars after coming out of the chorus I dropped the "FM Sub Bass" to just leave the other two bass sounds playing, and I faded those out over the next 8 bars as well, to stop at the same point the stabs did.

Drum-wise I dropped to just kick, snare, clap, 808 hats, "Beachball Loop," and "Dirty Electro" for the first 16 bars out of the last chorus. After this I stopped the 808 hats as well, and then after another 8 bars started to fade the "Beachball Loop" and "Dirty Electro" down to nothing over the final 8 bars. I added the "Crash," "Noise Crash," and "PB FX Stab" in a couple of places and copied over the "Reverse Piano" to end the track on. I did think about having a "Boom Kick" at the very end of the track, but in the end I decided I didn't need it. If you look at the way the parts gradually get removed in the arrangement window, it will give you a good indication of how it will sound without you even having to play it. As always, I played this whole section back a few times to make sure it was flowing the way I wanted, and once I had made a few final adjustments to the position and curve of the automation data, I was happy to move on to the final part that needed working on: the intro.

FIGURE 24.4
The arrangement with the DJ outro added almost completes the basic arrangement.

In some ways the intro would be a "mirror" of what happens at the end of the track, because both parts serve the same (or at least a similar) purpose: to give the DJ something that is easy to mix with. Of course it wouldn't be an exact mirror image, but would do many of the same kinds of things only in reverse. So to begin with, I started things in the same way as I did the first verse of the radio edit with the "Room Kick," "Top Kick," and "Beachball Loop" playing and with the "PB FX Stab" and "Noise Crash" sounds (but with level turned down quite a lot). After 8 bars I introduced the "Sub Kick" and added another crash/FX hit to emphasize this point. To make things transition nicely between the "Room Kick" and the "Sub Kick," I reduced the level of the "Room Kick" over the last bar before the change.

I am not a huge fan of having only drums running for *too* long at the beginning or end of tracks or remixes, so I copied over the "HPF Bass" part from the verse and faded the volume up to its current level from zero over the first 8 bars and then let it run at that level for another 8 bars. At bar 17 I wanted to introduce the "Fat Moog Bass" sound, but when I tried copying over the verse pattern it felt like too much had happened at once, so I edited the part a little and muted some of the notes so it now had a more "cut up" feel, which would serve as an introduction to the "full" part that would come in a little later. I let this "cut up" part run over the "HPF Bass" for the next 16 bars and then decided it was time to introduce the full bass line, so I reverted back to the original "Fat Moog Bass" part and also introduced the "FM Sub Bass" sound, at bar 33.

This was a good point to start picking up the pace a little with the drums and percussion, so I added in the "808 Hats" part. I didn't want to give too much away too soon on the percussion side, so I just left it at that for now. I also wanted to introduce the stabs at this point because they were a part of the first verse, which was now only 16 bars away. If I wanted them to come in subtly, I would have to start bringing them in here. I copied over the "Main ES2 Stabs" sound and drew in some automation on the "Stabs" bus to drop the Autofilter cutoff frequency all the way down and then open it up over the next 16 bars. As a means of hinting at what was to come and to emphasize the change, I dropped the "Sub Kick" for the last 2 bars before the drop into the first verse.

I know it seems like a really simple process, but these kinds of intros often are. This is a part of the track that *has* to be relatively simple because if it's too complicated and elaborate it will be difficult for a DJ to mix with. In addition, if people are listening to it in its intended destination (a club) then they ordinarily wouldn't be listening to this part of the track on its own anyway. That is sometimes a difficult thing to remember. When you listen to a club mix while you are working on it in the studio, you have to remember that you are hearing it in a very different way to how the general public will be hearing it. Things that seem overly simplified in the clinical studio environment could work perfectly in a club environment. With that in mind, this really worked for me as a club mix intro, but obviously I would need to work on some vocals to use in this part, so I saved the project and had a listen through the whole song as it was currently with the vocals in.

As I was listening through the vocals, my attention was drawn to the very first line ("Did you ever…"), and I had an idea to use just those three words as a vocal "effect" on the intro of the track. I would need to create quite a "spacey," delayed sound for this part and I wanted to slightly change the timing of the words as well. I created a new audio track and copied that line of vocal over, and then I cut it into the three separate words. I had something in mind for the rhythm I wanted, so I nudged the timing of each of the words around until it felt right. It was only a small change, but I felt that it suited the overall feel of the intro.

In terms of the effects for this vocal part, the choice was really quite simple. This type of vocal effect is something I use quite a lot, so the choice of plug-ins and starting points for the settings was pretty familiar to me. I started off by using a Channel EQ with the "Phone Filter Wide Band" preset, but on this occasion, I bypassed the low-pass filter to open up the top end of the sound. Next up was the Pitch Correction plug-in with the Response set to 0 ms, to give that heavily *autotuned* vocal sound. I followed this with an Ensemble plug-in to give a sense of width and soften the sound slightly. Then I added a Tape Delay with a 2-beat delay time and a Space Designer reverb with the "4.0s Swimming Pool" preset loaded. Finally in the chain was a Tremolo plug-in set up to give an autopan effect with a cycle length of 4 bars. I spent quite a while setting the wet/dry balance in the delay and reverb effects because I wanted something that was quite distant sounding, but at the same time, I still wanted it to be intelligible. I have to admit that I don't normally use the Tape Delay and Space Designer for this type of effect. When I am working with my full set of plug-ins at my disposal, I normally use the UAD RE-201 Space Echo for the delay part and the UAD EMT 140 for the reverb part, but the settings and plug-ins I had chosen here replicated the sound pretty well.

Now that the effects were right, I started placing this "Did you ever" vocal line where I felt it was needed. First I placed it at around bar 33, and then every 4 bars until bar 45. I lowered the volume of the early copies a little because they felt a bit loud compared to the music behind them. With them in place every 4 bars, I had to make a minor adjustment to the Tape Delay feedback amount so that the echoes of each line had just about died out by the time the next line was heard. It sounded good, so I copied over the same thing (every 4 bars) to the start of the breakdown and adjusted the volume accordingly. I really liked the effect, so I decided to use this vocal part and effect at the end of the track after the last chorus as a way of keeping a hint of the vocal going toward the end. I copied the parts over to the end of the track and placed them every 4 bars again, but this time I did a vocal fade-out to more or less nothing over 16 bars, so that the vocals sounded like they were receding off into the distance.

That was all I really wanted to do vocally, so I took a break, got something to eat (don't forget to do that!), and then came back about 20 minutes later to listen to the club mix from start to finish. Apart from a few small changes to some levels and EQ/filtering automation, things were very much as I hoped they would be. I know it might seem like I get things right "the first time" a lot, but remember that I have only been describing the "outline" of the process rather than every single little step and adjustment. It's also worth remembering that I have been doing this quite a while and experience certainly helps you to know what will and won't work, so it makes the decision-making process much easier! Now that I was happy, I went through and bounced the same versions I had for the radio edit (full vocal, instrumental, and a cappella), saved a "final" version of the file again, saved a copy, and renamed it so I could start working on the dub mix.

FIGURE 24.5
The final arrangement of the club mix, including the extra vocal snippets during the DJ intro and DJ outro.

For the dub mix, I wanted to keep the same musical arrangement but with less vocals and make those slightly more effected anyway. As I listened, I decided I really wanted to keep the vocals quite minimal. I don't always want to keep them as minimal as I had in mind to do here, but on this occasion, I thought that the music was strong enough to carry the track with hardly any vocals. That said, I really liked the "Did you ever" vocal I had used in the club mix intro, and wanted to make more use of that in the dub mix. I left everything the same up until the point where the first verse came in originally.

I liked the two lines "Did you ever stop believing" and "Did you ever need a reason" from the first verse, but if I was going to use them how I imagined, I would need to spread them out over a whole verse. So I cut those two lines out into separate regions and kept the first one where it was, while moving the second one to 8 bars later. I copied the end of each line over to the "Delay FX" track as well, to give a nice echo at the end, and copied the first word of each line onto the "Reverse Verb Vocal" track to give a nice sweep up into each line. Listening back, I thought the vocal sound wasn't quite right for the dub mix and wanted something a little more effected, so I created a copy of the original vocal track and copied these two lines on to the new track, and experimented with some different effects. The final effects set up for this track were an AVerb ("Long Subway Tunnel" preset, but tweaked a little) followed by an Autofilter (I chose the band-pass option, set cutoff to 68%, resonance to 11%, fatness to 100%, input distortion to 204%, output distortion to 60%, and adjusted the Main Out level until it was about the right volume).

The use of just those two lines made it feel much more like a dub mix, which, in my opinion, is generally quite repetitive. And it was that one fact that gave me an idea of something to try in the chorus section. In some cases I will leave the whole chorus of a song pretty much intact in a dub mix, but as I have already explained, in this case I wanted it to be much more of an instrumental mix with a few vocals placed strategically. That being the case, I chose to get rid of the chorus altogether and replace it with just those two vocal lines again. Fortunately, the melody worked just as well over the chorus as it did over the verse, so I didn't have to change anything at all except for an increase in level owing to the more powerful musical backing at this point. After I had applied the volume changes I was pleased with the result, but it needed a turnaround at the end. Seeing as I hadn't actually used the title of the song yet, I copied over the "Make it you" line from the end of the chorus onto this new sound and this worked perfectly as a turnaround, letting you know that it was the end of a section. In addition, the dub mix now bore at least some relation to the title of the track!

For the next section (the gap between chorus and verse and the second verse itself), I copied over the "Did you ever" vocals from the intro and ran that through the whole 16-bar section that led up to the second chorus. I did this because I wanted to have a break from the "Did you ever stop believing"/"Did you ever need a reason" section of the first verse and chorus. It also worked really well because it was a thinner sounding and more minimal vocal to go with the reduced dynamics of the music. When I got to the second chorus, I basically just reused the first one because, even though it was a dub mix, I still wanted some repetition and familiarity in the chorus. There might have been subtle differences in the phrasing between the first and second chorus in the original version of the track, but that wasn't so much of an issue here because of the inherent nature of dub mixes. By this point I had reached the breakdown section, so I decided to listen to the whole arrangement up until this point.

I felt that things were definitely flowing nicely and I really didn't want to introduce any new vocals this far into the mix, so during the breakdown I decided to just keep the intro "Did you ever" vocals running all the way through, rather than going to the original breakdown vocal ("When everything in my life..."). The main reason was obviously to keep the vocal minimal and sparse, but I also liked the way it seemed to give the musical build-up a little more emphasis anyway, because the vocals weren't detracting from the pitch bend and upward sweep of the effects used at the end. Once again, I needed to adjust the levels of these vocals to bring them into balance with the changed musical dynamics.

For the final chorus I was planning to use the same chorus vocal once again, to keep the repetition and familiarity there. There was one small issue though, because the original outro chorus had two lots of 12 bars, so I ended up with a 24-bar section (32 including the 8-bar instrumental section). As I didn't want to bring the vocals in at the beginning of the 32 bars, I therefore had 24 bars to accommodate what had become a 16-bar chorus. So I had the first chorus as is, but without the "Make it you" vocal at the end, and then just repeated the "Did you ever stop believing" line from the first half of the dub chorus, and then ended with the "Make it you" as I would normally on a dub chorus. In effect, I had one and a half dub choruses following the 8-bar instrumental section. It felt a tad strange, only having one and a half, but then it kept the 32-bar length overall for this section, which was my main goal.

With this dub mix, and especially because the vocal use was really quite minimal, I was in a way, more concerned about keeping the flow of the music working and then just dropping the vocal in places over the top. There are a number of DJs who really don't like playing vocal tracks because they are too "cheesy" and "commercial." This is one of the reasons why the market for a dub mix exists at all. In fact, this mix is closer to an instrumental mix than the vocal mix, so that would probably suit those DJs just fine. I took a quick break and then had a listen back at a lower level (because I didn't need to check bass levels or anything like that) while I was checking my emails. I did this because I wanted to listen on a more subconscious level, to see if the vocals gave enough of a hint of the track without being too obvious. After listening, I made a couple of minor changes to vocal levels, but that was basically it. I kept the vocal levels lower overall in this mix (compared to the club mix) because they were quite spacey anyway, and I wanted them to sound pretty distant.

Back in the day, dub mixes were often completely different productions, but these days that happens less and less. I think one factor is that the fees are so much lower and deadlines so much tighter that the remixer can't really justify doing what is, effectively, a second remix just for the purposes of a dub mix. As such, most dubs these days are just "dub vocal" versions of the main club mix. And that is what I wanted to do here. Had the client actually told me that they really wanted to put a lot of focus on the dub, then perhaps, had scheduling allowed, I might have changed the production as well as just the vocals. In this case, the client hadn't even asked for a dub but I wanted to do one anyway because I thought a dub would allow the music to come through a little more and I was really happy with the result, so the dub was more for my personal satisfaction than anything else! Hence me not spending ages working on a different production.

After a couple of final listens through, I was happy to commit this mix, so I bounced the same three mixes as I had for the radio edit and club mix, saved a "final" version, and then opened up a new Logic project and imported the three "full vocal" versions onto adjacent audio tracks, and prepared myself to work on the final stage of the process: mastering.

The very first thing I should reiterate is that I am not a mastering engineer by any means, so I simply use the mastering process as my attempt to

FIGURE 24.6
The final arrangement of the dub mix. Note the completely different use and placement of vocals near the top of the Arrange window.

get the best sounding masters I can. Sadly, and perhaps this is a fatal flaw in my own attempts at mastering, I usually do it (when it comes to remixing at least) on the same day I finish a remix. If that day is at the end of a three- or four-day process, then my ears have become so accustomed to the sounds and details and subtle nuances of the track that it is almost impossible for me to be entirely objective about how things sound. Even the space of a day or two might put enough distance between the memory of how the track sounded in my head and the reality of how it sounds on the day, but as sometimes happens, I didn't have the luxury of doing that if I was going to meet my deadline. So without further excuses and justifications, let's take a look at what I did.

For this part of the walkthrough I really wanted to use some more "heavy duty" plug-ins so I haven't included the Logic project file I created because you wouldn't necessarily have any of the plug-ins, but what I have done is taken screenshots of all of the settings I used (available on the website) for reference. Nonetheless, I will still describe the process and the reasons behind the decisions I made.

The first thing I wanted to address was the dynamics of the track, and that meant compression. I use a number of different plug-ins when I am mastering, and there isn't any particular detailed methodology in deciding which ones to use. I often start off (as I did in this case) with the excellent UAD Precision Series, and if things are sounding how I want them to, I can always try others. On this particular track the first plug-in in the mastering chain was the UAD Precision Multiband on the "Loudness" preset. I sometimes use this preset as a starting point because I find that it often adds a nice bit of punch to the track.

However, it can also make things a little too bass heavy or top heavy as well, so I often need to adjust a few things to keep it under control. As well as having all of the usual compression parameters available in each of its five bands, it also has a "Mix" control, which allows you to do parallel compression. The preset values sounded okay with this track, but I was sure I could get it sounding a bit better, so I played with the individual thresholds and ratios a little until I felt that I had a better balance, and then spent a while changing the gain of the individual bands and adjusting the mix control.

Once I had some good settings on the Precision Multiband I added a PSP Vintage Warmer to act as a single band "bus"-type compressor and to give a little warmth and drive. This plug-in has been highly recommended by a great number of people in the dance world because it can (when used carefully) add a real richness and warmth on pretty much anything you throw at it! In my case, it was set up subtly so it was giving no more than a few dB of gain reduction even on the loudest passages. I didn't really want the track to sound any more compressed after passing through the Vintage Warmer, just a little, well, warmer!

The next thing to do was to add some overall EQ to the track. I think that if I was being totally open, the biggest issue I had with working on this remix was getting the big stab sounds and the vocal balanced up against one another. This all-important midrange area of the track can cause major problems because there are often more things fighting for space in this frequency area than in any other. That is why I had applied quite a lot of EQ plug-ins at various stages, boosting different midrange frequencies and then having to adjust volume levels. This, then, was my last chance to try to get the sound I was hearing in my head. I started off using the UAD Precision EQ, but after spending a little while working with it I remembered that I had an unactivated demo of the UAD Manley Massive Passive EQ available. I have heard some wonderful things about this EQ from various colleagues in recent months, so I activated the demo and gave it a try. I have to say, I was very impressed with the overall smoothness of the tone and it wasn't long before I was convinced that this is a really great EQ and I can understand why so many people swear by it. After some big smiles about the amount I could actually push the different bands without it ever really sounding harsh, I settled down to the task at hand.

One of the great things about a truly good EQ (hardware or software) is that it enables you to get in there and manipulate the sound in a way that makes sense. Boosting one frequency seems to pull up (or down) adjacent ones in a way that appears to go beyond just mathematics, and takes on an *organic* quality. Perhaps that is what I had been missing all along in trying to get the midrange of this track balanced up with everything else. It certainly made the task easier, and it didn't take too long before I had finalized the settings for the EQ. I ended up using a 2 dB shelf boost at 68 Hz (for the weight in the kick), a 3 dB bell boost at around 3300 Hz with a medium bandwidth (to bring out the vocals and stabs a touch), and, finally, a 3.5 dB high shelf boost at 8200 Hz (for the presence and air on the hats and vocals).

FIGURE 24.7
Final EQ settings I used on the Universal Audio Manley Massive Passive EQ for the mastering of the track.

There is a great deal said about the benefits of tape saturation when mastering, and some people go as far as recording the stereo mixdown onto a real tape machine and then recording it back into a DAW to get that *real* "tape sound." I am not totally convinced there isn't an element of snake oil in all of this, at least in some cases, but I was certainly willing to give it a try. Sadly, I didn't have a tape machine to hand, so I decided to activate another plug-in demo that claims to give many of the benefits of the tape sound but in a digital plug-in. I am talking about the UAD EL7 Fatso Sr. Once I had loaded this up, I set the track playing and ran through a few of the presets to see what it was capable of. Many of them were far too juicy for what I wanted at this stage in the process, but I can totally see myself using them on individual sounds within a mix to add some drive/distortion at an early stage in the mix. In the end I did find one preset ("Tape Buss") that seemed fairly subtle, but at the same time noticeable (and beneficial!). The compression section seemed to be only taking off another couple of dB, which was good, because I prefer to do the mastering compression in a couple of different stages with different compressors, to get varying tonal characteristics and sounding, on the whole, more transparent than just one compressor working really hard. It was clear, even though I wasn't at all familiar with this plug-in, that it was doing more than simple compression, and I was totally convinced when I bypassed the plug-in to compare to the signal going into it. After a couple of bypass/active runs I was sold!

Finally I set up two UAD Precision Limiters. Even after all of the aforementioned effects, there was still quite a lot of headroom available, so I set them both up the same with 3 dB of gain and I would use just one for "Radio Master" (so there was still headroom for the broadcast compressor/finalizer they use to work without killing the dynamics of the track totally) and then for the "Club Master" I would add in the second one to get maximum level. And that was it. I took a short break just to give my ears one final rest before a run-through of all of the mixes in their "mastered" state, and then bounced down the "Radio Mastered" and "Club Mastered" versions of each of the three mixes. I was happy with the results, but they are undoubtedly not as good as a professional mastering house would get, hence me always keeping an unmastered version and sending that to the client as well, should they wish to get the track mastered at a later time.

The final step, after bouncing all of the different versions of the mixes, was to send 320 kbps MP3 versions (I chose the "Club Mastered" versions in this case) to the client for approval. Now all I had to do was sit back and wait for his comments. If he wanted any changes I would, within reason, do them as long as what he was asking for didn't compromise my "artistic vision" for this mix too much! And then, once I got approval from the client, I would use the submix buses (kick, loops, fx, bass, etc.) to make stems for each of the mixes, and would upload the full package (320 kbps MP3 versions of the mastered and unmastered mixes, 24-bit WAV versions of the mastered and unmastered mixes, and stems for all mixes) to the client using SendSpace or YouSendIt. If they requested everything to be sent as a physical copy on DVD then I would, of course, do that as well, but fewer and fewer people do that these days.

So in summary, what do I think of the remix? I am very happy with the result, especially given that I didn't have all my usual "tools" to hand, as I wanted to create the remix (at least up to the mastering stage) just using the built-in Logic tools. Are there things I would do differently? Yes of course, and I probably could have done, because, aside from the client's deadlines, I could easily have gone back over things after I had delivered the mix to him to make it sound better for the purposes of this book, but I didn't want to do that. It would have been very easy for me to make a bullet point list of how I created the perfect mix after spending a month doing it, and leaving out all the parts where I changed my mind and went back over things, but I don't think that would have been as realistic or, I hope, as helpful. I wanted to present this part of the book in as much of a real-world (warts and all) way, so you could see the *real* process rather than a glossy idealized version of it.

In hindsight and listening back, I would have liked to have spent a little more time on the mixdown and maybe try a few different versions with slightly different balance and emphasis, but that is partly the nature of the business. Deadlines are often tight and you have to get things done in time, so you aim to get the best result you can within the boundaries that are set for you. Leonardo da Vinci is quoted as saying "Art is never finished, only abandoned," and I think that is certainly true here. I can listen back to every remix I have ever done, and every original track I have ever produced, and make a list of things I would do differently now. But that is part of the nature of the beast. The last, and perhaps most useful, piece of advice I can give to you as a remixer is that you have to learn when to let go. In some ways this could be seen as a compromise and it probably is. Some of you will possibly read that as me saying that it only has to be "good enough," and that is also true. When you have the luxury of time you can possibly get closer to the ultimate version of any given track. But there comes a point when you're not making it any better, only different. And knowing when you reach that point is a skill that can't be taught, it has to be felt.

SECTION 4
The Business of Remixing

CHAPTER 25
Promotion

Let me open this chapter by saying that everything you have read (and hopefully learned) up to this point will only really benefit you if you take on board what I have to say in this and the next few chapters! All of the technical and creative expertise in the world will not do you any good without promotion and, later on, management or representation. You could be the best remixer in the world but if nobody knows that, you will find it hard to get any work! In an ideal world it would be easy to get management and have someone else do all the hard work of promoting you and your services, but the reality is that, unless you are very lucky, you will probably start out doing a lot of groundwork yourself. So with that in mind, let's take a look at what you can do yourself when you are starting out.

THE SHOWREEL

The whole point of a showreel is to actually *show* people what you can do rather than simply relying on you telling them how wonderful you are. The word *showreel* originated in the movie industry where it was basically a compilation of clips of work that the artist (be it actor, director, editor, etc.) has completed to show his or her talents. Many years ago a musician's showreel would probably have taken the form of a tape compilation; later it was a DAT (Digital Audio Tape) or CD, but today it is more often than not just a collection of MP3 (or sometimes WAV/AIFF) files that can be emailed to prospective clients or downloaded from the artist's website. It is usually accompanied by some kind of "biog" (biography), which gives a little more information about the artist in question.

The big problem with sending showreels to potential clients has always been that of getting noticed. Some record companies receive hundreds of demos each week, and even with the best of intentions, it is simply impossible for the person in charge to listen to each and every one of them. When demo DATs or CDs were the norm you were often advised to do something to make your showreel or demo stand out from the crowd. Given that most would be submitted in anonymous looking padded envelopes, it was often suggested to do something creative with the packaging to draw attention to your package. While this may have worked, there was also a fine line between drawing attention and screaming "I'm desperate...pick me!!!" The problem was that nobody really knew exactly where that line was; perhaps more confusingly, the position of that line would vary from company to company, from person to person, and even from day to day! There was no easy answer.

Fast forward to the present day. While the era of the MP3 showreel has undoubtedly made it easier, quicker, and more affordable to send out your showreel to potential clients, the same problems still exist. In fact, if anything they are much worse. Not only do you have to compete for attention with the possibly hundreds of other emails coming from those seeking work, you also have to avoid being trapped in a Junk mailbox with the risk of the email being "unread" and ending up disappearing off the bottom of someone's screen without ever being seen, let alone read. Perhaps an attention-grabbing subject line in your email could do the trick, but once again you run the risk of crossing that line. So how do you make a good impression and get your showreel listened to? Well, one possible answer comes from a decidedly "retro" idea. Are you ready for this? Okay, here it is: send a CD! Now I know it isn't as easy or as convenient and it takes more effort, but in a way, that is exactly the point!

Most tech-savvy people these days know that it is easy to send a personalized email to a "mailing list" and theoretically (assuming you have done your preparation) send out your showreel to hundreds (if not thousands) of music industry people in just a few minutes. There are (many) times when this is appropriate and perhaps even expected, but submitting showreels is one area where I would really recommend going the extra mile and making that extra effort. Aside from the fact that it shows that you are committed and dedicated because you have made the extra effort, it might just get somebody's attention too! The days of getting hundreds of CDs are probably gone. I would guess that most A&R people get a vast majority of their submissions via email. So instead of your showreel being maybe one of a hundred from that week, it might be one of ten, or perhaps even one of five! Certainly better odds than being one of potentially several hundred emails a week that have to be ploughed through. In fact, I was asked recently to send some tracks over to a record label on CD because my A&R contact there made pretty much this exact point, and asked to hear the tracks on CD because it was easier for him! Food for thought, at least.

In terms of what to include on your showreel, my recommendation is that you go for quality over quantity. When you are starting out you probably won't be able to namedrop too much, so the temptation is to include a lot of tracks, many of which the person you are sending it to might never have heard of, simply to show your versatility and to give the impression that you have done a lot of things. To be honest, I think you would be better off sending a showreel with your *best* three or four tracks rather than ten that you think are good but are hedging your bets a little because you don't know what they might like! Unfortunately, many times you will be judged by the worst track on your showreel, not the best. So bear that in mind when you are choosing what to put on there. As time goes on and you start getting some good paid work under your belt and your remixes start getting club play, radio play, and club chart exposure, you can start to include a few more, but I would never go above about six to eight tracks on a showreel unless you are incredibly diverse and are sending the showreel to a label or company that has a catalog of work as varied as the showreel you are sending.

Where possible, and when you get to the point where you have a good number of tracks to choose from, you should customize each showreel to the person or company you are sending it to. For example, while you, as a remixer, might be great at funky house, electro house, trance, and breakbeat, it is quite unlikely that the company you are sending it to will be interested in all of those genres. So do your homework about what the company or label might be interested in (perhaps check out their recent releases to see what kind of remixes they commissioned on those) and choose the tracks to include on your showreel based on that. Obviously if you don't have any tracks in the genre(s) they are interested in, the chances of you getting the gig are reduced, but it is certainly still worthwhile. Stranger things have happened!

One other thing to consider is whether to include the full tracks/remixes or just shorter "edits." This is a question I have wrestled with myself in the past, and with my years of experience, I can say with certainty that there is no right or wrong answer! You could argue the case both ways though. Full-length remixes would show that you understand how to build a club mix arrangement of a track and sustain interest and even build excitement and tension over a seven-minute or so period, but most people would just assume that you knew that anyway. Edits would give the sense of what the remix was about without the (undoubtedly very busy) listener having to sit through three minutes of "DJ Intro" and "DJ Ending." These edits would also show that you understand how to create a dynamic journey for your remixes in the three-minute or so window you have for most radio edits these days. As you start working with bigger and more commercial artists (if that is, indeed, part of your goal) the radio edit becomes increasingly important.

At this point I should probably deal with a question that some of you might have: "What if I am just starting out and haven't done any commissioned remixes yet?" The elusive first commissioned "official" remix is one of the hardest to get, because you have to ask people to take a chance on your abilities. So it is a good idea to have *something* to send them, even if it is only a bootleg or a remix that you have done using an a cappella vocal. It will show your technical skills and your creative ability and sense of originality. You won't get paid for doing these first few remixes but it will be time well spent if it helps you get that first gig and start building up your catalog of "official" remixes.

I realize that I make it sound quite easy: do a couple of unofficial remixes, do a bit of self-promotion, send a CD to a few labels, get your first official remix, do some more self-promotion, start building up your showreel, more self-promotion, start making contacts, get more paid work, and so on. I am sure that most of you will know by now that it really isn't as quick and easy as that, but that really is the fundamental process. Any one of the stages could take a long time, or a little (relatively speaking anyway) time, or you could end up being stuck at any given stage for quite some time. There are no guarantees of success or magic formula for this or for pretty much any other job in the world (music industry or otherwise), but if you have taken the time to build up your skills and honestly believe that you have what it takes to be able to do this for a living, then you have to keep working away on the back-end stuff, the boring stuff, the stuff we all need to do, every week or every month, when we would rather be in the studio working on something musical.

One final thing to mention before we move on is the issue of copyright. If you have completed a remix for a record label, in almost every circumstance (I can't actually think of any exceptions I have experienced personally) the record label will be the sole owner of the copyright of the remix whether you got paid or not. On the surface that would make it seem like you would be infringing their copyright if you were to distribute the product they owned, the remix. Even the process of sending out a showreel of remixes you had done would seem to fall under this umbrella of copyright infringement. Fortunately, it doesn't necessarily. In United States Copyright Law there is a doctrine called "Fair Use," which allows the use of otherwise copyrighted material as long as a number of factors can be satisfied. The fact that showreels are not-for-profit and will not have any effect on the market of the song you have remixed as they aren't being made available to the general public, would certainly go a long way toward satisfying those criteria. And while I am not in a position to offer legal advice in any way (please consult a solicitor if you have *any* legal questions), I have been unable to find a single case of a person being prosecuted for copyright infringement for including a remix/song/movie he or she has legitimately worked on as a part of a showreel. If you were to make that showreel (or any individual tracks) available to the general public, that would be a different matter as the sales of the track/movie could well be compromised by the free version you are making available.

Showreels are what you send to a record label/company when you are trying to convince them to give you the gig, but at some point you will presumably be hoping that people will start coming to *you* to remix their tracks rather than you having to chase them. That would normally start to happen when your profile gets high enough, but there are things you can do to maybe give things a little helping hand to move them along. I think it would be going too far to say that the music industry was actually *built* on hype, but there is certainly a lot of it around! So while you are waiting for your showreel to build up with commissioned work, you can certainly start getting a buzz building about you through the power of the Internet!

SOCIAL NETWORKING: A BRIEF HISTORY

Although the principles behind it began life much earlier, the world of social networking really began its meteoric rise to prominence back in 2003 with the launch of Friendster. This site was originally launched as a safe place for busy people to expand their network of friends and social contacts. It was a great success right from the start and the idea was soon adapted by the now legendary MySpace, which took the best ideas from Friendster and built upon those. Over the following years the MySpace user-base grew to over 100 million users and the site expanded from being a purely social and friend-based site to being a huge Internet portal for bands and music. It wasn't long before every band had its own MySpace page with photos, videos, and music tracks for people to listen to.

In many ways these MySpace band profile pages took over from the idea of a band having its own fan club. By having what amounted to a huge mailing list of "friends" (or fans in this context), bands could interact to some extent with the people most likely to buy their music. It offered the bands a (free) means of informing their fans of news, upcoming releases, and concerts, and offered the fans near instant access to that information and a means to interact directly with the band, or most likely an "appointed representative," more directly than ever possible before. It was, perhaps, this feeling of being able to send a message to the band directly that became one of the biggest attractions of MySpace to music fans.

Very soon after this, the inevitable slew of imitators (and a few innovators) came along and the world was suddenly bristling with sites like MySpace, and "social networking" became the new buzzword in both industry and society. You weren't anybody unless you had a MySpace account. So many bands and artists (and remixers!) hopped on the social networking bandwagon and started to build up a good amount of "friends." But herein lies the first problem. Websites like these have a tendency among a good number of people to encourage popularity contests, with people seeking to have as many friends as they can without being overly selective. As a result of this, it is almost certain that any band or artist on MySpace will have a not insignificant proportion of their "fans" who aren't really fans at all, so the numbers can be slightly misleading. Nonetheless, sending out a bulletin on MySpace takes the same amount of time whether it is to one person or to 100,000 so there is really no harm in having people in your friends list who aren't really fans, because they will simply ignore the bulletin you send out.

This ease of sending out bulletins, however, soon came to be abused like so many other things that are related to the Internet. Unscrupulous people soon started sending out messages that were nothing more than attempts to redirect the reader to a website to try to get them to buy or sign up for something, and MySpace very quickly became flooded with messages of this kind and lost quite a bit of popularity. By this time, however, there were already a number of other sites that, while perhaps not offering a dedicated "music" section of the site, did still offer most, if not all, of the same opportunities to bands and artists wishing to promote themselves. Sites such as LinkedIn and Bebo attracted large numbers of subscribers, but it was, eventually, Facebook that took the crown of the most popular social networking site.

Facebook offers a number of features similar to MySpace but without the dedicated "music" pages. There is also no music player (at least not "natively") on a Facebook page, which is a definite downside to Facebook. But the people behind Facebook were very clever in that they made the site almost infinitely extensible by the use of "apps," which add additional functionality to a Facebook page to those who actually want it. Facebook was still primarily a site set up for actual *personal* networking rather than *business* networking, in the sense that would be useful for people in the music industry. This was remedied with the addition of "fan pages." With a "fan page" set up you have pretty much all of the day-to-day functionality you would want as an artist, and everything you used to do on MySpace is now possible on your Facebook account. The only thing some people feel is lacking is the ability to customize the visual aspects of your page in the same way you can with MySpace. You can, with MySpace, have a page background and font color and styles that are in keeping with other branding you might have had elsewhere, whereas every Facebook page has the same overall appearance, making it more difficult to establish visual branding in the same way MySpace allowed you to. But this hasn't detracted people, and, as of writing this, Facebook now has over 500 million users worldwide.

The actual search methods and the whole process of adding friends or fans on Facebook makes it less likely to be abused in the same way MySpace was, but it does still suffer to a similar extent from the "popularity contest" effect MySpace has. This is an inconvenience, as I mentioned earlier, but it is, in reality, a fairly small price to pay for the benefits that can be had.

In addition to the social networking sites there are a couple of other things that fall broadly under the same heading: Twitter, SoundCloud, and YouTube. I am pretty sure that most of you will at the very least have heard about Twitter, and most likely, already have an account, so you probably already know the basic idea. I guess that it is probably closer to a blogging site than a social network, but as you can subscribe to someone's blog and become a "follower" and be listed as such, it does take on an element of social networking. Twitter has, in recent times, seen an explosion of popularity, with people Tweeting about anything from the mundane to the monumental, but to me at least, it is one of the best "value for time" social networks there is, simply because you don't get bogged down in complexity. Posts or tweets are very simple affairs, and there are applications for many smart phones now that enable you to tweet on the move. It's simple, short, direct, and to the point. Once you build up a list of followers, you can use Twitter as a means to either give them information or redirect them to another website for more information should you need to. You would probably still need either a Facebook or MySpace (or similar) page, or an actual website, to be able to give people anything of substance, but its "low maintenance" approach is what makes it a very good time investment, in my opinion.

SoundCloud is a fairly new kid on the block, and, like Twitter, it takes a different approach from the other social networking sites. In fact, some people might not really consider it a social network, and technically, they could be right, but the fact that, like Twitter, you can have people who subscribe to your account and who can then receive messages and information from you puts it at least loosely into the social networking category for me. SoundCloud is focused on one thing and one thing only: Sound! SoundCloud isn't concerned what mood you are in or if you have some photos from your last crazy night out, all it is about is music. It is, in essence, a glorified online media player and file storage location for the music producer or artist. You can upload tracks and you can specify if they are or are not downloadable, you can upload tracks as you go along (subject to certain limits), and each has its own dedicated URL, which you can then email to people so they can go to the relevant page and either stream or download the music. Some nice touches are the ability for listeners to post comments on the tracks and to actually attach them to a particular point on the track that's playing. They could, effectively, attach a comment to a certain part of a song and therefore make their comment more clear and relevant, seeing as you already know which part of the song they are referring to. I think it is a useful site, but I am yet to be fully convinced about it purely because (apart from the ability to attach comments to the track in particular places) I really don't see what it offers that can't be achieved with a Twitter account and either your own website for uploading your files to, or alternatively, an account with something like SendSpace (*www.sendspace.com*) or YouSendIt (*www.yousendit.com*) to upload your files to.

Finally we should probably mention YouTube. Once again, and like both SoundCloud and Twitter, it isn't really a social network in the traditional sense, but as I stated for SoundCloud, the fact that other users can "subscribe" to your "Channel" means there are social networking elements coming into play. In quite a few ways, the uses to us as remixers (and possibly artists ourselves) are quite similar to those of SoundCloud, but with a few differences, one very obvious downside, and a few extra benefits. The obvious downside is that YouTube is a *video* site and you can't actually upload MP3 or WAV/AIFF files for people to listen to, at least not directly. It is pretty simple to make a "video" (using rudimentary and often free video software for both Mac and PC) if all you want to achieve is to get your music up on to YouTube to be available for people to listen to. All you need is a JPEG or BMP image (perhaps your logo if you have one) and the MP3 or WAV/AIFF audio file, and then, in not much more than a few minutes, you can create a "video" that is simply a static image with your audio playing. So it won't be very interesting for people to watch, but that might not be too much of a concern for you. Once you have created this "video," you can upload it, and assuming that you enter in all of the information and *tags* properly, it will soon be searchable by the people who watch the estimated 2 billion videos per day. Not only will it be searchable, but if you are clever with your tags, it will also show up in the "Related Videos" section, so you should get some views from people who might not have searched for your track directly but have come across it by simply browsing through the videos and related video links.

One of the main benefits of YouTube is the sheer volume of users it has. You still need to work hard to get the tags right so your video has the best chance of showing up in search listings, but there is potential, *huge* potential so it's definitely worth looking into. In fact, according to some advice I was given by Eddie Gordon (of Media2Radio fame, *www.media2radio.com*), YouTube has now become the first place people will go to online to search for music! Quite a big statement to make and quite an important one to listen to if you are looking for ways to promote yourself. One quick word of advice though: You remember we spoke about showreels a little earlier and about "fair use?" Do you remember me saying that showreels were generally okay because you weren't making them available to the public? Well YouTube is a *very* public place, so if you were uploading remixes you had done on to YouTube as a means of having an online showreel, you would be well-advised to talk to any record labels you have worked for and ask their permission before you actually upload anything. Most of them will be fine with it, but will probably ask that you don't upload anything until they have given you permission to, so that it doesn't interfere with their promotional plans or disrupt their sales. And, as always in situations like these, full notification of the owners of the copyright to the track is always a very good idea.

Next we look at why and how social networking sites are useful to a remixer such as yourself.

USING SOCIAL NETWORKING TO YOUR ADVANTAGE

I mentioned earlier in this chapter that there are things you can do when you are starting out (or even looking to develop your remixing career further) that will help you to build "profile," and will, eventually, help make it easier for you to get work. The music industry is very competitive and there are so many people out in the world competing for a relatively small amount of work, that anything you can do to help yourself get a competitive edge is hugely beneficial to you in the long run. Social networking sites can help you in this respect by allowing you to build up fans (or followers/friends) and giving you a means of direct contact with them. As a remixer, however, you will probably end up using the sites slightly differently from the way in which a recording artist, singer, or band would, simply because the part of the industry we are working in has a different dynamic and mindset, which we need to take into consideration.

The fundamental difference is that as a remixer, you want to appeal to A&R people, DJs, and the general public, while a recording artist or band would probably be more interested in promoting themselves to the general public, as they will be the ones who ultimately buy their music. You see, as a remixer you are not reliant on heavy sales and tour and merchandising income, as many of today's recording artists are, but you *are* reliant on heavy support from certain key DJs. There are a few reasons for this, the first of which is that, on most pop singers and acts, the remixes may well be included on the CD/download package, but they will, in most cases, contribute a very small part of the overall sales figures for that track. So sales and revenue aren't usually a motivating factor in pop artists having remixes done of their track. The simple fact is that club music and club culture are here to stay and *huge* numbers of people go to clubs on the weekends. By getting remixes done, record labels are able to get their artists' songs heard in an environment where they wouldn't otherwise be heard. That is probably the main reason for having club remixes of pop records these days. Remixing has become a part of the furniture in the music industry today, and A&R people at most major labels almost take it for granted that remixes will be done if the track is at all appropriate for remixing (and in some cases even if the track *isn't* appropriate!); the decision comes down to a business decision at that point: who...and how much?

The simple answer to the "how much?" question is..."as little as possible!" Record labels are businesses and they exist to make money. Because of that, they won't spend money unnecessarily and without being able to fully justify it. That's why, if you don't really have any major remixing achievements to your name in the past, you will be hard-pushed to get any decent amount of money out of them. But there is still money to be made once you are established and have a bit of a track record. To give you some examples, as I write this I know of at least a few remixers who can command $16,000 for a single remix, but that really does represent the absolute peak at this time. Ten years ago that figure would have been closer to $40,000, but the music industry has suffered a substantial decline since then. To be honest though, if you did get yourself in to that position, $16,000 for a single remix isn't *that* bad...is it? On a more realistic level, it is quite possible that after you have started getting some bigger remixes, you can start getting fees of $800 to $1600 for a remix, and then working your way up from there. There are very few remixers who will get $16,000 for a remix; a few who will get $8,000; quite a few who will get $3000–$5,000; many who will get $1600; and a *lot* who will get between $800 and $1200. When you start out, be prepared to do them for even less than that, or possibly for your first remixes, have to do them for free just to build trust, establish your brand, and get something together to put on your showreel.

When talking about the "who," there are a massive number of things to factor in: the size of the record label, popularity of the artist, success of the last single, who remixed the last single, who out of all of the remixers of the last single seemed to deliver the most well-received remix, what style the possible remixers are, what style is *de rigueur* at the moment, and many others.

The size of the record label and the popularity of the artist will have an impact on the decision as to which remixers are considered, for largely budgetary reasons. Put simply, a bigger artist on a bigger record label will have a bigger budget for remixers, which (normally) will mean bigger remixers. That doesn't, however, put you out of the running if you are just starting out because there is always a chance that a label will commission a lesser-known remixer to work on the remix package for a much smaller fee, just to add a different "flavor" to the existing mixes without adding too much to the budget.

It also happens that, quite often, if a remixer does a particularly good remix for an artist then the artist or label will ask for that remixer to work on their next track as well. In fact, sometimes this can go on over a period of several releases spread over a couple of years or longer. There is a lot to be said for trying to build up a good relationship with whomever is in charge of commissioning the remixes, because with a professional, helpful, and courteous attitude, and, of course, great work, you can get some good repeat business from labels without having to do the sales pitch each and every time you are looking for work. Bear in mind, though, that styles and tastes do change, and unless you are changing with that to some extent, there is only a certain amount of time you can ride off the back of your previous success with any particular artist or label.

Most record labels will want to get the track played in a variety of different clubs, which means they will commission remixes across a variety of genres (or subgenres, at least) so the other remixers they have already commissioned will play a part in who they choose to finish up the package. If they already have, for example, a couple of house mixes, then they might look to another genre for the final mix rather than commissioning somebody to deliver a remix similar to something they already have. A part of this will sometimes involve getting remixes from up-and-coming remixers in genres that are just becoming popular, so if what you are doing is a little different, and in some way, at least, cutting edge you might have a better chance of getting the gig than if you were competing against the established "big boys" in another genre. The flip side of that position is that what you are doing might be considered too "underground" or risky, so you might actually reduce your chances of getting the gig compared to doing something safer.

The truth is that you really shouldn't chase any particular style just to try to get work, because that will rarely succeed. Try to do something that really resonates with you, you feel comfortable doing, and you *enjoy* doing! You should definitely push yourself and try to develop your own unique sound, but you should do it in your own way rather than trying to jump on the latest bandwagon simply to try to steal a piece of the action. By doing something you are really into, you will be able to inject an excitement and dynamism into your work, which along with some good self-promotion will go a good way toward getting you established and well on your way to better paid work.

If this is pretty much how the industry works, then why bother with the whole social networking thing? Well, as I implied earlier, how much you could get paid is at least partly dependent on your "profile." Sure, you can raise your profile by remixing some big records and getting your name in the club charts as a remixer. That is traditionally how it is done. If you are a DJ as well, then getting regular DJ bookings and working in some bigger gigs every now and then will also help. But social networking is about *networking*, and in many cases, who you know can help open doors and get you work that you might not have otherwise gotten.

You may be able to track down some people who work for record labels on Facebook, but if you send them a friend request and they have no idea who you are, it stands a good chance of not being accepted. However, DJs might be more open to having new people on their friends lists and often a polite request asking them if they would be happy for you to send them new material you are working on will be met with a positive reply. After all, DJs are all about the music they play, and if they can get brand new music free of charge by being part of your promo list then, unless your music really isn't up to standard (and you wouldn't be sending it out to people like that if it was, would you?) or it really isn't the kind of music they play, they will probably be more than happy to do so. Of course you do need to do at least a little homework here to find out what they *do* play, but that shouldn't really be too hard to do.

The temptation, when approaching both record labels and DJs, is to go for the biggest names first. This makes sense, because like everybody else in this industry, you are aiming for the top! And while I would never discourage you from aiming high, I would also advise you to remember that there are a lot of smaller labels and lesser-known DJs that might still be able to offer you work or support for the work you are already doing. Becoming a success isn't *all* about getting work for the big artists and labels, because there are great songs released on smaller labels as well and it might be easier to convince a smaller label to take a chance on you if you are a relative unknown.

So by using Facebook, MySpace, Twitter, and whichever is the newest and coolest social network at the time you are reading this to build a good collection of contacts, by making sure that you keep those contacts updated with what you are doing, keeping your name constantly in their mind (without being a nuisance of course and *no* spamming!), you can get a level of recognition among DJs that might take much longer to do through more conventional means. I have made several references throughout this book to the importance of DJs to remixers, and I will go into more depth in a Chapter 27 on the specifics of this symbiotic relationship, but for now just remember that, in the world of the remixer, the DJs are quite possibly the most important group of people when it comes to you being successful or not.

Having said that, there is no reason why you can't start building a fan base for your work among the general public as well. Whether you have written your own material or have purely been remixing, you can start putting your work up on MySpace or Facebook or other sites, and once it is available and people start to listen to it, you will find that some of them will like it and send you a friend request, become a "follower," or whatever the format is for the particular site.

You see, what the Internet has done is provide a level playing field, well, almost level at least. Social networking sites, sites such as Beatport and iTunes, and the power of search engines, have given every single music producer, remixer, and artist the same tools, the same chance of exposure, and the same accessibility and availability as every other. This is the theory, at least. Unfortunately, while everybody has the same opportunities, some people are in a better situation to take advantage of them. So the more successful you are, the more likely you are to be in a situation where you can employ someone to spend time doing all the promotion work for you. When you are starting out, and even as your career begins to develop, there are so many things you have to do yourself (for financial reasons), that it is often hard to find the right balance between the fun stuff (writing, programming, remixing, experimenting) and the necessary stuff (accounts, paperwork, promotion), and this can vary from week to week depending on how much time you have available. So self-discipline and being able to stay focused are very important too.

The biggest problem facing *any* artist, established or otherwise, is that of public awareness. You could have the best material in the world but if nobody knows about it, nobody will buy it. Or, in the case of a remixer, nobody will employ you to do remixes. That is really the key point of using these social networking sites. Where a band or singer will use those kinds of sites to inform the record-buying public that they have a new song coming out, you as a remixer can use the same principles to inform people within the industry of what you have been working on and to sell your service (rather than your product) to them. The bigger your contact list through these sites—assuming you have been adding people who could potentially be useful to you rather than just adding people randomly—the more chance you have of someone getting in touch with you to do some work. But are these social networking sites enough?

BEYOND SOCIAL NETWORKING

Even with all the media attention that sites such as Facebook get on a regular basis, there are still other avenues to explore. Let's have a brief look at other online promotional strategies.

Ten years ago, when I started professionally in the music industry, the Internet was, while not exactly a novelty, nowhere near as much of an important tool as it is today. But times move quickly and it wasn't long before having your own website became essential if you were to be taken seriously in the music industry. Several years ago it was seen as a brilliant leap forward that you could place photos, your biography, discography, and other information online so you didn't have to keep sending it to people in the mail. And the fact that you could also have clips of your music available for people to listen to instantly was quite remarkable! Artists spent a lot of time setting up websites to promote themselves, and it stayed that way for quite a while. Then the social networking sites came along and, after a while at least, have practically negated the need for an artist website.

Today, more people are likely to search for an artist on MySpace, Facebook, or YouTube than search for their artist's website. In fact, it has gotten to the point where many dance music artists and producers only have an artist website so they can have links to their MySpace, Facebook, Twitter, and SoundCloud pages (along with any others they might have). Some of these websites might contain a few pictures, a basic biography, and a contact page, but otherwise they serve more as a portal to access all of the other pages, as well as, presumably, giving them a recognizable email address that is less generic than a Hotmail or Gmail account.

Do you really need any more than that on your artist website? In my opinion, yes you do. The fact remains that for all the options and facilities that social networking sites give you, there are always going to be things you would do differently in an ideal world. MySpace is visually quite customizable, but in terms of what you can and can't include on your page, it is relatively limited. Facebook is also quite limited in terms of what you can include, but it is also visually quite bland. With a website of your own it can be customized to your heart's (and perhaps your bank account's) content. It can be laid out exactly how you want visually, it can incorporate as little or as much information as you want, and it can include as many features as you want.

For example, if you were a DJ and a remixer you could have a page on your website that looks like a world map with "pins" in the map to represent the gigs you have coming up. Each "pin" would have information attached to it, which became visible when the user hovered the mouse over the pin. You could even make them "clickable," so that if the user clicked on them it would open up a new browser window and load the venue's website. Try doing that on a social networking site!

Another option would be to have a secure (password protected) area of your website where you could permanently store MP3 versions of all the tracks and remixes you have done. This is something I actually do myself so that, should anybody (artist or record label) ask me if I have samples they can listen to, I have instant access to everything and can pick out what I feel is most relevant to them without having to upload the files. I simply go through my list of direct links to each of the files and paste them all into an email with login (username and password) information, and they are ready to go. The beauty of doing things this way is that, with a bit of preparation, it is possible to access from pretty much anywhere in the world. If you have a laptop or a mobile phone that has email facilities, you can easily email yourself an up-to-date list of all the links, so that even if you were on a train somewhere, if someone requested a particular track, you could simply cut and paste the relevant link(s) from your list document and email them.

It might seem a little unnecessary to be worried about responding that quickly to a possible work inquiry but, if that ever occurs to you, just remember that there are probably at least 10, maybe 20, maybe 100 other people out there fishing for exactly the same job you are being considered for. Time is of the essence. Plus it will give you an air of professionalism. Being prepared, being organized, having everything you need to hand immediately, *will*, while not strictly relevant to being a remixer, show that there is a good chance you will be reliable and will get things done on time, and not give the client any cause for grief and stress further down the line.

So there are just a few things you could incorporate into an "artist" website that would be difficult (perhaps even impossible) to integrate directly into one of your social networking pages. I am sure there are many others relevant to your situation, and with web technology developing at a head-spinning rate, there are sure to be plenty more developments over the coming years that will have the possibility of being useful to the remixer.

If you are one of the unfortunate majority of people who have to work on all of this stuff yourself and you struggle to update all of the various different sites, and if you asked me which you should concentrate on, I would probably advise you to prioritize the social networking sites simply because of the access it gives you to both your fans among the general public and the DJs, artists, and record label people who will, ultimately, help to make or break your career. I still do believe that an artist website gives other people the sense of something more professional about you.

Speaking of online forums, those are definitely something you should spend a little bit of time focusing on as well. There are a growing number of them appearing online, with some being more general dance music based and others being much more genre/subgenre-based, but what they all have in common is being a source of invaluable information about the genre itself and about who, within that scene, is getting the best feedback and why. Forums like these are a remarkable research tool if you are looking to find out what is going on in any given branch of dance music.

They can also be used as a means of promoting yourself. Most of them are fairly easy to sign up to and you would be well-advised to sign up to a few, at least, and then you will be able to post any news you have up on the forum where you know there is a *very* good chance that the people reading will like the music you make, and the style of remixes you are working on.

From past experience, I've learned that it pays to be enthusiastic in your posts on these kinds of forums, but not to go *too* far with things where you risk coming off as arrogant. "Quietly confident" is a good way to explain it! Another consideration is your choice of username. Choose a username that is (or at least closely resembles) your "professional name" so the people on the forum know who you are. I once signed up to a forum as "Soul Seekerz" and got a lot of positive messages from people saying how nice it was of the actual artist to join the forum and take time to talk to the "fans" directly.

Once again, and in summary, the various social networking (and related) sites that exist, along with music forums, are very useful tools for making sure people know what you are doing and, among industry people, raise the level of awareness of who you are and what you do to the point where you are in people's minds without necessarily having had a huge hit record.

THE PERSONAL TOUCH

All of the things mentioned so far in this chapter have one thing in common: they are all online activities. And the reason I have spent so long describing them is that they are vitally important in the marketplace today. So much business is conducted online that to neglect it as a means of promoting yourself would be an unforgivable mistake. Having said that, there is still really no substitute for *actual* face-to-face contact. It might not be easy, depending on your location, to get out to clubs where there are "big name" DJs playing, and it may be even more difficult to actually get to the offices of the various record labels (even assuming that you could manage to get an appointment to see the relevant people). But, if it *is* possible, I thoroughly recommend taking whatever opportunities you can find to have face-to-face meetings with people, because, as much as they recognize your name, your email address, or even your Facebook page, people are far more likely to remember you if they have met you in person and have gotten on well with you.

This will obviously be much easier to do if you are a DJ who is actively out and regularly playing, because you will probably meet quite a few people through the natural course of your DJing work without really having to try that hard. But even if you're not a DJ, there are still ways of getting to meet these people. For example, if there was a particular DJ you liked, who was doing quite well and who you thought might be interested in playing the kind of music you're making, the first thing you could do would be to try to find her on MySpace or Facebook and send her a message saying that you really like her work and that you are a producer/remixer (as appropriate) and you would like to have her on your friends list. Assuming that she accepted, you could then send her a brief note of thanks before waiting a little while, perhaps a week or two.

At that point you could send her another message asking if she would be okay with you sending her promos of any tracks you work on, and an email address to send the promos to. Again, if she is happy to do this, the next step would be to keep checking back on her pages or artist website for any information on gig dates. If you saw one near you, you could get in contact. I would try sending a message through the social networking page where you first got in touch with her; this route is probably preferable since you don't want to annoy her by bombarding her actual email inbox with loads of messages—it's about being able to be a fan while still respecting boundaries and personal space. Let her know that you are looking forward to coming to the gig and seeing her play. If the gig is a long way in the future, a well-timed newsletter shortly before the gig could help to bring you back to the front of her "record box," so to speak.

When you go to the gig, try to get a chance to speak to her while she's are there. Whether you do get a chance to actually speak to her depends on many factors, not least of which is the club itself. Some clubs are *very* unhappy about any of the "punters" (and to the club, that is *exactly* who are you, just another "punter") getting to speak to the DJ, whereas others are more open about it. Even if the club is happy to let people talk to the DJ, you might not get an opportunity because the chances are that she will already have her entourage in the DJ box, but like so many things in this business, if you *don't* chase any chances then the opportunities will not chase you! And if you don't get a chance to talk to this particular DJ on the night itself, you still have that communications channel open with her through her MySpace or Facebook account. A quick message a day or two after the show saying how much you enjoyed the atmosphere in general and the show itself would almost certainly be met enthusiastically, and might lead to the start of a more regular and more ongoing dialog with this person.

SUMMARY

In this chapter we have looked at just a few of the strategies you can use for self-promotion when you are starting out or in that early phase of your career. Of course, there are many more. What we are trying to achieve here is, essentially, marketing! And that is a skill set all its own. Sure, some of those skills have been diluted these days, but if you feel you could use a little help in this respect, there are some excellent books on the market that deal with marketing, sales, and promotion in general, and some that are more optimized toward online marketing and S.E.O (Search Engine Optimization) techniques that can help put your websites closer to the top of search engine listings—never be a bad thing!

My final word on self-promotion is to not be afraid to see how other people do it! Think of a remixer who is in a position you would like to be in. It would be ideal if that person was in a genre that was either the same, or at least, quite similar to the genre(s) you work in, because marketing techniques can vary for different market sectors. See if you can get some inspiration from the way other remixers handle their promotion.

In the next chapter we look at the pros and cons of having a manager, agent, or promoter of your own.

CHAPTER 26
Management and Representation

In the last chapter we looked at what could more or less be called the "DIY" approach to promotion, and the way in which nearly all of us will start out doing things. But, as time passes and your profile as a remixer grows, you may well start looking for management or representation to take over some of the administrative workload, to push you toward new (and possibly bigger) clients you simply didn't have access to before, and to take your career to the next level. A good manager might also be able to provide guidance and feedback on ways in which you could develop your sound to make yourself more successful, and provide a valuable second opinion on the work you are doing.

All of this comes at a price, of course. Most managers take a cut of around 20% of your earnings. While that might seem like a lot, if they're getting you a great deal more work, it really isn't. On a practical level, you will probably make that money back just from the time you save in trying to get work and in the administrative side of the promotion. You can also factor in the fact that you will be getting pushed to clients you might not have had access to before, something definitely worth paying for. Finally, there is the *cachet* that comes with actually having management. Although getting a manager won't suddenly make you a better remixer overnight, if your manager is well-established and represents other established and respected artists, you will most likely be seen differently by the industry. It isn't necessarily fair, and it isn't really "right," but that is pretty much how it is.

You see, there are many extremely talented people out there who have not yet had their "break" and there are also other people who are perhaps not quite as talented but who have known the right people or just been in the right place at the right time, and are already on the road to success. But don't let that discourage you. The simple fact is that talent and ability *will* always make itself known and it will rise to the top as a result. Having a manager is a way of accelerating that, and in some respects, there is a glass ceiling for the unmanaged remixer. But there are always exceptions to any rule; some people make it quite a way up before they get management, whereas others get management fairly early on. The music industry these days is far more accessible than it was, say, ten years ago, and there are far more opportunities for the determined and ambitious producer/remixer than there was back then.

Let's assume that management is something you are hoping to find at some point, and let's take a look at what you might be looking for in a manager and, conversely, what a manager might be looking for in the artists he represents. My personal checklist for a manager is someone who can fulfill most (if not all) of the following criteria:

- Can take care of most of the administrative part of my work
- Will actively promote me and my services
- Can open doors for me that I wouldn't be able to do myself
- Can work with me to develop my sound
- Can keep me informed of any changes in the industry that I need to be made aware of
- Will be brutally honest with me when I ask his opinion of my work
- Will be professional in his dealings with other people and with me
- And most importantly, is someone I can trust *completely* to do what is best for me and my career

Without at least a strong majority of those qualities, I wouldn't be interested. The simple fact is, if you are going to be working with someone on an "exclusive" basis (where he is the only person representing you), you need to know that he is doing the best for you, because contractually, you won't be able to employ anyone else to fill in any "holes" in what he is doing. Your personal list might be a little different from mine.

For example, I am quite happy taking care of my own websites and social networking sites, but to others, it might be important that their manager takes care of that as well (if not personally then at least arranges for someone to do it). Other people may not want to hand over all of the administrative part of the work to their manager, perhaps because they want to have total control of that to avoid any mistakes. Needs and requirements vary, but you should definitely think about what you would need from a manager before you find yourself in that situation. Many people get caught out in life by finding themselves in circumstances they're not prepared for, and then being forced to make a quick decision. If you make the wrong decision, you could be trapped in a bad situation for a long time! That's why it's always good to at least have your list of important criteria building in the back of your mind, because you never know when a manager might approach you.

What else might you be looking for in a manager? Well, a good thing to consider, for a couple of reasons, is who else he represents. On the one hand it makes sense to look for a manager who represents other producers or remixers who are in vaguely the same genre as you, because he will have industry contacts relevant to what you do. But this is very much a double-edged sword, because if he already represents people who are in the same general area as you are in terms of style, you might find yourself competing directly with more established artists for your manager's time and attention, a situation likely to end in tears.

This isn't so much of a problem if your potential manager looks after quite a few different people spread across a number of (often related) genres, because there might only be a couple of other remixers on his roster who do what you do. But if that manager represented *only* remixers in a narrow genre, I would listen for those alarm bells (unless he only looked after very few people).

This brings me nicely to the question of numbers. It is definitely worth taking into consideration how many people your potential manager looks after. If the numbers are into double digits and there is only one person (rather than a management team), I would at least want some kind of reassurance that I wouldn't simply end up at the bottom of a big pile, picking up the scraps of work the other remixers didn't want or weren't available to do. The same argument can be had about who to sign your records to if you produce your own material. There is a good argument for signing to a smaller label rather than the biggest one, because there is always a risk that if you sign your track to a big label with a lot of artists you may well never really get much attention from them, whereas a smaller label could work much harder for you and make you feel much more valued as an artist.

Common sense comes into play here. The preceding statement has a lot of common sense going for it but, even then, you still need some credibility from a smaller manager (or record label, in the previous example). You wouldn't sign away your talent to someone, even if he offered you a better deal, and even if you felt he would give you more attention, unless you felt that he had the connections, the infrastructure, and the ability to take you forward in your career. So be aware that good intentions, while admirable, don't necessarily mean good results. And, as I have already mentioned, the music industry uses a *lot* of hype, so most of the people working with any degree of success in the industry have at least graduated their "Bulls**t 101" class and are able to fluently "talk up" their business/service/product and be very convincing about it. Reader beware: anybody who promises you anything in the music business is probably talking out of a place where the sun doesn't shine, because it is, in my experience, impossible to promise *anything* in the music industry, especially now that it is in such a state of constant change. Don't dismiss everyone as talking complete nonsense, and, of course, give people a chance to back their words up with something more substantial, but don't be taken in by words alone.

It might sound like a contradiction, but think about this: you might actually *benefit* from having a manager who could sell ice to Eskimos, because that means he will probably be very persuasive in selling your services to potential clients as well. Of course, you would have feel confident he is being truthful and that no false claims are being made, but a manager who can speak convincingly is a very useful person to have on your team!

One final thing I would like to mention before I look at things from the manager's point of view is "exclusivity." Most managers will, if they choose to represent you, want to do so *exclusively*, which means that they are your only manager and only they get you work. Pretty straightforward really, but there are potential complications. The range of artists each manager represents will vary. Some will only look after, for example, hip-hop producers and remixers, others may only look after house producers and remixers, while others still may look after trance, techno, drum 'n' bass *and* dubstep producers and remixers. If you only work under one name in one genre, this really isn't an issue for you as long as you are with a manger who works in your genre. But what if, as is the case with me, you have different projects under different names in quite different genres, but your manager only really works in one or two of those genres?

Well, you have options, of course. What you could do is have a deal with Manager A that is an "exclusive" management deal but *only* for Project A, and then you could have another "exclusive" deal with Manager B for Project B, and then have Project C and D, which you are developing, not represented by any manager. This is what I do and this is what works for me. It does have a downside though, in that the two managers don't have a day-to-day dialog going and there have been times when they have both booked me for work over the same few days and I have had to do some manic calling around to try to reschedule things. There is a lot to be said for having a fully exclusive arrangement with your manager whereby he looks after all of your projects. This is much simpler, though unlikely to be the best solution unless you happen to find a manager who is comfortable, experienced, and connected in many different genres.

Some managers are happy to work on a nonexclusive basis, where they aim to get you work but allow you to have other people trying to get you work as well. This might *sound* like a better option; after all, five people trying to get you work is better than one, no? Well actually, not necessarily. The problem occurs when these managers overlap. If you had one manager for the U.S.A., one for the U.K. and Western Europe, one for Eastern Europe and Russia, and one for the Far East and Australia, then you would probably be okay, subject, of course, to the scheduling problems I highlighted earlier. But if you had a number of different managers all working the same territories, there is a distinct chance that a record label or production company could receive a few emails or calls over a short space of time from different people, each claiming to represent you. At that point, they may just feel confused and unsure about who to call and could well choose someone else to do the remix, just to avoid the complications and potential "politics" of what is clearly a complex situation.

The relationship between a manager and an artist is complex. It isn't as simple as choosing a manager you want and then hiring him to work for you in the same way you might choose a decorator! Although *technically* an artist employs a manager, rather than a manager employing an artist, it is still very much a two-way street in terms of the requirements for working together. Most managers are very selective about which artists they will represent and will need to see some pretty compelling evidence to support your interest in working with them. But do managers have a "criteria list" like the one outlined previously, even if it is a less formalized one? How do they make their decisions about who to accept and who not to accept onto their roster? To find out more, I asked one of the most respected remixer managers in the U.K., Matt Waterhouse, for his thoughts and experiences in the remix management world. Here is what he had to say.

If you received an unsolicited demo reel from a producer/remixer would you give it a listen and consider putting him forward for work?

Matt Waterhouse: Definitely. But my primary aim is to manage my roster long-term, so I would only add if I felt that the remixer complemented the teams I already have on board and more so if they fit a niche that the existing remix teams do not cover.

What qualities do you think a remixer needs to have to prosper in today's market?

Flexibility to adapt to the changes in the main music styles that fluctuate over the years.

Do you have a checklist that you use for any potential new artists on your roster?

The remixes would need to be to a very high standard and work for both commercial clubs and dance radio alike.

How involved do you get in shaping the sound of your artists?

100%. Digital Dog and Cutmore are two teams I took on who had never worked in the industry before, and we have developed their sounds to a level that is as good as any team I look after.

Any advice you can give to somebody wanting to get into the industry?

Be passionate about the music you are making and be open to constructive criticism… after all, the manager is always right!

As you can see, Matt Waterhouse has fairly strict criteria for selecting artists he wants to work with. It is also interesting to learn that he is quite willing to seek out someone he feels he wants to work with, who isn't already established. So, it isn't necessarily true that you will have to go and seek out a manager. There may be times that they will come to you!

If you are having trouble finding management, or if you simply don't feel that you are ready to commit yourself fully into an exclusive management arrangement, there is another option you might like to pursue. There are quite a few club promotion companies around these days that specialize in doing promotional mailouts of new records to club DJs. In theory, these companies have a purely logistical role and just deliver the tracks to the DJs either physically (CDs through the mail) or digitally (by having the tracks available as MP3s to download on their "members only" websites). Sometimes, however, these companies take a more active role with the record labels and are often asked to help source remixers for a given project, because they have a great deal of experience and knowledge about who is doing well on the remixing scene at any given time. Is this another potential route into success as a remixer?

Like every other option we have explored so far, it is far from easy, because the people in charge of these companies are just as selective as managers as to who they recommend. The only difference is that, because their main focus is promotion rather than management, they might be a little more approachable and open to working with new people. In addition, they are likely to be less genre-focused than a manager would be and far less likely to want to tie you into any kind of formal or exclusive deal. In many ways, any dealings you have with people like this should be more like dealing with a broker who simply takes a "finder's fee," rather than anything structured like a management deal.

To get a better perspective on this, I spoke to a good friend of mine, Brad LeBeau, from Pro-Motion in New York, and asked him similar questions to see if his take on the situation was any different from Matt Waterhouse, the manager.

Do you, as a promoter, have any influence with your label clients as to who to use for remixes?

Yes I do.

If you received an unsolicited demo reel from a producer/remixer would you give it a listen and consider putting them forward for work?

Yes I would, but would opt to let them work on a new/upcoming artist before trying them out on a larger project/established talent.

What qualities do you think a remixer needs to have to prosper in today's market?

- Complete command of programming skills as well as remixing talents
- Patience in working with a prospective client
- Confidence
- Determination

Is there any advice you can give to someone wanting to get in on the action?

With the advent of technology, aspiring producers/remixers who were once unable to work without the funds necessary to get into the studio are now afforded an equal opportunity. Because ownership is essential, I urge aspiring remixers to take the plunge and buy their own equipment and practice, practice, practice. Don't be shy! Approach those they respect in the industry, ask questions, and absorb as much knowledge as they can!

It's interesting to see that a lot of the attitudes are very similar, despite their different perspectives and the fact that they're based on opposite sides of the Atlantic. Of course, you will probably have noticed some differences in attitude and expectations between the two, but this is to be expected because the music industry (and in particular the dance music part of the music industry) is actually quite different in the two countries.

Looking back over what these industry professionals had to say, I think there is a lot to learn and a lot to do before you should even think about approaching anybody to represent you. Taking into account that, as I have said already, every manager, every promoter, and indeed every remixer will have different criteria, I think we can still summarize a few key points from what you have read here as a good checklist, for when you're ready to consider if you're personally and professionally ready to look for management.

- Have you already gotten at least some profile and experience?
- Are you getting enough work to convince a manager that you are worth taking on?
- Being completely honest with yourself, do you have something that sets you apart from other people trying to get in to remixing? (This could be a stylistic edge or simply a matter of quality.)
- Can you work quickly and efficiently?
- Are you prepared to accept that you might need to make changes to your work in order to satisfy the client?
- Are you prepared to accept that you might be required to work late or weekends on occasion, in order to meet deadlines?

If you have answered "yes" to all of the above questions, it might be worth your while trying to find management. Obviously, the further along in your career you are, the easier it will be to find management. Unfortunately, success breeds success and you might get to a point where you are successful enough to get a manager, but at this point, you might feel that you don't need one because you are already getting enough work. At that point it becomes a purely practical decision based on how much you feel your time is worth. Would you rather spend a few (or many) hours each week promoting yourself and tracking down new clients to work for, or would you rather give up a percentage of your income and have someone else do that for you? In most cases people will choose to have a manager, because in addition to freeing up a lot of time, a good manager will already have a good client base that he will simply plug you into and will, hopefully, start getting you work.

I have been lucky enough to have great management for the past six years and I wouldn't change that for anything. Yes, I have given up 20% of my income for that period, but what my manager has done for me professionally simply can't have a price tag attached to it. He has brought me in a ridiculous amount of work, and offered me a great deal of advice and guidance on how to take my remixing to the next level. He has been there with me throughout the process of almost constant reinvention. Moreover, and because of the sheer volume of work he has gotten throughout his professional career, he is always a reliable and dependable second opinion on anything I am working on (remixes and original productions). Would I be where I am today without him? I can't say for sure, but if I am being totally honest, I doubt it. And that is possibly the best recommendation I can give for having a manager.

In the previous chapter I mentioned how important DJs were to the success (or failure) of a remixer, so in the next chapter we take a closer look at this strange relationship and try to find out how this developed, what is involved in this unique relationship, and what you can do to make it work for you.

CHAPTER 27

The Importance of the DJ

The world of the remixer is different from that of the artist in many ways, but fundamentally similar in one. That similarity is that the aim of what you do is to sell records. Here, however, things change. The first and most fundamental difference is that, as a remixer, your job is to help to sell *somebody else's* records. I have already mentioned the importance of being respectful to the original artist and song, because the remix you do should be as much (if not more) about the original song than about you as a producer/remixer. I have also mentioned that the benefits (aside from the obvious one of the remix fee) of being able to be promoted indirectly by the record label that commissioned the remix can be very valuable as well. But what I haven't really spoken much about is the market to which you, as a remixer, will be appealing.

THE DJ AS THE CONSUMER

A singer or band aims their marketing and promotion at the general public, as it's members of the general public who buys their records. With a remix, things are slightly different. There are times when the remixes will be included on the final CD or download, but there are just as many times when the remixes are commissioned purely for club promotion purposes. In either case, you, as the remixer, have a different market you are appealing to: the DJ. There will always be some members of the public who like the remixes and actively seek out the remixes to buy, but I am almost certain that a majority of the sales of the remix are specifically made to DJs wanting to play the track in the clubs and on radio stations. This is why I talked much earlier in the book about the importance of making the tracks easy for DJs to mix into and out of. If you are aiming purely at the general public, you could start and finish your tracks however you liked (within reason of course) because it wouldn't matter to someone playing the CD whether your track would be easy to mix into. But given the majority of the market are DJs, you have to consider that when making your tracks.

Just as singers or bands often have fans who will buy most of their records and support them and go to their concerts, we as remixers can have DJ "fans" who like what we do and actively seek out new remixes and original tracks by us if they like what we do. This is just another part of using the marketing of the tracks you remix to your benefit. You might do a remix for a well-known artist and not really get a great remix fee for it, but if it is marketed and promoted well, it could raise awareness of what you do among DJs, who could then become fans of yours, actively support the work you do, and when you release your own songs, they would most probably buy and play those as well. *That* is where you get the indirect benefit, because on your own material you will be earning money from the sales.

Of course, even just having your name "out there" as a well-known and respected remixer can have other benefits, because you could, once you have enough of a profile, start getting people coming to you wanting you to work for them (or with them) without you having to be constantly going after work proactively. If this does start to happen, I recommend that you continue being proactive, and don't just sit back, relax, and wait for the work to start coming in. If you have management, they will always be actively marketing you anyway; but if people come to *you* then it puts you in a stronger position in terms of negotiations.

THE DJ AS THE PRODUCT TESTER

Another way in which DJs can be very important to us as remixers (and producers) is in respect of the feedback we can get from them. None of us, no matter how motivated and determined, can do as much research as we probably should. And by research I mean being out there in the clubs listening to the music, seeing what is working and getting the crowd going the most, and listening for developments in the music and changes in style that start creeping through so you can keep one step ahead of any new sounds that start to become popular. If we could go out every night of the week to a different club in a different city, of course we would be seeing and hearing enough, but then we wouldn't have the time to actually make any music! A network of DJs who you know and talk to regularly (not necessarily in person, maybe through social networking sites) can be a massive help in this respect. By doing things this way, you can easily get more of an idea of what is happening globally. After all, it's a big world out there and every single DJ and club-goer is a potential customer, so if you can get someone else to do your research for you in many different countries at the same time, I don't see any possible reason why you shouldn't jump at the chance.

Having said that, you do need to be realistic about it and be prepared for massively conflicting views from different DJs. Music is, after all, a very personal thing. It is a personal reaction to a personal story, so there is no end to the number of different ways in which that story can make someone feel. It isn't always easy to try to make sense of wildly different feedback from different DJs, nor does the fact that their reactions could change from one week to the next help make things any clearer. But after a while you get pretty good at figuring out what the general consensus seems to be, while simultaneously picking out any specific information that could help you.

For example, there is often a very different reaction to a song or genre in different geographical areas. What goes down well in London might not be liked at all in Manchester. Tracks that are rocking New York might clear a dancefloor in L.A. A remix that makes a club erupt in Singapore might get nothing more than looks of mild confusion in Rome. There may well be deeper reasons for this, but I don't think that it is necessary, as a remixer, to fully understand *why* all of this happens. It is useful to be able to spot patterns and have some kind of mental snapshot of what the club scene is like in different countries around the world, because it can definitely help you to adapt the remix you are doing to the intended market. For example, if I know I am doing a remix for a U.S. label that is intended (mainly at least) for the U.S. market, I will do things slightly differently. I wouldn't change my style completely but it might have a slightly different angle for U.S. remixes compared to U.K. ones.

In addition to the general "crowd reaction" type feedback a DJ can give you, there is a good chance that many of them are producers or remixers these days as well, so they might even be able to give you information in a more detailed and technical way about things that might improve the mix. However, be warned: if you ask 20 DJs you will get 20 different suggestions as to how to improve the track. You can't possibly make all of the changes, because many of them will be at odds with each other, but you will probably see a generalized pattern start to form. The ones that have been suggested the most times are the ones that the track is actually most likely to benefit from.

There is, of course, always the option that you do a few alternate mixes for different DJs or markets. I am not talking about an alternate mix that is a completely new production (although that *is* possible of course), but more a mix that is tailored slightly to a particular DJ's comments or feedback. It would be impossible to do this for every DJ who gives you feedback, but if you felt that the DJ was important enough or if you knew him very well and trusted his opinion, and if you had the time of course, then it might be worth it, especially if the changes he thought might help are fairly minor production issues. Building relationships with DJs in this way and listening to their input, taking the time to acknowledge it at least, whether or not you actually take their advice, and, of course, putting them on your "promo list," will be of great help to you. There will always be some DJs who just take all of the MP3s you send them and never bother getting back to you. This does seem a little unfair seeing as you are providing them with free music, but this kind of thing is to be expected no matter what industry you are in.

THE DJ AS THE COLLABORATOR

If having a list of DJs you can get feedback from is very useful, then actually collaborating with one is even more useful. Not only do you get the DJ feedback more or less instantly when they play out at clubs, you also get the wealth of experience of how different tracks go down in different clubs. That doesn't mean your list of DJ contacts is worthless now, because one DJ alone won't be able to give the sheer variety of club reaction reports that a number of DJs will, but seeing as they are a collaborator, whatever feedback you *do* get is likely to be much more in-depth and specific. I don't think that the feedback is the most important part of collaborating with a DJ, I think that particular role is held by the fact that a DJ can "road test" your own tracks or remixes *before* you finish them up (all assuming that you have time to do so, and this is much more relevant to remixers than producers). This way you can make any minor changes before the masters are delivered to the record label/client.

You would hope that any *minor* changes made after the A&R department had initially heard the mix would simply be signed off on, but sadly, something as simple as changing the level of a drum part by 1 or 2 dB might be enough to turn a "yes we like it" into a "no we don't." Sometimes (quite rarely though I should add) a simple change (which you feel is an improvement) can be enough to have the client tell you to change it back to how it was before. Fortunately, if you save new versions of the project each day (I often do it more than once, each with a new filename) and don't simply overwrite the same file repeatedly, such changes will be easy enough. What is more difficult are the times when you honestly believe that the changes you have made for the "new" version make it sound better, but the client doesn't agree and prefers the older version. Unless you feel that the old version doesn't represent you as a producer/remixer very well (which is unlikely if we are talking about subtle changes), you should probably accept that the client has her own preferences and, ultimately, she is the one who has to be happy with the remix. I don't see that as compromising your artistic integrity, although some might.

There is another key benefit of working with a DJ, and that is you can get all of your tracks and remixes heavily supported and prioritized wherever and whenever the DJ is playing out. It's another form of free publicity that shouldn't be overlooked. Given the emphasis I have tried to give throughout this book to the fact that promotion and publicity are very important—perhaps even the single *most* important thing in your career, rightly or wrongly—it should be clear that this is a very valuable benefit. The biggest problem for any up-and-coming artist, and even some more established ones, is how to raise awareness of what you do. By working with DJs you have an instant delivery mechanism for your product, and given that the DJ community itself can be quite a close-knit community, the chances are that if your DJ collaborator plays the track at a club, there will be other DJs there who might like it, and might ask for a copy of it. Think of it as a form of *viral marketing* for the real world and the 3D people, and you aren't far wrong!

Finally, working with DJs gives you an instant insight into the smaller details of the remix that might save you a lot of time. While you might be perfectly capable of figuring out a great arrangement that builds and drops in the right way and at the right times, it might take you a little while to get there, whereas your friend "DJ Collabor8or" will probably have a much more instinctive approach to the whole process. As we should all know by now, time is rarely the friend of the remixer, so anything you can do to save time without compromising quality is a real lifesaver.

THE DJ AS... YOURSELF!

You have seen the ways in which a DJ can be a useful ally, but there is always the option of becoming a DJ yourself. If this is the case, you get many of the benefits outlined previously along with the possibility of additional income. I don't mean to imply that earning money as a DJ is an easy option. Far from it, in fact. Getting work as a DJ is just as hard as getting work as a remixer, perhaps even harder in some ways. There is a huge amount of competition for DJ bookings, and breaking through into that market can take a lot of effort. Many of the same problems arise (getting an agent/manager and then them being able to convince the promoter or venue to take a chance on you), but there is some light at the end of the tunnel. If you have managed to build a name for yourself as a remixer/producer, that *can* open doors for you as a DJ, although you shouldn't assume that it necessarily will. Even if it does, you can't demand large fees until you have proven yourself.

Also, not everybody who is a good remixer/producer will necessarily be a good DJ. While the two disciplines are related, they are also very different. Confidence is a big issue, because someone who feels comfortable in a studio environment where he can take his time to get things right might not feel comfortable up on stage in front of a potentially very large crowd where things have to be right immediately. To some extent, this confidence comes with practice: the more gigs you do the more you will feel confident about your abilities to get things right, but to some, just the idea of getting up on stage at all will make them feel uneasy.

There is also the time issue. Being a DJ, especially if you are fairly successful and are being asked to play in different countries, can take up a lot of time. That might be time you want to spend in the studio or time you want to spend with your family. Either way, you might find that it simply isn't possible for practical reasons. The good news is that, in this situation, you can still work with someone else who is a DJ so you can get the best of both worlds.

There is another factor here as well, and this is one I am reluctant to talk about because it is a less-than-positive aspect of the music industry these days, but I feel I should bring it up in the spirit of full disclosure. I am sure we are all more than aware of illegal file downloads. This affects producers more directly than remixers because the sales of their own records will be negatively affected, but it does even affect remixers because even the biggest labels with larger budgets have had their profits cut because of illegal downloads. As a result, less profits means lower budgets to spend on getting remixes in the first place. So where does DJing fit into all of this? Well, many record labels are now putting more of an emphasis on gigs as a good source of income for their artists, because a ticket for a gig can't be downloaded illegally. It is possible as an electronic music producer to actually get gigs, but because of the nature of what we do, actual "live" performances are quite a big deal and can be very hard to pull off. So the nearest equivalent for many would be a DJ set. In the same way that a ticket to a gig can't be downloaded illegally, neither can the admission to a club. And because of this, DJ gigs can, if you can get them, be a good source of income.

Of course, with the global economy being the way it is at the time of writing, even DJ fees have taken a tumble in recent years, because the clubgoers don't have as much money to spend, and hence, club attendance is down. But this "live" performance aspect is still far more secure than simply selling music.

I have seen a number of people posting on forums stating that they only really make tracks these days to raise their profile to try to get DJ work. Essentially, they don't expect to earn anything from selling records and pretty much give them away as some kind of "loss leader," just to get their name out there and on everybody's lips. At that point, they try to use that profile to boost their DJ work. It's a pretty bad state of affairs really, because historically, an artist would have gigs/live shows as a means of promoting their recorded material. But in recent times it seems that things have turned pretty much a full 180 degrees, so that now the music has almost become a promotional tool for the live performance. Hopefully things will change in the future, but for now at least, DJ work remains a very viable supplement to any production work or remixing you might do. Although many purist DJs hate the idea, laptop DJing is now here to stay, so it has, in some of the technical aspects at least, become easier than ever for a producer/remixer to become a gigging DJ.

SUMMARY

Since the beginning of dance music as we know it, the role of the DJ has always been an important one. At the beginning, the DJ and the producer/remixer were often different people, but as time has gone on, more and more DJs have started producing their own tracks and more and more producers have started DJing. On some levels I would prefer that the two were separate, as, without being elitist in any way, I believe that the "product" in both industries would be better for it. The people who are best at DJing would be DJing and the people who are better producers would be producing; there would be less competition in both areas. However, I am also a realist so I totally understand why it has happened the way it has.

There are a great number of practical advantages to having the DJ input in a production or remix, and someone coming from a production background could, perhaps, bring a more dynamic performance aspect to DJing. I think that one of the biggest reasons this crossover has happened is, as with so many things, money. Put simply, why would you want to split any money earned from sales of your productions or any remix fees with someone else if you didn't have to? And likewise, why would a DJ want to split any DJ fees with a collaborator? It might sound selfish, and there are advantages to collaborating that can in many cases outweigh the financial loss, but it is also a case of practicality. As fees for both areas have dropped in recent years, there is less money to go around. Consequently, having to split an ever-diminishing fee becomes more of an issue, as most people have a minimum amount they need to earn to continue being able to work.

Let's say that you needed to bring in $1600 a month. If the total income from your productions, remixes, and DJ gigs with your partner was $6500 a month, you could afford to split the money (assuming that both sides were pulling their weight of course!) and still be earning enough. However, if the total amount you were bringing in was only $2400 a month, you simply couldn't afford to split the money and still keep on doing what you are doing. Even with the best intentions in the world, at that point things have changed, so that it is now practically impossible to continue. But if you were working alone, you would be bringing in enough to continue. I know it sounds rather selfish and cutthroat, but I am just being honest. I know a large number of people who have had to drop their collaborator, and have even taken a reduction in fees (DJing and remixing), because even after the reduction, they still come out better off than when they were sharing the fees.

I don't want to encourage you one way or the other, but I do think it only fair that I bring that aspect of the business to your attention. Whichever you consider the best option, I hope that it has become clear how important a DJ (either a partner/collaborator or yourself) is to your work as a remixer.

CHAPTER 28
Legal and Contracts

Although as a remixer you might not have to deal with all the legal intricacies of a full-blown record deal, there are still some contractual issues you might come across during your career. I am obviously *not* qualified to give legal advice as such and you should definitely consult a lawyer or attorney with any legal questions. The points made in this chapter are merely to make you aware of some of the more obvious and commonplace things you should look for in remix agreements. So with that caveat out of the way, let's get started.

The first thing I should mention is that, while remix agreements exist to protect the interests of the record label (or artist), and perhaps to a lesser degree, the rights of the remixer, it is far from certain that you will be supplied with one for every remix. In fact, if I had to make a guess, I would probably say that at least 50% of the remixes I have done in my career *haven't* had a remix agreement provided. Only the bigger labels provide them as a matter of course; the smaller labels (or artists themselves if you are working for them directly) tend not to provide them. It's fair to assume that one of the reasons for this is that larger labels will often have their own legal departments, so they aren't saddled with the cost of having to get separate documents drawn up each time. Another option is that a label might simply get a pro-forma remix agreement drawn up by a lawyer and then use this each time.

The absence of a remix agreement doesn't mean you are totally unprotected. Assuming that you have some kind of written communication from the client and that it wasn't all agreed completely verbally (which is very rare, if not unheard of), there will be some implied rights and obligations on both sides. I will list some of the fundamentals of most remix agreements here so you know just what you are getting into.

OWNERSHIP OF MASTERS

This is perhaps one of the most important things you need to be aware of. It may well seem obvious to you, but you need to know that by doing the remix (and assuming it is accepted by the client) you are transferring the full rights of the sound recording of your remix to the client. What that means in real terms is that the full remix and all component parts, including any recordings of live instruments, are owned by the client, and they have the right to use them as they wish, without exception, and in perpetuity (meaning forever). It is also common to have a clause in the agreement that also states that the client is under no obligation to actually use the remix you have provided. This might seem counter-intuitive if they have actually paid you for the remix, but things like this do happen from time to time and you can't let it get to you. It might seem unreasonable, and you may feel like the remix you have given them is your best work and be keen for it to be heard, but by signing the remix agreement, if this clause is present, you are agreeing to that possibility so there is nothing you can really do in that situation.

A part of this same clause might also include you giving the client the right to use your name (your professional name, at least, which may, or may not, be your real name), your image or photo, and any biographical information you provide to them in connection with the usage of your remix. I can't see any reason why you wouldn't want this, because aside from the financial benefits of remixing, the secondary—or perhaps primary, depending on your motivation—reason for doing these remixes is to promote yourself. Nonetheless, you should be aware that you are agreeing to this, so if you have any issue with this you need to check any remix agreement for this clause and, if necessary, request that it be removed. Of course, requesting doesn't mean getting!

PUBLISHING RIGHTS

Normally, if you are writing your own music or contributing to the writing of a piece of music, you will retain a publishing interest in the work. If you are unfamiliar with the concept of publishing, perhaps the simplest explanation I can give you is that you get a publishing share when you contribute to the *writing* of a song (either the musical or lyrical part). You don't necessarily have to be involved in the production, recording, or performance, but, of course, it doesn't preclude that either. You could, for example, write some lyrics on a piece of paper and hand it to a music producer friend which then becomes the basis of a hit record. Your involvement was purely writing those words, but you would be entitled to a publishing share of the record. Publishing income isn't derived from record sales (or download sales), but is based on the performances of the song, either at a music venue or broadcast in some manner.

When you are working on a remix it is entirely possible that you completely rewrite the musical part of the song and use none of the original musical parts or even the original chord sequence. In this case you might think you are entitled to a share of the publishing; sadly, I have yet to come across a remix agreement that allocates a share of publishing (even a *pro rata* share) to the remixer. There has always been a clause in the agreement that states that the remixer relinquishes all claims to the publishing rights of the masters. There have been many occasions in the past where a remix has actually become the most popular version of a song and has, in fact, made the song a hit. In that instance you might feel like you should be entitled to a share of the publishing, and perhaps, on a moral level at least, you might be; but again, this clause is standard, so even in the absence of a remix agreement, I think it would be very hard to dispute this "standard practice" term.

PAYMENT

So now for the one most people will be most interested in: payment. This is actually quite straightforward in most cases. I would say that a vast majority of remixes are on a "flat fee" basis, which means that you get paid a single fee for the remix with no further payments or royalties in the future. Like so many things in the music business, there will be times when you might be asked to do a remix for a percentage of profits, for a smaller fixed fee, and a royalty later on. In fact, there are many possible options, but these tend to happen more when you are dealing with either a much smaller label or directly with the artist herself. Bigger labels tend to stick with the "flat fee" basis.

There is one other thing to look out for, or possibly even for you to request, which is the provision of a small royalty payable in the event your remix becomes the "lead mix." As I mentioned just now, it happens quite often that a remix becomes the version of a track that is the most popular, which gets all of the radio play, and which, ultimately, is the reason for the success of the record. A good example of this is the Todd Terry remix of Everything But The Girl's "Missing," which is probably the version of the song most people assume is the original version, but is, in fact, a remix. This is what I mean by "lead mix." In that instance it does sometimes happen that a small royalty is payable, but first, that is increasingly rare, and second, you have the burden of proof when it comes to defining "lead mix" and establishing that your remix was, in fact, just that. If we are talking about a record that becomes a huge hit, if you have that clause in the remix agreement it is definitely worth pursuing.

OTHER THINGS TO CONSIDER

While those are the main things to look out for in a remix agreement, there are other things you need to be aware of that might catch you out if you're not paying attention. One that I have often come across relates to exactly what the label is expecting you to provide to them. There may be a clause in the agreement that specifies that the client requires any or all of the following: club mix/extended mix, dub mix, radio edit, TV edit, mixshow edit, instrumental, and a cappella. In addition it may specify in the remix agreement that you have to provide stems, which can be a painstaking and time-consuming job in itself if you are required to provide separate stems for each version of the remix you completed. But if that is what the remix agreement specifies, then that is what you have to provide. The difficult thing is that you will normally be provided with a remix agreement only once the remix has been accepted by the client, at which point it is really too late to argue, and the only choice you have is to either do the "extra" work and get paid, or give up on all the work you have already done. For the most part, it is pretty standard to be expected to provide a club mix, dub mix, and radio edit, and instrumental and a cappella versions are fairly easy to run off, so that shouldn't take up too much extra time. The only thing I would try to avoid is having to provide stems. Even if there is a clause in the remix agreement that states that you won't always be asked to provide them, if the clause exists, the client has a right to ask.

Another common clause in more in-depth remix agreements is one that indemnifies the client against any potential legal action, by stating that you have not used any samples in your remix. The definition of "samples" in this context is unauthorized usage of another musical work in whole or in part; it *doesn't* means samples in the sense of those included on a sample CD or collection or similar—unless the license agreement of the sample CD/collection specifies that you need to get permission before using any of the samples in any commercial recording. It is incredibly rare but, believe it or not, occasionally the odd one crops up that requires you to get permission! Any original recordings (stems or single parts) provided by the label are, of course, usable. If you did want to use a sample from another piece of music, it would be down to you to get all the required permissions and authorizations and provide these to the client *prior* to them accepting the remix.

Another common inclusion is to have some kind of credit for your work. This is where you can request a written credit along the lines of "Remix and additional production by Simon Langford." This might seem obvious because your name (or the name of the project you are working under) will already be on the remix anyway, but you can also ask for additional information to be included such as "Remix and additional production by Soul Seekerz (www.myspace.com/soulseekerzmusic)" or you may wish to include an email address or, as I usually do, my manager's contact information. Don't get too carried away because the label probably won't include an essay-length credit if you write one. Also, you should know that usually the clause includes a "get out" for the label, which states that they only have to use "reasonable endeavors" to include it. And finally, with an increasing proportion of sales being in the form of digital downloads, the days of having a credit included on the sleeve or disc itself are numbered. The actual information included on a digital download can vary from provider to provider, and this means there is no guarantee that even if the label does its best to include your credit, the download provider will actually do so. There is also the option of including your credit in the metadata for the MP3, but again, that is very much dependent on both the download provider *and* the label doing everything they should in the way they should, and is far from guaranteed.

Some contracts will stipulate that you should seek legal advice before signing the agreement, and in theory, I would have to agree. But the reality of the situation is that the cost of seeking legal advice could eat up a large percentage of the actual remix fee, so most remixers simply don't bother. Many of the remix agreements I have been sent are worded almost exactly the same, and I wouldn't be surprised if they were all based on some kind of public domain "pro forma" agreement available somewhere. After you have seen quite a few of them, you get used to looking over them yourself for the most important points you are expecting to see. I am obliged to tell you that this is not a substitute for professional legal advice, clearly, but I also feel obliged to let you know the reality of the situation for myself and for most of the remixers I know.

SOME ODDITIES

If you ever take the time to fully read the remix agreements you are sent, and you should of course, there might be a few eye-opening moments from time to time. One that I came across recently was a clause that stated that the masters of the remix had to be delivered, by hand, to the A&R Director (or his equivalent) at the main office of the company. Of course, I didn't actually do this, and I wasn't asked to, but the fact remains that I could, in theory, have been asked to do just that and I wouldn't have had any grounds to argue, given that I had signed the remix agreement. Similarly, I have also seen a clause that requests the masters be delivered on tape to the label! I imagine this is probably just the result of a rather old remix agreement that hasn't been updated in a while, as I doubt very many remixers would even have access to a reel-to-reel tape recorder. In the event that you do see something like this in a remix agreement, the chances are that you won't actually be required to do it. But if you are in any doubt at all, give the client a call or send them an email to double check.

SUMMARY

I know it might not seem like we have covered a lot of things here, but these really are the fundamental points of most of the remixing agreements I have come across. The biggest difference between the simpler and more concise agreements and the longer agreements is purely one of obfuscation designed to make things unnecessarily complex. The longer ones serve no purpose, as when you finally manage to figure out what is being said, it is a simple point that probably could have been made in a much more concise way. Additionally, some remix agreements will expect you to provide additional information, such as names of producers/remixers involved, addresses of studios where the remix took place, and other similar kinds of thing. While I don't believe this is truly necessary, it certainly doesn't do any harm. I only get that request on remix agreements from one major client, and the information requested is the same each time, so I have that particular "appendix" stored as a PDF file on my computer, which I simply print out whenever I get a remix agreement from them.

In terms of legal and contractual issues, the life of a remix is certainly much simpler than that of an original song, because the paperwork involved is much simpler and much less convoluted. But it might pay you to get your lawyer to have a look over one of the more involved remix agreements, and take the time to explain everything to you in layman's terms if you are having difficulty understanding anything. As I said, they are for the most part largely similar, so if you can get your head around one then most of the others will seem more intelligible.

Index

C

D

M

N

O

P

S

T